Signal Processing and Artificial Intelligence Technology for High-End Equipment Fault Diagnosis

Signal Processing and Artificial Intelligence Technology for High-End Equipment Fault Diagnosis

Guest Editors

Bingyan Chen
Yao Cheng

MDPI

Basel • Beijing • Wuhan • Barcelona • Belgrade • Novi Sad • Cluj • Manchester

Guest Editors
Bingyan Chen
Centre for Efficiency and Performance Engineering
University of Huddersfield
Huddersfield
UK

Yao Cheng
State Key Laboratory of Rail Transit Vehicle System
Southwest Jiaotong University
Chengdu
China

Editorial Office
MDPI AG
Grosspeteranlage 5
4052 Basel, Switzerland

This is a reprint of the Special Issue, published open access by the journal *Machines* (ISSN 2075-1702), freely accessible at: https://www.mdpi.com/journal/machines/special_issues/65LUC20871.

For citation purposes, cite each article independently as indicated on the article page online and as indicated below:

Lastname, A.A.; Lastname, B.B. Article Title. *Journal Name* **Year**, *Volume Number*, Page Range.

ISBN 978-3-7258-4961-1 (Hbk)
ISBN 978-3-7258-4962-8 (PDF)
https://doi.org/10.3390/books978-3-7258-4962-8

Contents

About the Editors

Bingyan Chen

Bingyan Chen received his Bachelor's degree from the School of Mechatronics Engineering, Zhongyuan University of Technology, Zhengzhou, China, in 2017, and his PhD degree from the State Key Laboratory of Rail Transit Vehicle System (formerly State Key Laboratory of Traction Power), Southwest Jiaotong University, Chengdu, China, in 2023. He is currently a Research Fellow at the Centre for Efficiency and Performance Engineering, School of Computing and Engineering, University of Huddersfield, Huddersfield, United Kingdom. His research interests include non-stationary signal processing, machine learning, machine health monitoring, and rotating machinery fault diagnosis. His research work focuses on the basic theory and application of health management (condition monitoring and fault diagnosis) of rotating machinery systems. His doctoral dissertation was awarded the Outstanding Doctoral Dissertation Award of Southwest Jiaotong University in 2024. He has participated in more than 12 competitive research projects in this field, contributing to knowledge advancement and innovation in condition monitoring and fault diagnosis of rotating machinery, especially rolling element bearings. He has published 27 articles in internationally renowned SCI-indexed journals (3 of which are ESI highly cited papers) and 12 articles in EI-indexed journals and international conferences to disseminate his scientific research outcomes. He is also one of the editors of two international conference proceedings.

Yao Cheng

Yao Cheng received his Bachelor's degree in Probability Theory and Mathematical Statistics and his PhD degree in Vehicle Operation Engineering from Southwest Jiaotong University, Chengdu, China, in 2013 and 2019, respectively. He is currently an Assistant Researcher at the State Key Laboratory of Rail Transit Vehicle System (formerly State Key Laboratory of Traction Power), Southwest Jiaotong University, Chengdu, China. His current research interests include intelligent fault diagnosis, dynamic signal processing, and machine learning. His research focuses on intelligent analysis of rail vehicle dynamics and digital operation and maintenance. He has been awarded nearly 20 research projects from the National Natural Science Foundation of China, Sichuan Provincial Department of Science and Technology, national key laboratories, and enterprises, contributing advanced knowledge and innovations in high-speed train transmission system dynamics models, multi-source data-driven intelligent monitoring and diagnosis algorithms, and the fusion of physical mechanisms and multi-source data. He has published 42 SCI-indexed papers in internationally renowned journals in the fields of multi-source data fusion, mechanical vibration signal analysis, drive system dynamics, and intelligent fault diagnosis, including 25 papers as the first or corresponding author and 4 highly cited ESI papers. He has been granted eight invention patents. He has chaired several academic conference sessions and given invited presentations. He also serves as a Young Editorial Board Member of several journals and serves as a Guest Editor for several SCI-indexed journals.

Preface

We are pleased to present this Reprint of 13 research papers on signal processing and artificial intelligence technologies for high-end equipment fault diagnosis. This Reprint aims to capture the latest advances in the development and application of novel signal processing and artificial intelligence methods for the purposes of condition monitoring and fault diagnosis of advanced industrial machinery systems.

Mechanical equipment are widely used in various industrial fields, significantly improving production efficiency and people's living standards. However, due to various harsh operating conditions and environments, mechanical systems in various industrial fields often face the severe challenge of key component failure. Critical industrial sectors place stringent requirements on the safety and reliability of mechanical systems to avoid unexpected failures or downtime. Timely fault diagnosis and early warning systems are effective means of preventing catastrophic accidents, and the application of advanced sensing, monitoring, and diagnostic technologies can effectively improve the accuracy of machinery health monitoring and fault diagnosis.

The research papers in this Reprint cover mechanical system dynamics modeling and fault mechanism analysis, signal processing and machine learning-enabled condition monitoring, fault diagnosis and remaining useful life prediction for mechanical systems, and the application of various sensing technologies, including vibration, temperature, thermal imaging, electrical signals, and force or torque. Application areas for these technologies include the railway, wind power, machinery manufacturing, and aerospace industries, with high-end machinery including high-speed trains, wind turbines, aero engines, machine tools, and industrial robots. The intended audience includes researchers from universities and research institutions, as well as engineers from industry, working on machine health monitoring and fault diagnosis.

We would like to thank all the authors contributing to this Special Issue for their valuable research and sharing spirit, as well as the peer reviewers and editorial office staff for their unwavering support in producing this Special Issue. We hope that these research papers will provide valuable insights into the mechanisms and principles of machine condition monitoring and fault diagnosis, and guide the practical industrial applications of machinery health management. We also hope that they will serve as valuable references for similar research objectives in other fields and as inspiration for further research.

Bingyan Chen and Yao Cheng
Guest Editors

machines

MDPI

Article

Analytical and Experimental Investigation of Nonlinear Dynamic Characteristics of Hydrodynamic Bearings for Oil Film Instability Detection

Yang Chen [1], Zequn Zhao [1], Hao Zhang [1,*], Xin Li [2] and Zhanqun Shi [1]

1 School of Mechanical Engineering, Hebei University of Technology, Tianjin 300401, China
2 National Key Laboratory of Diesel Engine Turbocharging Technology, China North Engine Research Institute, Tianjin 300401, China
* Correspondence: zhanghao@hebut.edu.cn

Abstract: Nonlinear vibration phenomena, such as oil whirl and oil whip, are common indicators of oil film instability in hydrodynamic bearings and are key signs of potential faults in rotating machinery. Excessive vibrations caused by oil film instability can accelerate bearing wear and lead to the failure of the rotating system. This paper presents a model for nonlinear dynamic coefficients, aimed at providing a quantitative approach for monitoring and predicting oil film instability. The impact of operational parameters and perturbation values on both linear and nonlinear stiffness and damping coefficients is investigated. Simulation results and experimental rotor vibration signals demonstrate that the nonlinear dynamic coefficient model effectively characterizes oil film instability and accurately predicts rotor trajectory, while traditional linear models are only applicable under low-speed and small-disturbance conditions. Compared to traditional analytical models and numerical solutions, the nonlinear dynamic coefficients have higher accuracy and efficiency and can reliably identify the onset frequency of oil film instability. This study clarifies the relationship between nonlinear dynamic coefficients and rotor dynamic response, laying a theoretical foundation for the monitoring and prediction of oil film instability.

Keywords: hydrodynamic bearing; nonlinear dynamic coefficients; oil film instability; condition monitoring

1. Introduction

Journal bearings, as key components, are widely used in rotating machinery due to their simple structure, strong load capacity, and long service life. As a crucial component of rotor systems, the bearing's dynamic characteristics directly affect the overall dynamic behaviors and safe operation of the system [1–3]. Oil film instability usually occurs under low eccentricity and a high rotational speed [4]. The violent vibration with a sudden increase in amplitude can easily lead to bearing wear and rotor rubbing. Although classical linearization models (such as the short bearing theory) can partially explain the oil film characteristics under steady-state conditions, it is difficult for them to accurately predict the transient instability process caused by the coupling and lag effects of time-varying bearing stiffness and damping. Therefore, accurately determining bearing dynamic coefficients, especially the nonlinear characteristics of oil film instability, is vital for the design and maintenance management of rotor systems [5].

The lubrication characteristics of hydrodynamic bearings were typically studied by directly integrating and solving the Reynolds equation, which incurred high computational costs. To address this issue, researchers introduced the concept of dynamic characteristic parameters. Lune [6] proposed a four-linear-stiffness-coefficient model by using the perturbation method, which has been widely applied. However, the linear dynamic coefficients are inadequate for analyzing and calculating the strongly nonlinear behaviors of hydrodynamic bearings [7]. To address this problem, scholars have initially proposed the concept of nonlinear dynamic characteristics. Czolczynsk [8] determined the nonlinear coefficients of gas bearings via the orbit method. Nevertheless, due to the intrinsic errors associated with the method, its application remains limited in practice. Nishimura et al. [9] explored rotor nonlinear behaviors supported by hydrodynamic bearings using the short bearing theory, considering unbalance and rotor configuration. Choy et al. [10] analytically expressed the nonlinear oil film force as an odd-power series of journal perturbation displacement, based on the short bearing approximation. Chu et al. [11] used a quasi-static nonlinear model to estimate the bearing's dynamic coefficients. Adiletta et al. [12] applied the long bearing theory to correct approximate theories and analyze the effects of a wide range of parameters on the response of a rigid rotor. However, the short bearing model neglects axial pressure flow, whereas the infinitely long bearing model ignores end leakage effects, so the calculation results of these models all have errors to a certain extent.

The past decade has seen significant improvements in the computational methods for nonlinear dynamic models and nonlinear dynamic characteristic parameters. Andréas and Santiago [13] conducted experiments to evaluate the dynamic response of a large orbit motion with hydrodynamic bearings under a single-frequency excitation force. Meruane et al. [14] proposed a framework that accurately identified dynamic parameters and determined the nonlinear transient response under various operating conditions. Tian et al. [15] developed a finite element model of a TC rotor and calculated the nonlinear bearing forces. Ying et al. [16] compared rotor nonlinear response with and without base excitation at different rotational speeds. Dakel et al. [17] established an innovative Timoshenko beam-based finite element model to analyze dynamic behavior in rigidly supported mechanical systems. Weimin et al. [18] introduced a technique that utilizes partial derivatives to computationally derive the nonlinear dynamic coefficients of journal bearings without prior knowledge of the shaft orbit. Smolík et al. [19] numerically investigated the nonlinear behavior of bearings within the framework of multibody dynamic simulations of turbochargers. Dyk et al. [20] enhanced the analytical representation of oil film nonlinearity by employing a modified polynomial model. Chasalevris et al. [21] considered lemon-bore and partial-arc bearings as two main bearing profiles in the nonlinear impedance analysis. Building on thermohydrodynamic coupling principles, Li [22] formulated a novel computational model that elucidates the nonlinear dynamic behavior of turbocharger rotor–bearing systems. This foundational work has stimulated further investigations into the stability thresholds and complex nonlinear phenomena associated with floating ring bearing configurations under operational conditions [23–25].

As rotating machinery advances toward higher speeds and greater precision, the nonlinear behaviors and instability effects induced by oil film forces have gained considerable importance. Dyk and Rendl [26–28] introduced a new lubrication groove design and derived analytical expressions for dynamic coefficients, accurately assessing the nonlinear characteristics and stability threshold. The nonlinear force parametrization using the perturbation method paradigm was developed in References [29,30]. EI-Sayed et al. [31,32] conducted a bifurcation analysis using second-order nonlinear coefficients. However, the model showed errors at low Sommerfeld numbers due to the simplification of the oil film forces.

In summary, simplified linear stiffness and damping coefficients cannot evaluate complex nonlinear behaviors. The analytical solution based on the short bearing model is only applicable to full circular bearings with a length-to-diameter ratio far greater than 1, although the convergence speed is fast [33,34]. Numerical solutions for nonlinear oil film forces lead to problems of large computational load and difficult convergence [35]. In addition, although there have been a large number of studies on the nonlinear dynamics of rotor–bearing systems, research on applying nonlinear dynamic coefficients to predict the instability of oil films is still insufficient. Therefore, the main contributions of the proposed method in this paper are summarized as follows:

(1) The proposed nonlinear dynamic coefficient model enables to evaluation of the nonlinear dynamic characteristics of the rotor–bearing system and solves the limitation of the linear model.
(2) Compared to traditional analytical models and numerical solutions, the nonlinear stiffness and damping coefficient model significantly saves computation time while accurately expressing the transient nonlinear characteristics of journal bearings.
(3) The simulation results and experiment signals both show that the nonlinear dynamic coefficient model can effectively characterize oil film instability and accurately predict the rotor trajectory.

The remainder of this paper is organized as follows: Section 2 presents the lubrication model for finite-length hydrodynamic bearings, deriving the analytical expressions for nonlinear stiffness and damping coefficients using the second-order perturbation theory. Section 3 validates the computational results of linear and nonlinear coefficients against the published data. Then, the shaft trajectory, oil whirl, and oil whip phenomena are investigated based on both linear and nonlinear coefficient models. Section 4 verified the theoretical analysis through the experimental displacement signals. Section 5 presents the conclusion of this study.

2. Methodology

2.1. Lubrication Model of Hydrodynamic Bearing

Figure 1 is a schematic diagram of the geometric structure and working principle of the full circular hydrodynamic bearing. O_b, O_j denote the geometric centers of the bearing and journal, respectively; R denotes the bearing radius and c denotes the radial clearance; e, θ represent the eccentricity and attitude angle, respectively, and φ denotes the angle that the journal rotates.

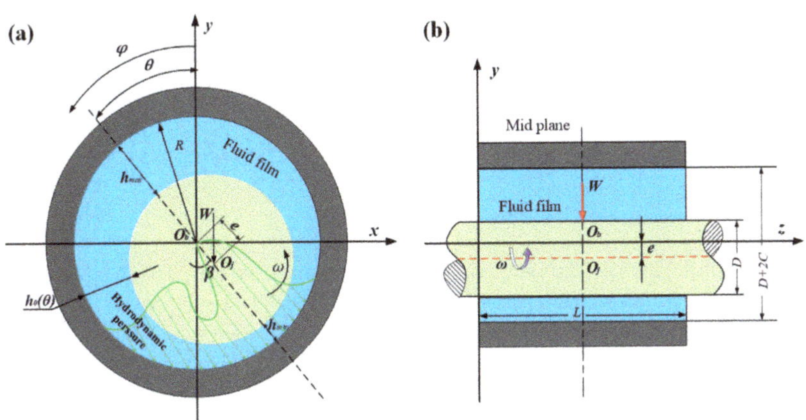

Figure 1. Schematic diagram of the hydrodynamic bearing: (**a**) front view; (**b**) side view.

Within the classical lubrication theory, the lubricant is assumed to be an incompressible Newtonian fluid undergoing laminar flow (the shear stress has a linear relationship with shear rate). However, under extremely high rotational speed conditions, the Reynolds number may exceed the critical value (Re > 2000), and the turbulence correction factor and inertia effect need to be introduced [36]. Under extremely high pressure and temperature, the characteristics of non-Newtonian fluids—as well as the viscosity–temperature and viscosity–pressure effects of lubricating oil—must be taken into consideration [37].

This paper focuses on the conventional working condition (Re < 2000), so the Newtonian fluid and laminar flow assumptions are adopted. The hydrodynamic within the oil film thickness is mathematically described by the Reynolds equation:

$$\frac{\partial}{\partial x}\left(\frac{\rho h^3}{12\eta}\frac{\partial p}{\partial x}\right)+\frac{\partial}{\partial z}\left(\frac{\rho h^3}{12\eta}\frac{\partial p}{\partial z}\right)=\frac{\omega R}{2}\left(\frac{\rho\partial h}{\partial x}\right)+\frac{\rho\partial h}{\partial t} \tag{1}$$

where x, z are coordinate representations in the circumferential and axial directions; ω is the rotation frequency of the journal bearing; η is the lubricant viscosity; and the oil film pressure p is dependent on the oil film thickness h, which can be expressed as follows:

$$h(\varphi,z)=c\times(1+\varepsilon\times\cos(\varphi-\theta)) \tag{2}$$

We define the dimensionless forms of each variable:

$$H=\frac{h}{c}\quad \varphi=\frac{x}{R}\quad Z=\frac{z}{L}(0\leq Z\leq 1)\quad P=\frac{p}{p^*}\left(p^*=\eta\omega R^2/c^2\right) \tag{3}$$

Then, the dimensionless form of Equation (1) can be written as follows:

$$\frac{\partial}{\partial\varphi}\left(H^3\frac{\partial P}{\partial\varphi}\right)+\left(\frac{R}{L}\right)^2\frac{\partial}{\partial Z}\left(H^3\frac{\partial P}{\partial Z}\right)=6p^*\left(\frac{\partial H}{\partial\varphi}\right) \tag{4}$$

H represents the dimensionless fluid film thickness. P represents the dimensionless pressure. φ and Z represent the dimensionless axial and vertical coordinates, respectively. φ_1 denotes the oil film rupture's start position and φ_2 denotes the oil film rupture's end position. The Reynolds boundary condition is used here with the pressure starting point at $P(\varphi=\varphi_1,Z)=0$ and the pressure endpoint at $P(\varphi=\varphi_2,Z)=0$, $\left.\frac{\partial P}{\partial\varphi}\right|_{\varphi=\varphi 2}=0$.

By integrating the oil film pressure distribution over the load-bearing zone of hydrodynamic bearings, the hydrodynamic bearing forces can be mathematically derived. The vertical component F_x, axial component F_y, and resultant force F from the direct integration of the Reynolds equation are as follows:

$$\begin{cases} F_X=\int_{-1}^{1}\int_{\varphi_1}^{\varphi_2}P\sin\varphi d\varphi dZ \\ F_X=\int_{-1}^{1}\int_{\varphi_1}^{\varphi_2}-P\cos\varphi d\varphi dZ \\ F=\sqrt{F_X^2+F_Y^2} \end{cases} \tag{5}$$

Since the Reynolds equation is an elliptic partial differential equation, obtaining an exact analytical solution is challenging. As a result, the finite difference method is commonly employed to solve the pressure distribution. Figure 2 illustrates that the circumferential direction of the entire oil film region is discretized into m points, and the axial direction is discretized into n points. The difference quotient of adjacent nodes is used to approximate the partial derivative values in different directions. The discrete form of Equation (4) is derived as follows:

$$\begin{aligned}
&\frac{\partial}{\partial\theta}(H^3\frac{\partial P}{\partial\theta}) = \frac{H^3_{i+0.5,j}(P_{i+1,j}-P_{i,j})-H^3_{i-0.5,j}(P_{i,j}-P_{i-1,j})}{\Delta\theta^2}\\
&(\frac{R}{L})^2\frac{\partial}{\partial Z}(H^3\frac{\partial P}{\partial Z}) = (\frac{R}{L})^2\frac{H^3_{i,j+0.5}(P_{i,j+1}-P_{i,j})-H^3_{i,j-0.5}(P_{i,j}-P_{i-1,j})}{\Delta Z^2}\\
&6\frac{\partial(H)}{\partial\theta} = 6\frac{H_{i,j}-H_{i-1,j}}{\Delta\theta}\\
&12\frac{\partial(H)}{\partial\tau} = 12\frac{(H)^{\tau}-(H)^{\tau-1}}{\partial\tau}
\end{aligned} \tag{6}$$

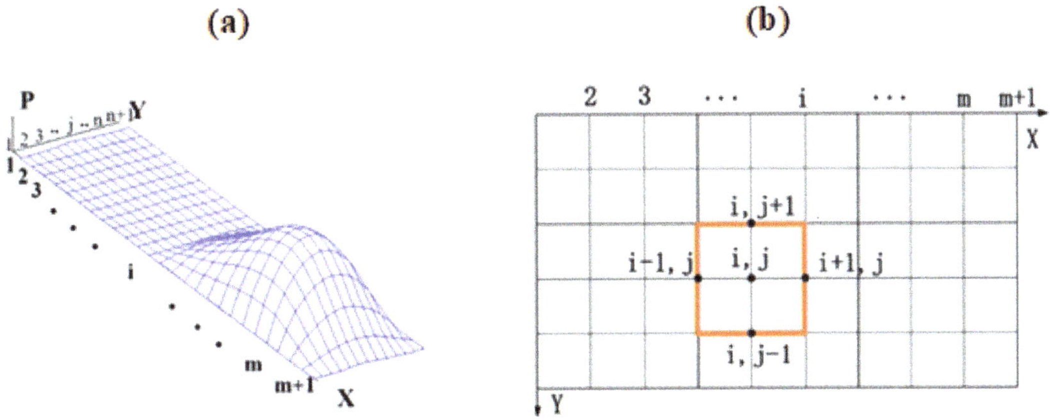

Figure 2. Schematic of the hydrodynamic bearing mesh; (**a**) mesh coordinates of the oil film region; (**b**) finite difference method.

Finally, the calculation formula in discrete form is as follows:

$$P_{i,j} = \frac{AP_{i-1,j} + BP_{i+1,j} + CP_{i,j-1} + DP_{i,j+1} - F_{i,j}}{E_{i,j}} \tag{7}$$

The calculation coefficients A, B, C, D, E, F are, respectively, expressed as follows:

$$\begin{aligned}
&A = H^3_{i-0.5,j}B = H^3_{i+0.5,j}\\
&C = (\frac{D}{L}\cdot\frac{\Delta\theta}{\Delta Z})^2H^3_{i,j-0.5}D = (\frac{D}{L}\cdot\frac{\Delta\theta}{\Delta Z})^2H^3_{i,j+0.5}\\
&E = A+B+C+D\\
&F = 6\Delta\theta(H_{i+0.5} - H_{i+0.5})
\end{aligned} \tag{8}$$

The coupled equations are solved using a relaxation iteration scheme to accelerate the convergence speed, and the relaxation factor is taken as 1–2, which is applicable for the selection of most lubrication issues [38]. The specific selected value is related to the mesh grid density and boundary conditions. The iterative process terminates when successive approximations satisfy the prescribed convergence tolerance, $\delta = 10^{-6}$. To determine whether sufficient accuracy has been achieved in each iteration result, the following relative convergence criterion is adopted to determine whether to abort the iteration process:

$$P_{i,j}^k = \beta(P_{i,j}^k - P_{i,j}^{k-1}) + P_{i,j}^{k-1} \tag{9}$$

$$\frac{\sum_{i=2}^{m}\sum_{j=2}^{n}\left|P_{i,j}^k - P_{i,j}^{k-1}\right|}{\sum_{i=2}^{m}\sum_{j=2}^{n}\left|P_{i,j}^k\right|} \le \delta \tag{10}$$

2.2. Modeling of Nonlinear Stiffness and Damping Coefficients

The rotor equilibrium position under different operational conditions is determined as the premise for evaluating the lubrication condition. For the hydrodynamic lubrication condition, according to the small perturbation theory [39], the oil film thickness and hydrodynamic pressure are expressed as follows:

$$\begin{aligned} &h = h_0 + \Delta x \sin\varphi - \Delta y \cos\varphi \\ &p = p_0 + \Delta p = p_0 + \tfrac{\partial p}{\partial x}\Delta x + \tfrac{\partial p}{\partial y}\Delta y + \tfrac{\partial p}{\partial x'}\Delta x' + \tfrac{\partial p}{\partial y'}\Delta y' \end{aligned} \tag{11}$$

Substituting Equation (10) into Equation (4) yields the dimensionless perturbation expression for the Reynolds equation:

$$\begin{aligned} &\tfrac{\partial}{\partial\varphi}(H^3\tfrac{\partial P_i'}{\partial\varphi}) + (\tfrac{R}{L})^2\tfrac{\partial}{\partial Z}(H^3\tfrac{\partial P_i'}{\partial Z}) = \\ &\left\{\begin{array}{ll} 6\cos\varphi - 3\frac{\sin\varphi}{H}\frac{\partial H}{\partial\theta} - H^3\frac{\partial P_0}{\partial\varphi}\frac{\partial}{\partial\varphi}(\frac{\sin\varphi}{\partial\varphi}) & P_i' = \frac{\partial P}{\partial X} \\ 6\sin\varphi - 3\frac{\cos\varphi}{H}\frac{\partial H}{\partial\varphi} - H^3\frac{\partial P_0}{\partial\varphi}\frac{\partial}{\partial\varphi}(\frac{\cos\varphi}{\partial\varphi}) & P_i' = \frac{\partial P}{\partial Y} \\ 12\sin\varphi & P_i' = \frac{\partial P}{\partial X'} \\ 12\cos\varphi & P_i' = \frac{\partial P}{\partial Y'} \end{array}\right\} \end{aligned} \tag{12}$$

Once the perturbation pressures are calculated, the linearized stiffness and damping coefficients are integrated by the following:

$$\begin{aligned} &\left.\begin{array}{l} K_{XX} \\ K_{YX} \end{array}\right\} = -\int_0^1\int_{\varphi_1}^{\varphi_2}\frac{\partial P}{\partial X}\left\{\begin{array}{c} \sin\varphi \\ -\cos\varphi \end{array}\right\}d\varphi dZ, \quad \left.\begin{array}{l} K_{XY} \\ K_{YY} \end{array}\right\} = -\int_0^1\int_{\varphi_1}^{\varphi_2}\frac{\partial P}{\partial Y}\left\{\begin{array}{c} \sin\varphi \\ -\cos\varphi \end{array}\right\}d\varphi dZ \\ &\left.\begin{array}{l} C_{XX} \\ C_{YX} \end{array}\right\} = -\int_0^1\int_{\varphi_1}^{\varphi_2}\frac{\partial P}{\partial X'}\left\{\begin{array}{c} \sin\varphi \\ -\cos\varphi \end{array}\right\}Rd\varphi dZ, \quad \left.\begin{array}{l} C_{XY} \\ C_{YY} \end{array}\right\} = -\int_0^1\int_{\varphi_1}^{\varphi_2}\frac{\partial P}{\partial Y'}\left\{\begin{array}{c} \sin\varphi \\ -\cos\varphi \end{array}\right\}Rd\varphi dZ \end{aligned} \tag{13}$$

The eight stiffness and damping coefficients are calculated by integrating P_i' over the oil film region. The relationship between normalized values and real values is converted by $K_b = k_b/W$, $C_b = c_b * \omega * c/W$, respectively.

The linear coefficients are based on the first-order Taylor expansion. Correspondingly, the solution principle for the nonlinear coefficients is based on applying the second-order Taylor expansion to the rotor's disturbance displacement and velocity at the equilibrium position.

The expression of the dimensionless pressure distribution P using the second-order expansion for the Taylor series is as follows:

$$\begin{aligned} P = {} & P_0 + P_X\Delta X + P_Y\Delta Y + P_{X'}\Delta X' + P_{Y'}\Delta Y' + \tfrac{1}{2}P_{XX}\Delta X^2 + \tfrac{1}{2}P_{YY}\Delta Y^2 \\ & + P_{XY}\Delta X\Delta Y + P_{YX}\Delta Y\Delta X + P_{XX'}\Delta X\Delta X' + P_{XY'}\Delta X\Delta Y' + + P_{YX'}\Delta Y\Delta X' P_{YY'}\Delta Y\Delta Y' \end{aligned} \tag{14}$$

Substituting Equation (13) into Equation (4) yields the dimensionless perturbation expression for the Reynolds equation:

$$
\begin{array}{l}
\frac{\partial}{\partial \varphi}(H^3 \frac{\partial P}{\partial \varphi}) + \frac{\partial}{\partial Z}(H^3 \frac{\partial P}{\partial Z}) = \\
\frac{\partial H_{XX}}{\partial \varphi} - \frac{\partial}{\partial \varphi}[6H_0^2 H_X \frac{\partial P_X}{\partial \varphi} + (6H_0^2 H_X + 3H_0^2 H_{XX}) \frac{\partial P_0}{\partial \varphi}] \\
-\frac{\partial}{\partial Z}[6H_0^2 H_X \frac{\partial P_X}{\partial Z} + (6H_0^2 H_X + 3H_0^2 H_{XX}) \frac{\partial P_0}{\partial Z}], \qquad P = P_{XX} \\
\frac{\partial H_{YY}}{\partial \varphi} - \frac{\partial}{\partial \varphi}[6H_0^2 H_Y \frac{\partial P_Y}{\partial \varphi} + (6H_0^2 H_Y + 3H_0^2 H_{YY}) \frac{\partial P_0}{\partial \varphi}] \\
-\frac{\partial}{\partial Z}[6H_0^2 H_Y \frac{\partial P_Y}{\partial Z} + (6H_0^2 H_Y + 3H_0^2 H_{YY}) \frac{\partial P_0}{\partial Z}], \qquad P = P_{YY} \\
\frac{\partial H_{XY}}{\partial \varphi} - \frac{\partial}{\partial \varphi}[3H_0^2 H_Y \frac{\partial P_X}{\partial \varphi} + 3H_0^2 H_X \frac{\partial P_Y}{\partial \varphi} + (6H_0 H_X H_Y + 3H_0^2 H_{YY}) \frac{\partial P_0}{\partial \varphi}] \\
-\frac{\partial}{\partial Z}[3H_0^2 H_Y \frac{\partial P_X}{\partial Z} + 3H_0^2 H_X \frac{\partial P_Y}{\partial Z} + (6H_0 H_X H_Y + 3H_0^2 H_{YY}) \frac{\partial P_0}{\partial Z}], \quad P = P_{XY} \\
\frac{\partial H_{YX}}{\partial \varphi} - \frac{\partial}{\partial \varphi}[3H_0^2 H_X \frac{\partial P_Y}{\partial \varphi} + 3H_0^2 H_Y \frac{\partial P_X}{\partial \varphi} + (6H_0 H_Y H_X + 3H_0^2 H_{XX}) \frac{\partial P_0}{\partial \varphi}] \\
-\frac{\partial}{\partial Z}[3H_0^2 H_X \frac{\partial P_Y}{\partial Z} + 3H_0^2 H_Y \frac{\partial P_X}{\partial Z} + (6H_0 H_Y H_X + 3H_0^2 H_{XX}) \frac{\partial P_0}{\partial Z}], \quad P = P_{YX} \\
2H_{XX} - \frac{\partial}{\partial \varphi}(3H_0^2 H_X \frac{\partial P_{\dot{X}}}{\partial \varphi}) - \frac{\partial}{\partial Z}(3H_0^2 H_X \frac{\partial P_{\dot{X}}}{\partial Z}), \quad P = P_{XX'} \\
2H_{XX} - \frac{\partial}{\partial \varphi}(3H_0^2 H_X \frac{\partial P_{\dot{Y}}}{\partial \varphi}) - \frac{\partial}{\partial Z}(3H_0^2 H_X \frac{\partial P_{\dot{Y}}}{\partial Z}), \quad P = P_{XY'} \\
2H_{XY} - \frac{\partial}{\partial \varphi}(3H_0^2 H_Y \frac{\partial P_{\dot{X}}}{\partial \varphi}) - \frac{\partial}{\partial Z}(3H_0^2 H_X \frac{\partial P_{\dot{X}}}{\partial Z}), \quad P = P_{YX'} \\
2H_{YY} - \frac{\partial}{\partial \varphi}(3H_0^2 H_Y \frac{\partial P_{\dot{Y}}}{\partial \varphi}) - \frac{\partial}{\partial Z}(3H_0^2 H_Y \frac{\partial P_{\dot{Y}}}{\partial Z}), \quad = P_{YY'}
\end{array}
\tag{15}
$$

Each perturbation pressure $P_{XX}, P_{YY}, P_{XY}, P_{YX}, P_{XX'}, P_{XY'}, P_{YX'}, P_{YY'}$ corresponds to the solution of each respective perturbation equation.

The nonlinear stiffness and damping coefficients are defined as the second-order approximation, ignoring all perturbation terms higher than the second order. Nonlinear stiffness is the second-order partial derivative of the reaction force with respect to displacement, and damping is the second-order partial derivative of the reaction force with respect to velocity. The detailed terms of each coefficient are as follows:

$$
\begin{array}{l}
K_{XXX} = \frac{\partial^2 F_X}{\partial X^2}, K_{XYY} = \frac{\partial^2 F_X}{\partial Y^2}, K_{XXY} = \frac{\partial^2 F_X}{\partial X \partial Y'}, K_{XYX} = \frac{\partial^2 F_X}{\partial X \partial Y'} \\
K_{YYY} = \frac{\partial^2 F_Y}{\partial Y^2}, K_{YXX} = \frac{\partial^2 F_Y}{\partial X^2}, K_{YYX} = \frac{\partial^2 F_Y}{\partial X \partial Y'}, K_{YXY} = \frac{\partial^2 F_Y}{\partial X \partial Y'} \\
C_{XXX} = \frac{\partial^2 F_X}{\partial X \partial X'}, C_{XXY} = \frac{\partial^2 F_X}{\partial X \partial Y'}, C_{XYX} = \frac{\partial^2 F_Y}{\partial X \partial X'}, C_{XYY} = \frac{\partial^2 F_X}{\partial Y \partial Y'} \\
C_{YXX} = \frac{\partial^2 F_Y}{\partial X \partial X'}, C_{YXY} = \frac{\partial^2 F_Y}{\partial X \partial Y'}, C_{YYX} = \frac{\partial^2 F_Y}{\partial Y \partial X'}, C_{YYY} = \frac{\partial^2 F_Y}{\partial Y \partial Y'}
\end{array}
\tag{16}
$$

where $K_{IJO} = k_{ijo}c^2/F$, $C_{IJO} = c_{ijo}c^2\omega/F$ are the dimensionless forms of nonlinear coefficients, and i, j, o correspond to X, Y, respectively.

The sixteen coefficients are derived by integrating the disturbed oil film pressure distribution:

$$
\begin{cases}
K_{X\zeta\eta} = \int_{-1}^{1} \int_{\varphi_1}^{\varphi_2} -P_{\zeta\eta} \sin \varphi d\varphi dZ \\
K_{Y\zeta\eta} = \int_{-1}^{1} \int_{\varphi_1}^{\varphi_2} -P_{\zeta\eta} \cos \varphi d\varphi dZ \\
C_{X\zeta\eta} = \int_{-1}^{1} \int_{\varphi_1}^{\varphi_2} -P_{\zeta\eta'} \sin \varphi d\varphi dZ \\
C_{Y\zeta\eta} = \int_{-1}^{1} \int_{\varphi_1}^{\varphi_2} -P_{\zeta\eta'} \sin \varphi d\varphi dZ
\end{cases}
\tag{17}
$$

Substituting Equations (12) and (15) into Equation (17) yields the following expansion of the equivalent force F_ζ for the dimensionless oil film component force expressed by the stiffness and damping coefficient:

$$
\begin{array}{l}
F_\zeta = F_{\zeta 0} + K_{\zeta X}\Delta X + K_{\zeta Y}\Delta Y + C_{\zeta Y}\Delta X' + C_{\zeta Y}\Delta Y' + \frac{1}{2}K_{\zeta XX}\Delta X^2 + \frac{1}{2}K_{\zeta YY}\Delta Y^2 \\
\quad +K_{\zeta XY}\Delta X\Delta Y + C_{\zeta XX}\Delta X\Delta X' + C_{\zeta XY}\Delta X\Delta Y' + C_{\zeta YX}\Delta Y\Delta X' + C_{\zeta YY}\Delta Y\Delta Y'
\end{array}
\tag{18}
$$

2.3. Dynamic Model of Rotor Motion

To study the rotor dynamic response under the linear and nonlinear coefficient model, the dynamic model of the rotor–bearing system was established. As shown in Figure 3b, the equation of motion of the rotor–bearing system can be expressed as follows:

$$\begin{cases} M\ddot{X} = F_x + Me_u\omega^2\cos(\omega t) \\ M\ddot{Y} = F_y + Me_u\omega^2\sin(\omega t) - W \end{cases} \tag{19}$$

where e_u, ω, t represent the unbalanced eccentricity, rotational speed of the journal, and time, respectively. F_x, F_y denote the bearing forces in the respective directions, obtained using the linear and nonlinear dynamic coefficients, respectively. W denotes the gravity and radial load in the vertical direction.

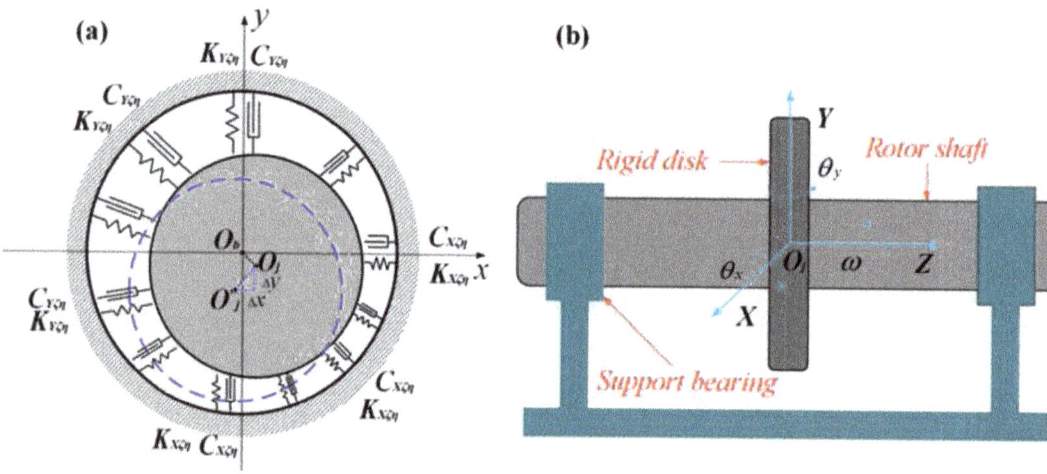

Figure 3. Nonlinear dynamic coefficient model: (**a**) nonlinear stiffness and damping coefficients; (**b**) scheme of rotor–bearing system.

The rotor motion equation is iteratively solved using the Runge–Kutta numerical integration to determine the rotor's position and velocity at each time step. A stable rotor vibration response is obtained through a certain number of iterations. The complete flow of the proposed nonlinear stiffness and damping coefficient calculation model, along with its validation method, is shown in Figure 4, with the specific simulation parameters listed in Table 1.

Table 1. Simulation parameters of the rotor–bearing system.

Bearing Parameter	Specification	Rotor Parameter	Specification
Bearing length L (mm)	20	Shaft radius R_1 (mm)	30
Bearing diameter D (mm)	20	Shaft length l (mm)	450
Radial clearance c (μm)	100	Density ρ (kg/m^3)	7850
Viscosity η (Pa·s)	0.0135	Load W (N)	500
Sommerfeld number S	0.01~1	Eccentricity ε	0.1~0.9

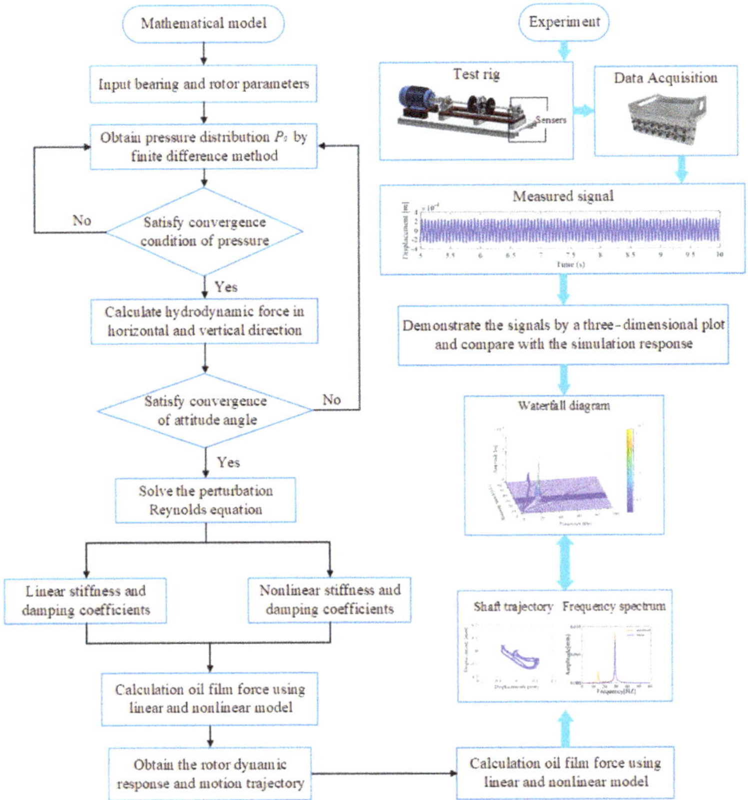

Figure 4. Flowchart of the simulation process and experimental verification.

3. Simulation Results and Discussion

3.1. Perturbation Analysis

In this section, the effects of both linear and nonlinear parameters on the perturbation force are investigated. The linear and nonlinear coefficients are derived by applying perturbations of varying magnitudes. Three different magnitudes of displacement and velocity perturbations are considered to evaluate the variation in the bearing dynamic coefficients.

Generally speaking, a larger grid number results in higher calculation accuracy and significantly increases calculation time. Therefore, in the numerical calculation process, an appropriate number of grids should be selected to ensure high calculation efficiency and convergence accuracy. Figure 5 shows that the changing trend in the maximum oil film pressure slows down significantly, and the relative error is close to 0.1% when the number of circumferential grids is 440 and the number of axial grids is 200. Thus, a grid size of 440×200 is selected for the following numerical calculation.

The first perturbation value is $\Delta X = 0.002, \Delta X' = 0.002, \Delta Y = 0.002, \Delta Y' = 0.002$. The oil film force decreases with the increasing Sommerfeld number when $S < 0.1$, while the oil film force increases with the increasing Sommerfeld number when $S > 0.1$. The results are displayed in Figure 6, indicating that the evaluated perturbed forces are approximately the same between the linear and nonlinear models at low perturbations.

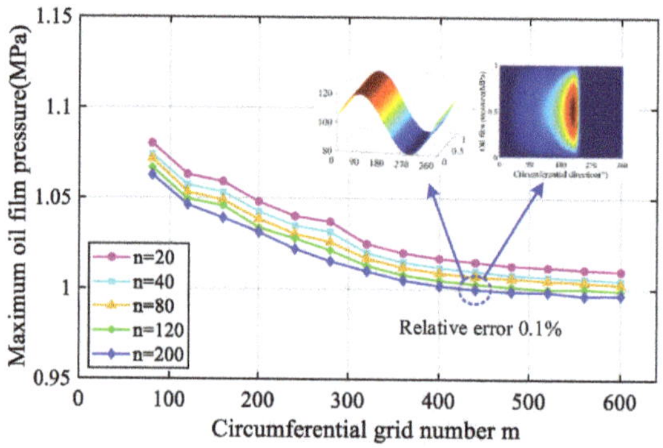

Figure 5. Verification of grid convergence accuracy.

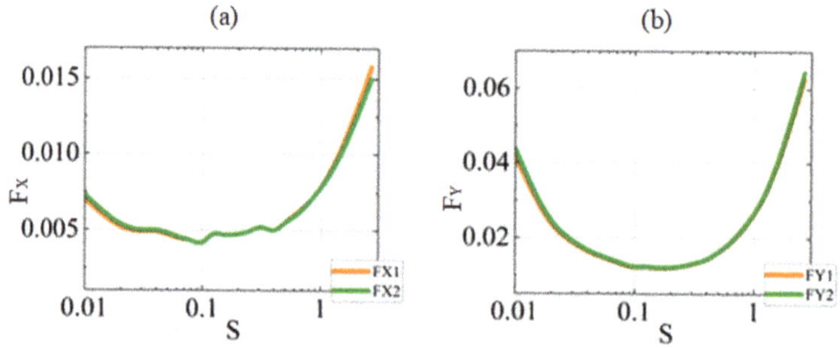

Figure 6. Oil film forces with linear and nonlinear dynamic coefficients under perturbation value $\Delta X = 0.002, \Delta X' = 0.002, \Delta Y = 0.002, \Delta Y' = 0.002$: (**a**) horizontal force; (**b**) vertical force.

The second perturbation value is $\Delta X = 0.02, \Delta X' = 0.02, \Delta Y = 0.02, \Delta Y' = 0.02$. The variation trend of the perturbed force with the Sommerfeld number in Figure 7 is the same as that in Figure 6. However, there are small deviations in the absolute force values between the two methods at moderate perturbation levels. The forces based on second-order coefficients are higher than those based on first-order coefficients.

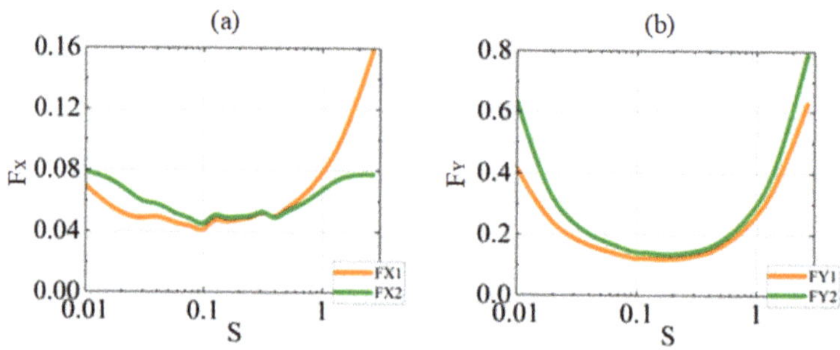

Figure 7. Oil film forces with linear and nonlinear dynamic coefficients under perturbation value $\Delta X = 0.02, \Delta X' = 0.02, \Delta Y = 0.02, \Delta Y' = 0.02$: (**a**) horizontal force; (**b**) vertical force.

The third perturbation value is $\Delta X = 0.2, \Delta X' = 0.2, \Delta Y = 0.2, \Delta Y' = 0.2$, as illustrated in Figure 8. The absolute value of the perturbation reaction force of the oil film decreases with the Sommerfeld number when $S < 0.1$. The perturbation force amplitude presents an increasing progression with operational speed as the Sommerfeld number increases when $S > 0.1$, while the linear oil film force has a flat trend. As depicted in

Figure 8a,b, the reaction forces evaluated using the two methods are already significantly different at high perturbations. Significant differences are found at both low and high Sommerfeld numbers. The absolute values of the forces based on the second-order coefficients are generally much higher than the absolute values of the forces based on the first-order coefficients when $S < 0.1$ and $S > 0.6$.

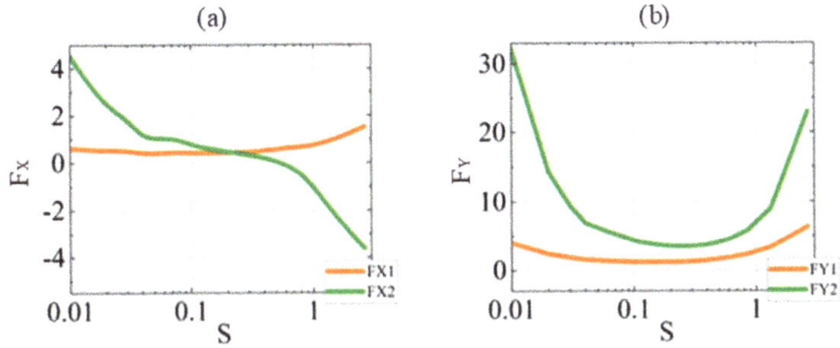

Figure 8. Oil film forces with linear and nonlinear dynamic coefficients under perturbation value $\Delta X = 0.2, \Delta X' = 0.2, \Delta Y = 0.2, \Delta Y' = 0.2$: (**a**) horizontal force; (**b**) vertical force.

The analysis in this section shows a significant deviation between the forces evaluated based on linear and nonlinear coefficients at large perturbation values.

However, with the increase in the perturbation value, the difference in the reaction force calculated by the linear and nonlinear coefficient models gradually becomes larger with the intensified nonlinear effects of the oil film.

3.2. Results of Nonlinear Stiffness and Damping Coefficients

This section presents a comparison of the calculated linear and nonlinear stiffness-damping coefficients with the results from the published Ref. [29]. Figure 9 illustrates the eight first-order stiffness and damping coefficients. A total of 32 selected nonlinear coefficients were compared, and they were distributed in the following manner. The second-order stiffness and damping coefficients are plotted in Figures 10 and 11. The solid line in the figure represents the results modeled in this paper, while the results of Ref. [29] are represented by the dashed line. I, J, O correspond to X and Y, respectively. In contrast, the second-order coefficients reveal significant discrepancies in the cross-coupled stiffness terms.

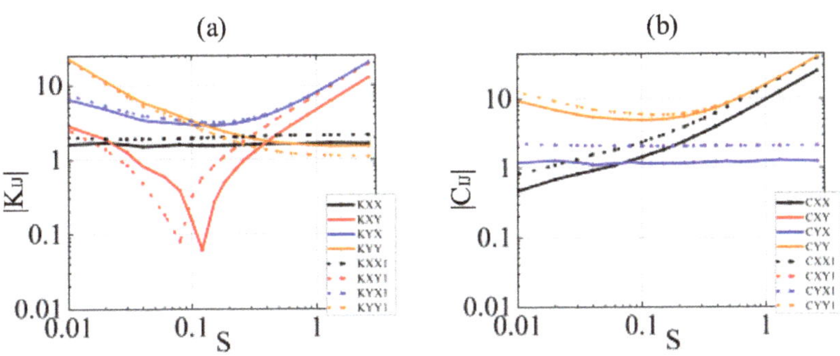

Figure 9. Comparison results of the linear dynamic coefficients: (**a**) stiffness coefficients; (**b**) damping coefficients.

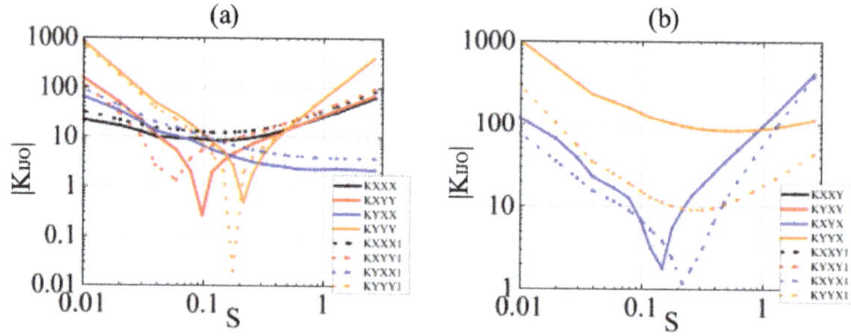

Figure 10. Comparison results of the nonlinear stiffness coefficients: (**a**) stiffness coefficients; (**b**) damping coefficients.

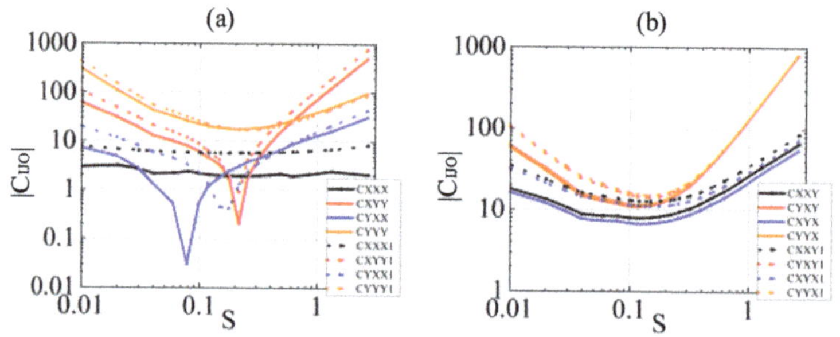

Figure 11. Comparison results of the nonlinear damping coefficients: (**a**) stiffness coefficients; (**b**) damping coefficients.

The simulation results in this paper show good consistency with the literature, regardless of the linear or nonlinear dynamic coefficients. Moreover, the steady-state nonlinear oil film force in Ref. [29] is based on the short bearing theory, which assumes that the axial length of the bearing is much smaller than its diameter ($L/D \ll 1$), and the circumferential pressure gradient can be ignored. Therefore, the steady-state force of the finite-length bearing obtained through numerical integration in this paper is more accurate, especially for the second-order nonlinear cross-stiffness coefficients, which require higher convergence accuracy of the oil film force.

The variation trends of the linear and nonlinear dynamic coefficients with different eccentricities are described in Figures 12–14. In Figure 12a, the absolute values of the linear stiffness and damping coefficient are high when the eccentricity is $e < 0.2$ and $e > 0.85$, and tend to zero near the eccentricity $e = 0.5$. The absolute values of the nonlinear coefficients far exceed the linearized values when the eccentricity is $e < 0.2$ and $e > 0.85$, as illustrated in Figures 13 and 14. In particular, the nonlinear coefficients are almost thousands of times higher than the linear results as the eccentricity approaches 0.05 and 0.95.

Under low eccentricity (representing an extremely high rotational speed), stiffness and damping coefficients demonstrate significant variations along the orbital path. Additionally, reduced eccentricity amplifies the nonlinear characteristics of the shaft's orbital motion relative to high-eccentricity scenarios. Under the condition of high eccentricity (representing an extremely thin oil film), the offset between the shaft and the bearing is larger, and the amplitude of variation in the oil film characteristics increases significantly. The nonlinear dynamic coefficients change significantly, and the bearing's nonlinear effects become very strong, which is of great significance for the subsequent solution of the dynamic motion equation.

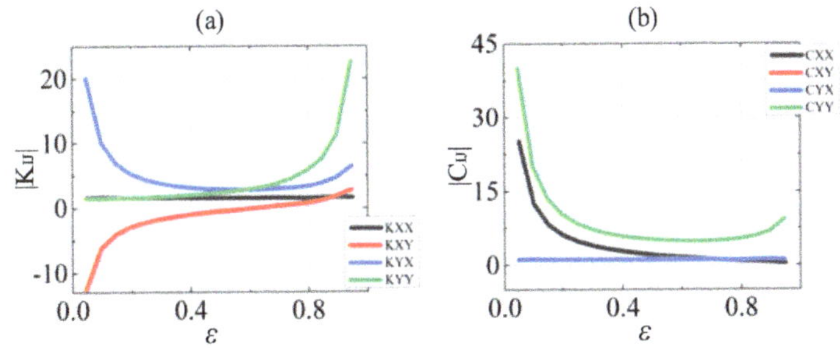

Figure 12. Linear dynamic coefficients with different eccentricities: (**a**) stiffness coefficients; (**b**) damping coefficients.

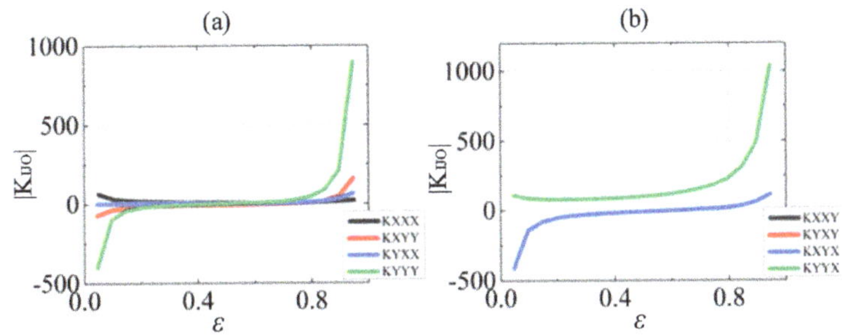

Figure 13. Nonlinear stiffness coefficients with different eccentricities: (**a**) stiffness coefficients; (**b**) damping coefficients.

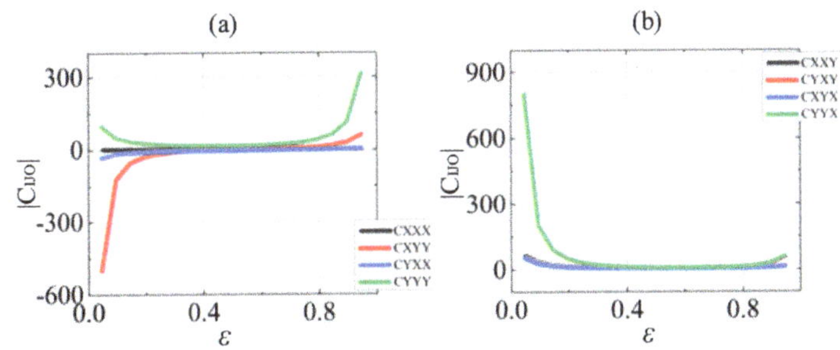

Figure 14. Nonlinear damping coefficients with different eccentricities: (**a**) stiffness coefficients; (**b**) damping coefficients.

3.3. Results of Nonlinear Motion Trajectory

A comparative evaluation of rotor-dynamic behavior and orbital motion characteristics is conducted in this section. When the journal bearing is subjected only to static loads in the vertical direction, the oil film force generated by the lubricant balances the external load. As the rotational speed increases, the eccentricity of the equilibrium position gradually reduces. Figure 15a illustrates the decrease in eccentricity from 0.95 to 0.17 as the rotational speed increases from 60 rpm to 3600 rpm. Figure 15b shows the polar plot of the bearing's equilibrium positions at various rotational speeds, where the journal rotates with a small eccentricity near the bearing's geometric center at 3600 rpm.

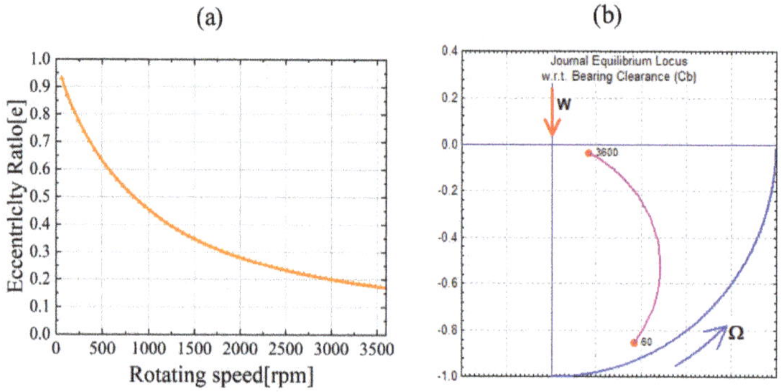

Figure 15. Schematic diagram of the rotor operational position: (**a**) eccentricity variation trend; (**b**) static equilibrium position.

The shaft trajectory is calculated in Figures 16 and 17. The rotor motion using linear coefficients only shows the change in the shaft center position with varying speeds at equilibrium positions, as shown in Figure 16a,b, without depicting the oil whirl features at different speeds. In contrast, the orbit calculated using nonlinear coefficients can represent the typical signatures of the oil whirl and whip, i.e., the shaft shows two nested rings or irregular shapes under real operating conditions in Figure 16c,d. When the rotational speed is below 20 Hz (i.e., when the eccentricity $e > 0.4$), the shape of the shaft orbit calculated using linear stiffness and damping coefficients is similar to that calculated using nonlinear coefficients, and the corresponding equilibrium position orbit is elliptical. However, the shaft centerline position computed with nonlinear stiffness and damping coefficients is more accurate. When the rotational speed exceeds 20 Hz, the orbit computed with nonlinear coefficients significantly differs from the linear results. The shaft orbit in Figure 17 presents a more complex orbit shape, which is closer to the actual shaft centerline orbit at different speeds under real operating conditions.

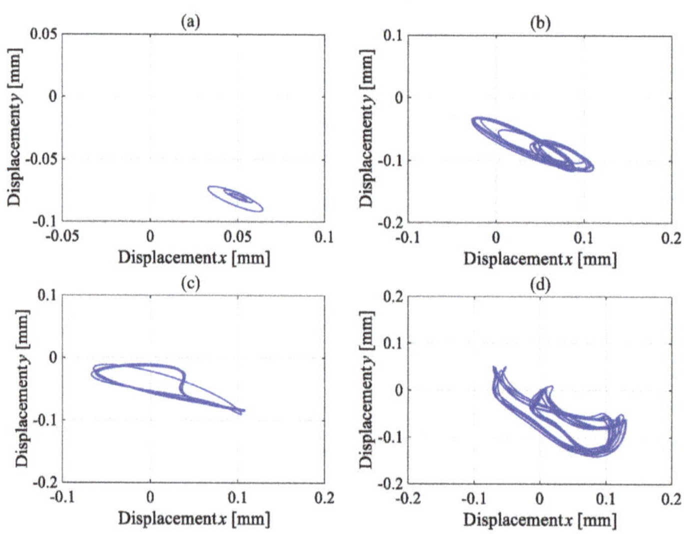

Figure 16. Shaft trajectory with linear and nonlinear coefficients at low rotational speed: (**a**) linear coefficients at speed of 6 Hz; (**b**) linear coefficients at speed of 14 Hz; (**c**) nonlinear coefficients at speed of 6 Hz; (**d**) nonlinear coefficients at speed of 14 Hz.

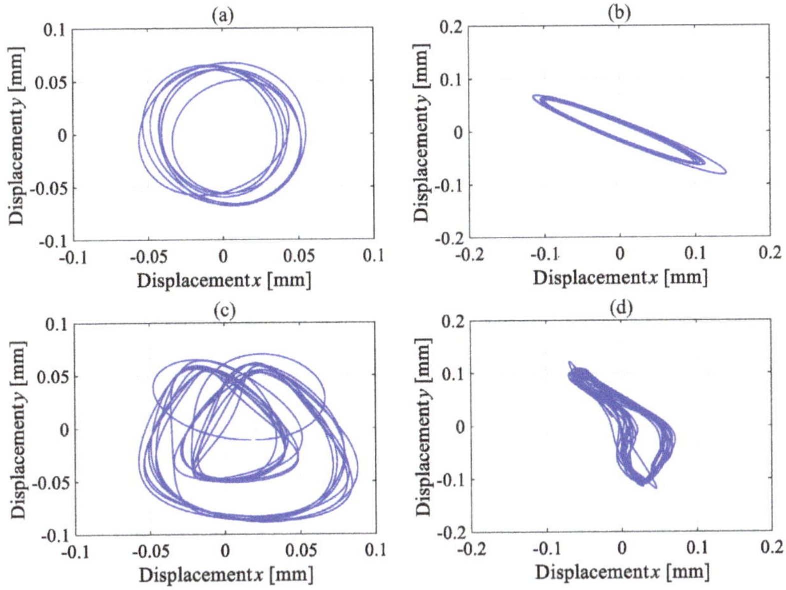

Figure 17. Shaft trajectory with linear and nonlinear coefficients at high rotational speed: (**a**) linear coefficients at speed of 30 Hz; (**b**) linear coefficients at speed of 50 Hz; (**c**) nonlinear coefficients at speed of 30 Hz; (**d**) nonlinear coefficients at speed of 50 Hz.

The frequency spectra of rotor displacement responses are depicted in Figures 18 and 19. At low rotation speeds of 6 Hz and 14 Hz, the rotor dynamic response contains only the rotational frequency, regardless of whether linear or nonlinear coefficients are used. However, when the rotational speed increases to higher speeds of 30 Hz and 50 Hz, the results of linear coefficients fail to represent sub-synchronous motion, i.e., oil whirl and oil whip, and there is no corresponding half frequency of the rotational speed in the spectrum. In contrast, the results of nonlinear coefficients can display the sub-synchronous whirl component at half the rotational frequency. In particular, the amplitude of sub-synchronous components becomes extremely large when the rotational speed is 50 Hz, which means the oil whip occurs.

Through the analysis in this section, it is evident that under fixed operating conditions, when the rotational speed exceeds a certain value (i.e., when the eccentricity is below a certain value), the rotor motion calculated by the nonlinear model can effectively represent the complex shaft centerline orbit at equilibrium positions corresponding to those speeds. The higher rotational speed implies greater disturbance and strong instability of the oil film. The nonlinear coefficient can predict and characterize the nonlinear characteristics in this case accurately. This capability addresses the complex nonlinear issues in hydrodynamic bearings that linear stiffness and damping coefficient models cannot capture.

It should be noted that the three-order dynamic coefficients have an impact near the critical instability threshold. Under extremely high eccentricity, the third-order coefficient has a significant influence on amplitude variation, and the effective stiffness of the system decreases, resulting in a reduced instability threshold. Moreover, high-order cross-coupling terms may excite subharmonic resonance in strongly nonlinear coupling systems, such as multi-span flexible rotors [40]. The second-order perturbation method adopted in this paper is able to characterize the main nonlinear behaviors, such as oil whirl and oil whip.

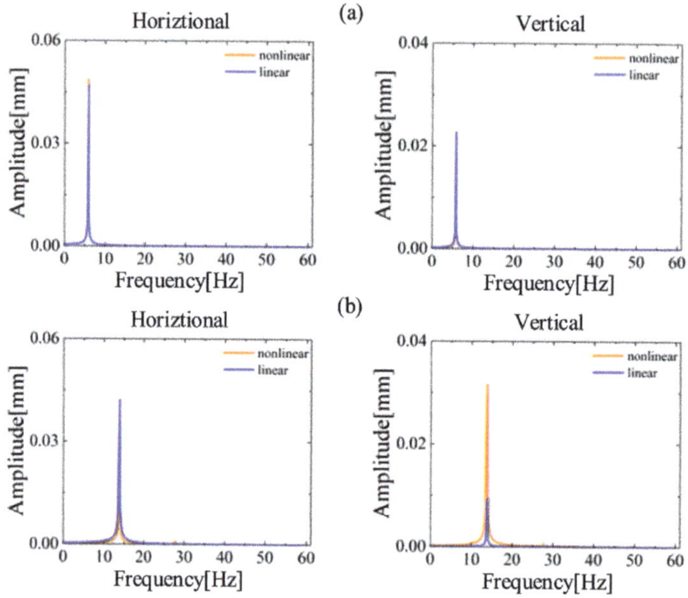

Figure 18. Frequency spectrum at low rotational speeds of 6 Hz and 14 Hz: (**a**) linear coefficients; (**b**) nonlinear coefficients.

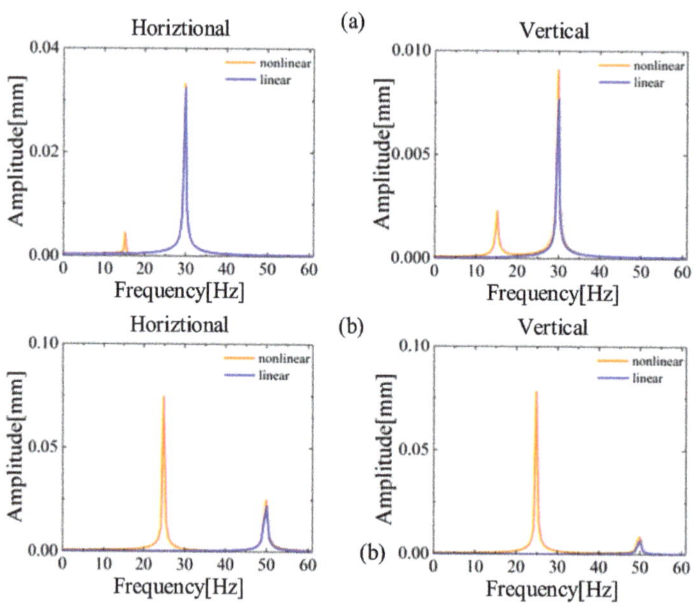

Figure 19. Frequency spectrum at high rotational speed of 30 Hz and 50 Hz: (**a**) linear coefficients; (**b**) nonlinear coefficients.

4. Experimental Verification

4.1. Test Rig Description

In this section, the simulation results are validated through experimental testing on a double-disk rotor–bearing system, as depicted in Figure 20. The test rig consists of two symmetrically mounted steel disks with a diameter of 12 mm, supported by two hydrodynamic journal bearings. The coupling is connected to a variable-frequency drive motor. High-precision laser sensors used to acquire displacement vibration (with a resolution of 0.1 mm) are fixed on brackets, measuring displacement in both the horizontal and vertical directions relative to the bearings.

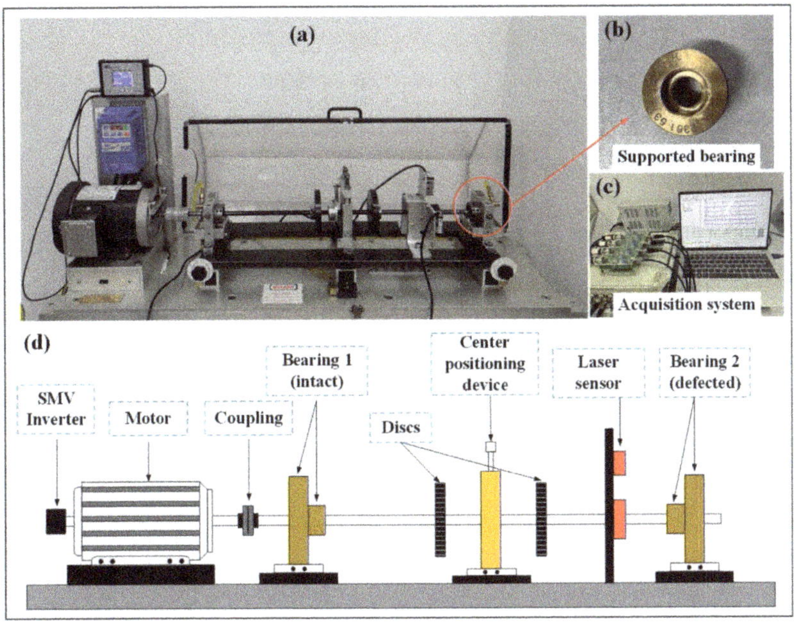

Figure 20. Experiment test rig configuration: (**a**) picture of test rig; (**b**) journal bearing; (**c**) acquisition system; (**d**) scheme of test rig configuration.

The laser detectors transduce optical signals into analog voltages first, which undergo high-speed digitization via a 16-bit data acquisition card to acquire the displacement signals. During the experiment, the synchronous run-up tests are conducted from 1 Hz (60 rpm) to 60 Hz (3600 rpm) in 1 Hz increments, and the sampling rate is set as 10 kHz to ensure the validity of the high-frequency range. Transient stabilization periods of 30 s are enforced before data recording to eliminate transient vibration instability.

4.2. Experimental Data Analysis

The experimental waterfall plots in Figure 21 (horizontal/vertical axes) reveal enhanced spectral complexity relative to numerical predictions, exhibiting multi-order vibration signatures dominated by synchronous (1.0×) and sub-synchronous (0.5×) components alongside higher harmonics. These additional frequency excitations stem from nonlinear dynamic interactions induced by inherent system imperfections, including shaft coupling misalignment effects and bearing clearance-induced parametric instabilities, which are not fully captured in the idealized simulation model. From the waterfall diagram, the variation frequency components of the displacement signals with the increasing rotational speed can be clearly observed.

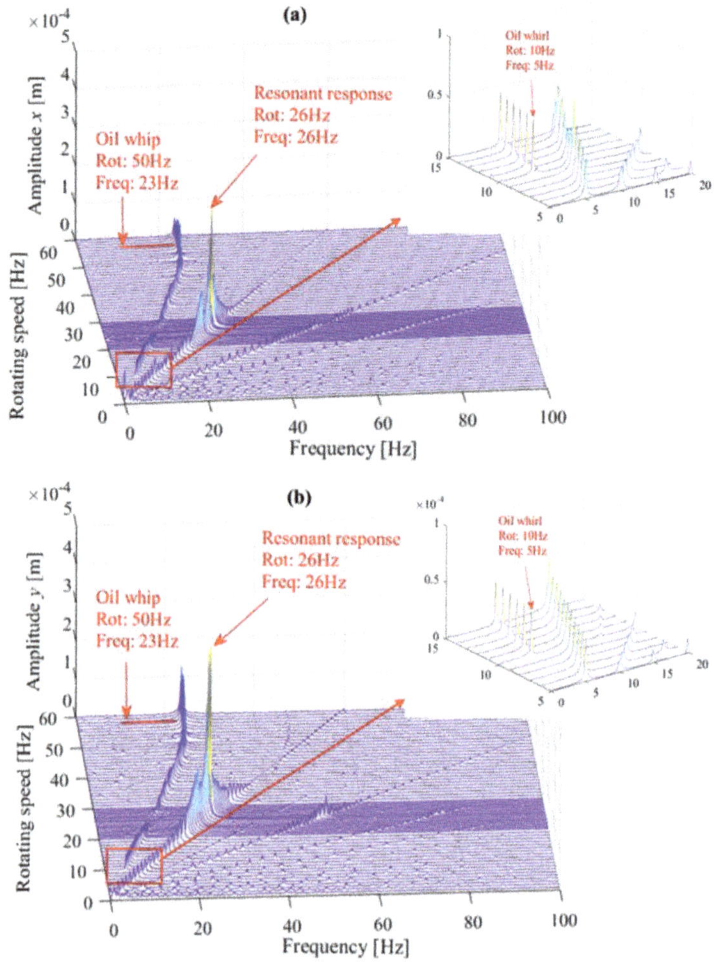

Figure 21. Waterfall diagram of displacement responses: (**a**) horizontal direction; (**b**) vertical direction.

At very low rotational speeds, the frequency components of the rotor's motion are primarily the rotational frequency and low-order harmonics. The oil whirl begins at around a rotational speed of 10 Hz. In this range, the rotor's vortex amplitude is relatively small, and the primary energy remains in the form of synchronous vibrations. The system's first-order critical rotational speed is approximately 26 Hz. The rotor vibrates intensely, and all frequency components other than the rotational frequency are suppressed. When the rotational speed continuously rises to twice the first-order critical rotational speed, oil whip occurs, and the vibration energy of the oil film instability is the dominant component of the whole system. The phenomenon of oil film instability confirms the calculation result of the nonlinear stiffness and damping coefficients.

Figures 22 and 23 display the detailed time domain waves, the shaft trajectory, and the frequency component distribution. The results at the low rotational speeds of 6 Hz are described in Figure 22a–c, and are characterized by an elliptical orbit with the main frequency component corresponding to the rotational speed. The rotor already begins to exhibit oil whirl at rotational speeds of 15 Hz and 30 Hz, as shown in Figures 22d–f and 23a–c, and the shaft orbit presents two nested circular ring shapes due to the existence of a half-rotational frequency characteristic of oil whirl. When the rotational speed is very high and the eccentricity is small, the shaft trajectories in the experimental results become very complex, as illustrated in Figure 23d–f. Oil whip already occurs when the rotational speed reaches 50 Hz. The experimental results align with the simulation results, showing that linear models are only suitable for low speeds, while nonlinear coefficients could accurately characterize nonlinear dynamic behaviors and predict the occurrence of oil film instability.

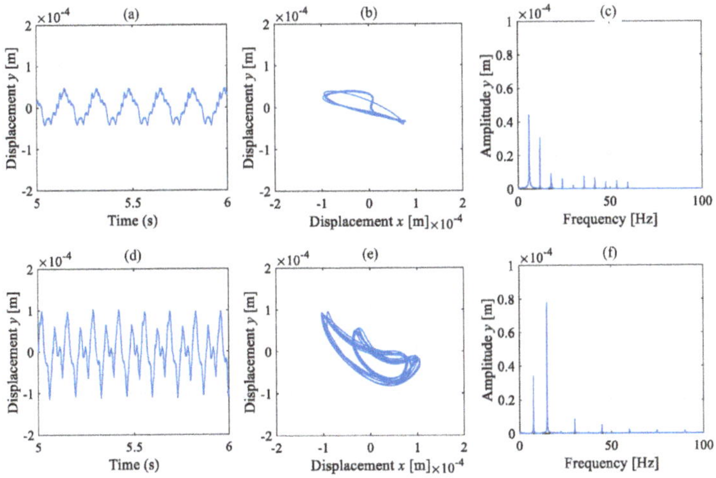

Figure 22. Experimental signals at different rotational speeds: (**a**) time domain signals at 6 Hz; (**b**) shaft trajectory at 6 Hz; (**c**) frequency spectrum at 6 Hz; (**d**) time domain signals at 15 Hz; (**e**) shaft trajectory at 15 Hz; (**f**) frequency spectrum at 15 Hz.

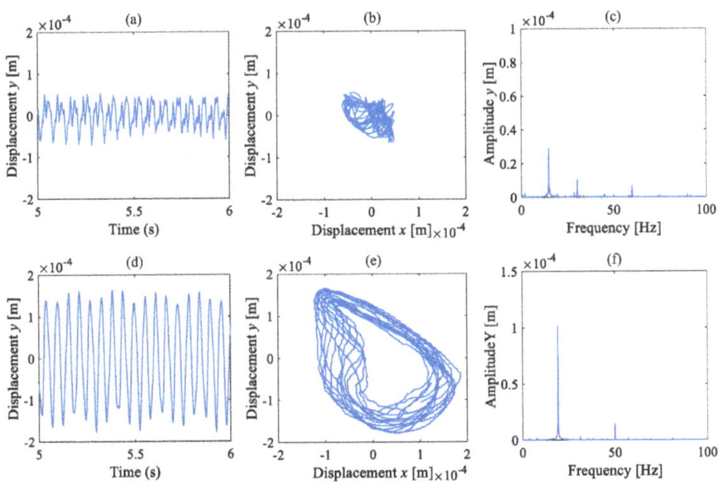

Figure 23. Experimental signals at different rotational speeds: (**a**) time domain signals at 30 Hz; (**b**) shaft trajectory at 30 Hz; (**c**) frequency spectrum at 30 Hz; (**d**) time domain signals at 50 Hz; (**e**) shaft trajectory at 50 Hz; (**f**) frequency spectrum at 50 Hz.

5. Conclusions

This study introduces a model for nonlinear dynamic coefficients using the second-order perturbation method. It examines the influence of various operational parameters on linear and nonlinear stiffness and damping coefficients. Finally, it evaluates the nonlinear dynamic response of a rotor–bearing system using both the linear and nonlinear coefficient models. It compares the predicted shaft trajectory and frequency spectrum with the experimental vibration signals. The conclusions of this paper are as follows:

(1) Compared to the linear coefficients, the nonlinear dynamic coefficients can evaluate the drastic changes in the perturbation reaction force when the Sommerfeld number is very high. The deviation value between the nonlinear and linear coefficient models becomes more significant when the perturbation is large.
(2) The nonlinear coefficient model is more accurate and can effectively capture complex rotor dynamic characteristics and complex motion trajectories under large perturbations and high speeds. The linear coefficients can only evaluate the rotor motion with small perturbations and low speeds.
(3) The experimental results confirm that the nonlinear coefficients can predict and assess significant nonlinear phenomena, such as oil whirl and oil whip. Additionally, the nonlinear dynamic coefficients are effective at identifying the onset frequency of oil film instability.

In conclusion, the quantification of nonlinear dynamic coefficients provides an effective theoretical tool for revealing the instability mechanism of oil films in high-speed rotating machinery. Firstly, this approach can guide the optimal design of bearing geometric parameters (such as clearance and lubricant viscosity), thereby suppressing the risk of instability at the source. Moreover, such nonlinear characteristic parameters can be embedded in the rotor dynamics simulation platform to accurately predict transient vibration behavior and provide condition monitoring strategies by capturing fault signatures such as harmonic frequencies. Furthermore, the proposed model lays a theoretical foundation for active control technologies (such as magnetic bearing compensation and adaptive fuel supply systems) as well as for the reliability design of high-end equipment.

Author Contributions: Conceptualization, Y.C. and Z.Z.; methodology, Y.C. and H.Z.; software, Z.Z.; validation, Y.C. and Z.Z.; investigation, Y.C. and Z.Z.; resources, X.L. and Z.S.; data curation, X.L. and Z.S.; writing—original draft preparation, Y.C.; writing—review and editing, Z.Z. and H.Z.; supervision, H.Z.; funding acquisition, H.Z. All authors have read and agreed to the published version of the manuscript.

Funding: This research was funded by the National Natural Science Foundation of China, grant number 52175085, and the Natural Science Foundation of Hebei Province, grant number E2022202185.

Data Availability Statement: Data are contained within the article.

Conflicts of Interest: The authors declare no conflicts of interest.

References

1. Oliveira, M.V.M.; Daniel, G.B. Vibrational signature of journal bearing oil starvation considering thermal effects and rotor unbalance variation. *Tribol. Int.* **2024**, *191*, 109132. [CrossRef]
2. Chen, Y.; Zhang, H.; Li, X.; Xiao, S.; Gu, F.; Shi, Z. Effects of Wear on Lubrication Performance and Vibration Signatures of Rotor System Supported by Hydrodynamic Bearings. *Lubricants* **2023**, *11*, 107. [CrossRef]
3. Sun, J.; Wang, L.; Li, J.; Li, F.; Li, J.; Lu, H. Online oil debris monitoring of rotating machinery: A detailed review of more than three decades. *Mech. Syst. Signal Process.* **2021**, *149*, 107341. [CrossRef]
4. De Castro, H.F.; Cavalca, K.L.; Nordmann, R. Whirl and whip instabilities in rotor-bearing system considering a nonlinear force model. *J. Sound Vib.* **2008**, *317*, 273–293. [CrossRef]
5. Ahmed, O.; El-Sayed, T.A.; Sayed, H. A Comparative Study of Theoretical and Experimental Analysis on Balanced and Unbalanced Rotors Supported by Oil- and Water-Lubricated Journal Bearings. *Machines* **2024**, *12*, 675. [CrossRef]
6. Lund, J. Review of the concept of dynamic coefficients for fluid film journal bearings. *J. Tribol.* **1987**, *109*, 37–41. [CrossRef]

7. Qiu, Z.; Tieu, A. The effect of perturbation amplitudes on eight force coefficients of journal bearings. *Tribol. Trans.* **1996**, *39*, 469–475. [CrossRef]
8. Czołczyński, K. How to obtain stiffness and damping coefficients of gas bearings. *Wear* **1996**, *201*, 265–275. [CrossRef]
9. Nishimura, A.; Inoue, T.; Watanabe, Y. Nonlinear analysis and characteristic variation of self-excited vibration in the vertical rotor system due to the flexible support of the journal bearing. *J. Vib. Acoust.* **2018**, *140*, 011016. [CrossRef]
10. Choy, F.; Braun, M.; Hu, Y. Nonlinear effects in a plain journal bearing: Part 1—Analytical study. *J. Tribol.* **1991**, *113*, 555–561. [CrossRef]
11. Chu, C.S.; Wood, K.L.; Busch-Vishniac, I.J. A nonlinear dynamic model with confidence bounds for hydrodynamic bearings. *J. Tribol.* **1998**, *120*, 595–604. [CrossRef]
12. Adiletta, G.; Guido, A.R.; Rossi, C. Chaotic motions of a rigid rotor in short journal bearings. *Nonlinear Dyn.* **1996**, *10*, 251–269. [CrossRef]
13. Andréas, L.S.; De Santiago, O. Identification of Journal Bearing Force Coefficients under High Dynamic Loading Centered Static Operation. *Tribol. Trans.* **2005**, *48*, 9–17. [CrossRef]
14. Meruane, V.; Pascual, R. Identification of nonlinear dynamic coefficients in plain journal bearings. *Tribol. Int.* **2008**, *41*, 743–754. [CrossRef]
15. Tian, L.; Wang, W.J.; Peng, Z.J. Dynamic behaviours of a full floating ring bearing supported turbocharger rotor with engine excitation. *J. Sound Vib.* **2011**, *330*, 4851–4874. [CrossRef]
16. Ying, G.; Meng, G.; Jing, J. Turbocharger rotor dynamics with foundation excitation. *Arch. Appl. Mech.* **2008**, *79*, 287–299. [CrossRef]
17. Dakel, M.; Baguet, S.; Dufour, R. Nonlinear dynamics of a support-excited flexible rotor with hydrodynamic journal bearings. *J. Sound Vib.* **2014**, *333*, 2774–2799. [CrossRef]
18. Weimin, W.; Lihua, Y.; Tiejun, W.; Lie, Y. Nonlinear dynamic coefficients prediction of journal bearings using partial derivative method. *Proc. Inst. Mech. Eng. Part J J. Eng. Tribol.* **2012**, *226*, 328–339. [CrossRef]
19. Smolík, L.; Hajžman, M.; Byrtus, M. Investigation of bearing clearance effects in dynamics of turbochargers. *Int. J. Mech. Sci.* **2017**, *127*, 62–72. [CrossRef]
20. Dyk, Š.; Smolík, L.; Hajžman, M. Effect of various analytical descriptions of hydrodynamic forces on dynamics of turbochargers supported by floating ring bearings. *Tribol. Int.* **2018**, *126*, 65–79. [CrossRef]
21. Chasalevris, A. Stability and Hopf bifurcations in rotor-bearing-foundation systems of turbines and generators. *Tribol. Int.* **2020**, *145*, 106154. [CrossRef]
22. Li, Y.; Liang, F.; Zhou, Y.; Ding, S.; Du, F.; Zhou, M.; Bi, J.; Cai, Y. Numerical and experimental investigation on thermohydrodynamic performance of turbocharger rotor-bearing system. *Appl. Therm. Eng.* **2017**, *121*, 27–38. [CrossRef]
23. Singh, A.; Gupta, T.C. Effect of rotating unbalance and engine excitations on the nonlinear dynamic response of turbocharger flexible rotor system supported on floating ring bearings. *Arch. Appl. Mech.* **2020**, *90*, 1117–1134. [CrossRef]
24. Conley, B.; Sadeghi, F. Experimental and Analytical Investigation of Turbocharger Whirl and Dynamics. *Tribol. Trans.* **2020**, *64*, 239–252. [CrossRef]
25. Zhang, Y.; Wang, W.; Wei, D.; Wang, G.; Xu, J.; Liu, K. Coupling analysis of tribological and dynamical behavior for a thermal turbulent fluid lubricated floating ring bearing-rotor system at ultra-high speeds. *Tribol. Int.* **2022**, *165*, 107325. [CrossRef]
26. Dyk, Š.; Rendl, J.; Byrtus, M.; Smolík, L. Dynamic coefficients and stability analysis of finite-length journal bearings considering approximate analytical solutions of the Reynolds equation. *Tribol. Int.* **2019**, *130*, 229–244. [CrossRef]
27. Dyk, Š.; Smolík, L.; Rendl, J. Predictive capability of various linearization approaches for floating-ring bearings in nonlinear dynamics of turbochargers. *Mech. Mach. Theory* **2020**, *149*, 103843. [CrossRef]
28. Smolík, L.; Dyk, Š. Towards efficient and vibration-reducing full-floating ring bearings in turbochargers. *Int. J. Mech. Sci.* **2020**, *175*, 105516. [CrossRef]
29. Miraskari, M.; Hemmati, F.; Gadala, M.S. Nonlinear Dynamics of Flexible Rotors Supported on Journal Bearings—Part I: Analytical Bearing Model. *J. Tribol.* **2018**, *140*, 021704. [CrossRef]
30. Miraskari, M.; Hemmati, F.; Gadala, M.S. Nonlinear Dynamics of Flexible Rotors Supported on Journal Bearings—Part II: Numerical Bearing Model. *J. Tribol.* **2018**, *140*, 021705. [CrossRef]
31. Sayed, H.; El-Sayed, T.A. A novel method to evaluate the journal bearing forces with application to flexible rotor model. *Tribol. Int.* **2022**, *173*, 107593. [CrossRef]
32. Sayed, H.; El-Sayed, T.A. Nonlinear dynamics and bifurcation analysis of journal bearings based on second order stiffness and damping coefficients. *Int. J. Non-Linear Mech.* **2022**, *142*, 103972. [CrossRef]
33. Ma, H.; Wang, X.; Niu, H.; Wen, B. Oil-film instability simulation in an overhung rotor system with flexible coupling misalignment. *Arch. Appl. Mech.* **2015**, *85*, 893–907. [CrossRef]
34. Hu, L.; Liu, Y.; Zhao, L.; Zhou, C. Nonlinear dynamic behaviors of circumferential rod fastening rotor under unbalanced pre-tightening force. *Arch. Appl. Mech.* **2016**, *86*, 1621–1631. [CrossRef]

35. Machado, T.H.; Storti, G.C. Nonlinear model for wear effects in hydrodynamic bearings applied to rotating systems. In *Nonlinear Dynamics of Structures, Systems and Devices: Proceedings of the First International Nonlinear Dynamics Conference (NODYCON 2019), Roma, Italy, 17–20 February 2019*; Springer International Publishing: Berlin/Heidelberg, Germany, 2019; Volume I, pp. 561–568.
36. Feng, H.; Jiang, S.; Ji, A. Investigations of the static and dynamic characteristics of water-lubricated hydrodynamic journal bearing considering turbulent, thermohydrodynamic and misaligned effects. *Tribol. Int.* **2019**, *130*, 245–260. [CrossRef]
37. Zhu, D.; Jane Wang, Q. Effect of roughness orientation on the elastohydrodynamic lubrication film thickness. *J. Tribol.* **2013**, *135*, 031501. [CrossRef]
38. Venner, C.H.; Lubrecht, A.A. (Eds.) *Multi-Level Methods in Lubrication*; Elsevier: Amsterdam, The Netherlands, 2000; Volume 37.
39. Ebrat, O.; Mourelatos, Z.P.; Vlahopoulos, N.; Vaidyanathan, K. Calculation of journal bearing dynamic characteristics including journal misalignment and bearing structural deformation©. *Tribol. Trans.* **2004**, *47*, 94–102. [CrossRef]
40. El-Sayed, T.A.; Sayed, H. Bifurcation analysis of rotor/bearing system using third-order journal bearing stiffness and damping coefficients. *Nonlinear Dyn.* **2021**, *107*, 123–151. [CrossRef]

machines

MDPI

Article

Research on Unbalanced Vibration Characteristics and Assembly Phase Angle Probability Distribution of Dual-Rotor System

Hui Li [1], Changzhi Shi [1], Xuejun Li [1,*], Mingfeng Li [2] and Jie Bian [3]

1 Key Laboratory of Rotor Vibration Monitoring and Diagnosis Technology for Machinery Industry, Foshan University, Foshan 528000, China; hnustlh06@163.com (H.L.); lhscz0110@163.com (C.S.)
2 Aecc South Industry Company Limited, Zhuzhou 412002, China; lmf18007330425@163.com
3 AVIC Hunan Power Machinery Research Institute, Zhuzhou 412002, China; hrbeu@163.com
* Correspondence: hnkjdxlxj@163.com

Abstract: This paper addresses the complex issue of vibration response characteristics resulting from the unbalanced assembly of the double rotors in the 31F aero-engine. The study investigates the vibration response behavior of the dual-rotor system through the adjustment of rotor assembly phase angle. Initially, a dynamic model of the four-disk, five-pivot dual-rotor system is established, with its natural frequencies and vibration modes verified. The influence of size and the position of the unbalance on the vibration amplitude in the dual-rotor system is analyzed. Additionally, the probability distribution of the assembly phase angles for both the compressor and turbine sections of the low-pressure rotor is examined. The results indicate that for the low-pressure rotor exhibiting excessive vibration, adjusting the assembly phase angle of the rotors' system's compressor or the turbine section by 180 degrees leads to a vibration qualification rate of 70.1435%. This finding is consistent with the observations from the field experience method used in the former Soviet Union. Finally, corresponding experimental verification is conducted.

Keywords: dual-rotor system; rotor unbalance; assembly phase angle; vibration amplitude; aero-engines

1. Introduction

The aero-engine rotor is a complex rotating system that operates under variable load conditions, with vibration issues accounting for over 70% of aero-engine failures [1]. Rotor imbalance often arises from manufacturing and assembly imperfections, such as mass eccentricity due to material inconsistencies, unbalanced couplings, or asymmetric keyways [2]. Unbalance excitation serves as the primary cause of rotor vibration [3], frequently resulting in premature repairs or even engine delivery constraints due to excessive vibration levels. This highlights the critical need for rigorous vibration management to enhance safety, reliability, lifespan, and performance. Achieving the dynamic balance of the rotor becomes an indispensable aspect within the engine's manufacturing and assembly process.

Dynamic modeling plays a crucial role in the investigation of aero-engine rotor balance, as it facilitates the identification of causes and effects related to rotor imbalance [4–6]. Lu et al. [7] have developed a comprehensive dynamic model for a dual-rotor system that incorporates nonlinear effects arising from ball bearings, thereby establishing the foundation for designing bearing–dual-rotor systems. Yu et al. [8] have constructed an advanced dynamic model for an aero-engine dual-rotor system, revealing that turbine fan blades not only introduce sudden imbalances but also lead to rotor–stator rubbing. Guan et al. [9] have extensively investigated the dynamic modeling of dual-rotor systems by considering gear meshing forces and uncovering nonlinear behavior resulting from gear-mesh-induced rubbing. Chen et al. [10] have successfully modeled a flexible rotor system with squeeze film dampers under dual-frequency excitations using multiple harmonic balance and Runge–Kutta methods to analyze periodic and quasi-periodic solutions of the system. Building on these modeling approaches, the next challenge is to address rotor imbalance, which has been a subject of various studies [11,12].

Addressing the issue of rotor imbalance, Kang et al. [13] investigated the rotational characteristics of dual rotors influenced by nonlinear bearings. Their simulation and experimental results demonstrated a strong correlation between the severity of rotor imbalance and the occurrence of reverse rotation. Zhang et al. [14] employed turbulent lubrication theory to develop a numerical model of a turbocharger rotor and bearing system, exploring how adjustments in imbalance and clearance can optimize system performance. Yang et al. [15] developed a dynamic model of a double-disk rotor system, accounting for multiple imbalances and multi-fixed-point friction impacts. Their study unveiled nonlinear dynamic behavior as rotational speed increased and analyzed the influence of unbalanced force direction on rotor vibration. Wang et al. [16] investigated how the direction of imbalance affects the dynamic behavior of a high-speed turbocharger double overhang rotor system. Song et al. [17] developed a bilateral imbalance prediction model for multi-stage assembled rotors, aimed at enhancing the initial assembly qualification rate and improving both the assembly accuracy and efficiency of the rotor system. While considerable research has examined the magnitude and location of imbalances [18], another key factor influencing rotor vibration is the assembly process, which can introduce additional imbalances [19].

In practical terms, the assembly of advanced engine rotors plays a crucial role in their design and manufacturing process. Effective control during assembly ensures that the rotor's imbalance remains within acceptable limits, which is essential for minimizing vibration in both the rotor and the entire engine [20,21]. Zhang et al. [22] optimized the assembly sequence by enhancing the algorithm for a multi-stage rotor assembly model, resulting in an impressive 84.22% reduction in residual imbalance. Yang et al. [23] introduced a linear assembly method that effectively mitigates error propagation through careful selection of measurement accuracy and the refinement of the assembly process. Zhang et al. [24] developed a geometric error propagation model specifically designed for multi-stage rotors to accurately calculate mass eccentricity vector and unbalanced forces, proposing an approximate expression model to optimize rotor assembly. In conclusion, these advancements in rotor assembly techniques and models have contributed to improving the efficiency of engine rotor systems [25].

The aero-engine rotor typically comprises a compressor rotor and a turbine rotor, which are integrated into a unified unit through coupling [26]. The turbine operates at high speeds, driven by high-temperature and high-pressure gas, providing power to the compressor. In practical rotor balancing, due to structural and assembly limitations, the compressor and turbine sections are balanced independently before being assembled as a whole. However, during high-speed rotation, this assembly introduces complex coupling imbalances. Consequently, even if the rotor is well-balanced on the balancer, it may still exhibit excessive vibration when installed in the engine. Investigating the impact of the assembly phase angle on overall rotor system vibration is an important research topic. Wang et al. [27] developed a dynamic model for a turboprop rotor and demonstrated that adjusting the phase combination of the unbalance can effectively mitigate excessive vibration in the turboprop engine. However, their study did not address the probability distribution of the assembly phase angle.

To address the challenges posed by the impact of the assembly phase angle, this paper is structured as follows: Section 2 establishes a dynamic model of a four-disk, five-support dual-rotor system using finite element analysis with theoretical validation and simulation verification of its accuracy. Section 3 conducts a dynamic balance analysis of the single rotor system while exploring the probability distribution associated with assembly phase angles. Experimental verification is presented in Section 4, followed by conclusions provided in Section 5.

The primary contribution of this study is the investigation of vibration control in dual-rotor systems by adjusting the assembly phase angle between the compressor and turbine sections of the low-pressure rotor. The results reveal that a 180° phase adjustment significantly improves vibration qualification rates, achieving over 70% alignment with field practices. Furthermore, this paper uniquely explores the probability distribution of assembly phase angles, providing a theoretical foundation that validates the practical methods previously applied in the former Soviet Union. These insights offer a deeper understanding of rotor dynamics and serve as a valuable guide for optimizing rotor assembly processes in aero-engine applications.

2. Dynamic Modeling of Dual-Rotor System

2.1. Model Simplification

The 31F aero-engine dual-rotor design is utilized in high-performance aircraft, where efficient power output and minimal vibration are of utmost importance. These dual-rotor systems find extensive application in military and high-speed civilian aircraft to ensure stability and reliability even under challenging operational conditions. The configuration of the 31F aero-engine dual-rotor system is depicted in Figure 1. This system comprises a low-pressure rotor and a high-pressure rotor, interconnected by an intermediate bearing. The low-pressure rotor consists of a low-pressure fan and a low-pressure turbine, which are connected via a sleeve bearing. Conversely, the high-pressure rotor encompasses multiple components including a nine-stage compressor disk, a one-stage sealing disk, a high-pressure turbine shaft, and a one-stage turbine disk. A rigid connection between the high-pressure turbine disk and the high-pressure compressor is established through the high-pressure turbine shaft to ensure robust interlinkage among these critical elements.

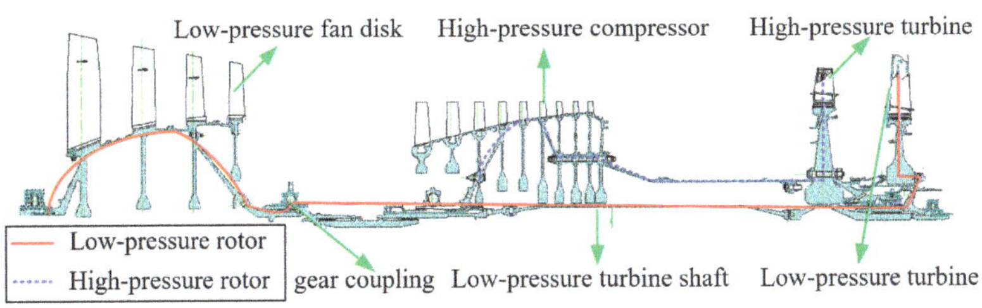

Figure 1. Structure of 31F aero-engine dual-rotor system.

By implementing rational structural reductions and conducting dynamic approximation analysis, the dual-rotor system of the 31F engine is simplified. A dynamic model of this system is established to represent a four-disk, five-support concentrated mass dual-rotor system. This model is utilized to investigate the coupling characteristics of unbalanced vibrations in the rotor system, as depicted in Figure 2.

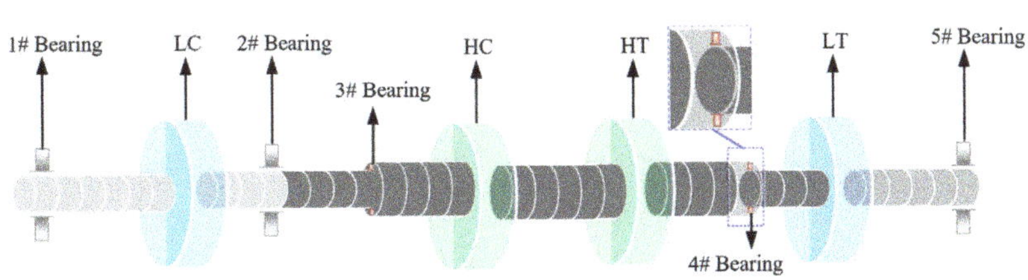

Figure 2. Simplified model of dual-rotor system.

The compressor and turbine disks are represented as rotating disks with equivalent masses and moments of inertia in the model. The four-stage fan in the low-pressure rotor is denoted as disk LC, while the low-pressure turbine disk is modeled as disk LT. The high-pressure rotor's nine-stage compressor disk is simplified to disk HC, and the high-pressure turbine disk is referred to as disk HT. This simplification reduces the actual multi-disk structure to a few-disk structure with a total of five supports. The low-pressure rotor consists of a low-pressure fan section and a low-pressure turbine rotor, with a support configuration of 1-1-1. The low-pressure fan section features two fulcrums and four-stage fans, where support 1# utilizes roller bearings for elastic support, while support 2# employs ball bearings for rigid support. The low-pressure turbine rotor includes one fulcrum at position 5#, which also uses ball bearings for elastic support. On the other hand, the high-pressure rotor adopts a support mode of 1-0-1, where supports 4# and 5# serve as rigid intermediate supports using roller bearings.

2.2. Dynamic Modeling Analysis

The Lagrange dynamic equation is expressed as [28]

$$\frac{\mathrm{d}}{\mathrm{d}t}\left(\frac{\partial T_{total}}{\partial \dot{\mathbf{q}}}\right) - \frac{\partial T_{total}}{\partial \mathbf{q}} + \frac{\partial V_{total}}{\partial \mathbf{q}} = \mathbf{F}_{total} \quad (1)$$

where T_{total} denotes the kinetic energy of the system, V_{total} denotes the potential energy, $\mathbf{F}_{total}$ denotes the external force, and $\mathbf{q}$ denotes the generalized coordinate vector of the system.

With the described simplifications, the finite element method models the rotor system to accurately compute the dynamic response at each node along the shaft, enabling effective comparison and validation against experimental results. The dynamic model of the dual-rotor system is illustrated in Figure 3. The low-pressure rotor system consists of 29 units and 30 nodes, with LC and LT positioned at the 5th and 28th nodes, respectively. The bearings are simplified as linear spring supports located at nodes 1, 9, and 30, with their stiffness represented by K_{b1}, K_{b2}, and K_{b3}. The high-pressure rotor system is divided into 11 units and 12 nodes, with HC and HT placed at the 3rd and 10th nodes, respectively. An intermediate bearing, denoted as K_{b4}, connects the 14 nodes of the low-pressure rotor and the 12 nodes of the high-pressure rotor.

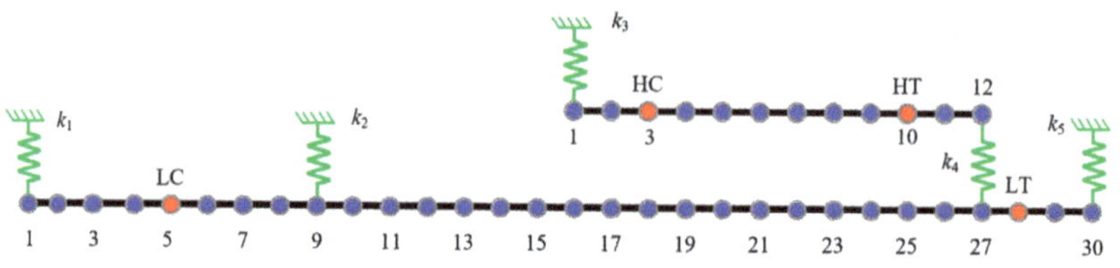

Figure 3. Dual-rotor dynamics model.

The dual-rotor system is discretized into a sequence of shaft units, turntable units, and elastic support stiffness units, resulting in the formation of the total mass matrix $\mathbf{M}_s$, stiffness matrix $\mathbf{K}_s$, damping matrix $\mathbf{C}_s$, gyroscopic matrix $\mathbf{G}_s$, and unbalanced force matrix $\mathbf{Q}_s$. The overall dynamic equation is then established in Equation (2), followed by the application of the Newmark-β numerical integration method to solve for the dynamic parameters.

$$\mathbf{M}\ddot{\mathbf{U}} + (\omega\mathbf{G} + \mathbf{C})\dot{\mathbf{U}} + \mathbf{K}\mathbf{U} = \mathbf{Q} \quad (2)$$

Drawing on the theory of rotating beam elements, each node in every segment of the shaft possesses three translational degrees of freedom and three rotational degrees of freedom [29]. The calibrated coordinate vector represents these degrees of freedom for any given node, as depicted in Figure 4a.

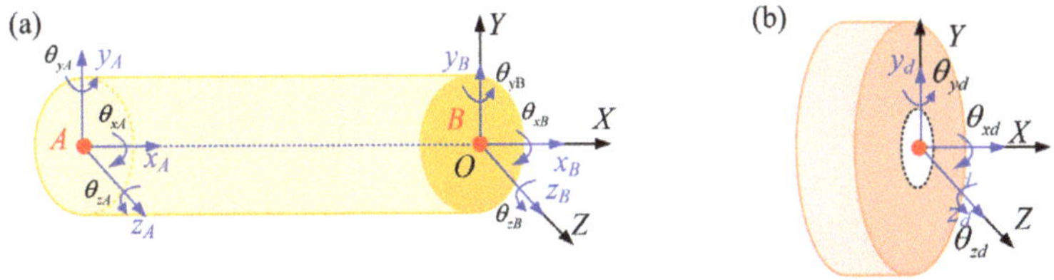

Figure 4. Element model: (**a**) shaft segment element, (**b**) the disk element.

The generalized displacement coordinates of any axis shaft segment element can be written as

$$\mathbf{u}_s = \begin{bmatrix} x_A & y_A & z_A & \theta_{xA} & \theta_{yA} & \theta_{zA} & x_B & y_B & z_B & \theta_{xB} & \theta_{yB} & \theta_{zB} \end{bmatrix}^T \tag{3}$$

The motion of each beam element in the global coordinate system OXYZ can be described as

$$\mathbf{M}_s\ddot{\mathbf{U}}_s + \omega\mathbf{G}_s\dot{\mathbf{U}}_s + \mathbf{K}_s\mathbf{U}_s = \mathbf{Q}_s \tag{4}$$

where $\mathbf{M}_s$, $\mathbf{G}_s$, and $\mathbf{K}_s$ represent the mass matrix, gyroscopic force matrix, and stiffness matrix of each shaft element, respectively. ω denotes the angular speed of the shaft element's rotation around the central axis, and $\mathbf{Q}_s$ represents the external excitation acting on the beam element.

The disk is typically treated as a rigid body exhibiting rotational effects, allowing it to be simplified into a concentrated mass and moment of inertia acting at its center of mass. The centroid node possesses six degrees of freedom: three translational displacements along the x_d, y_d, and z_d directions, and three rotational displacements about these axes. The simplified lumped mass element of the disk, therefore, includes these six degrees of freedom, which can be represented by the vector $u_d = \begin{bmatrix} x_d & y_d & z_d & \theta_{xd} & \theta_{yd} & \theta_{zd} \end{bmatrix}$. After the lumped mass simplification, the movement of the disk can be described as

$$\mathbf{M}_d\ddot{\mathbf{U}}_{di} + \omega\mathbf{G}_d\dot{\mathbf{U}}_{di} = \mathbf{Q}_d \tag{5}$$

where $\mathbf{U}_{di}$ represents the displacement of the disk element. $\mathbf{Q}_d$ is the external excitation of the disk element, $\mathbf{M}_d$ and $\mathbf{G}_d$ are the inertial matrix and gyroscopic force matrix of the disk element, respectively.

By disregarding the effects of bearing clearance, oil film thickness, and other factors, the five bearings are simplified into spring-damping units that act on the rotor fulcrum. The motion of the supporting units within the rotor system is described by

$$\mathbf{M}_b\ddot{\mathbf{U}}_{bi} + \mathbf{K}_b\mathbf{U}_{bi} = \mathbf{Q}_b \tag{6}$$

where $\mathbf{U}_{bi}$ represents the displacement of the support unit node in the dual-rotor system, $\mathbf{Q}_b$ denotes the external excitation force acting on this node, and $\mathbf{K}_b$ signifies the stiffness of the support element. In order to simplify calculations, this study focuses solely on radial stiffness and damping while neglecting cross stiffness and cross damping effects. Based on empirical data and experimental validation, the stiffness values of the spring elements in the rotor model are assigned as follows: $\mathbf{K}_{b1}$ = 0.9 × 10^7 N/m, $\mathbf{K}_{b2}$ = 2.5 × 10^8 N/m, $\mathbf{K}_{b3}$ = 0.3 × 10^7 N/m, $\mathbf{K}_{b4}$ = 2.5 × 10^8 N/m, and $\mathbf{K}_{b5}$ = 0.3 × 10^7 N/m.

The damping matrix of the dual-rotor system is set by Rayleigh damping, which is expressed as

$$\mathbf{C} = \alpha\mathbf{M} + \beta\mathbf{K} \tag{7}$$

where α and β are damping ratio coefficients, respectively, which can be written as

$$\begin{aligned} \alpha &= 2\left(\frac{\xi_{r2}}{\omega_{r2}} - \frac{\xi_{r1}}{\omega_{r1}}\right) / \left(\frac{1}{\omega_{r2}^2} - \frac{1}{\omega_{r1}^2}\right) \\ \beta &= 2(\xi_{r2}\omega_{r2} - \xi_{r1}\omega_{r1}) / \left(\omega_{r2}^2 - \omega_{r1}^2\right) \end{aligned} \tag{8}$$

where ξ_{r1} and ξ_{r2} represent the damping ratios, while ω_{r1} and ω_{r2} denote the first two natural frequencies of the dual-rotor system.

2.3. Model Verification

Using ADAMS 2020 simulation software, a simplified dual-rotor model is established as a four-disk, five-point rotor system, as shown in Figure 5. The compressor shaft and turbine shaft of the low-pressure rotor measure 381 mm and 923 mm in length, respectively, with an outer diameter of 30 mm each. The high-pressure shaft has an inner diameter of 40 mm, an outer diameter of 65 mm, and a length of 680 mm. Additionally, Table 1 provides the size parameters for the four equivalent disks, LC, LT, HC, and HT.

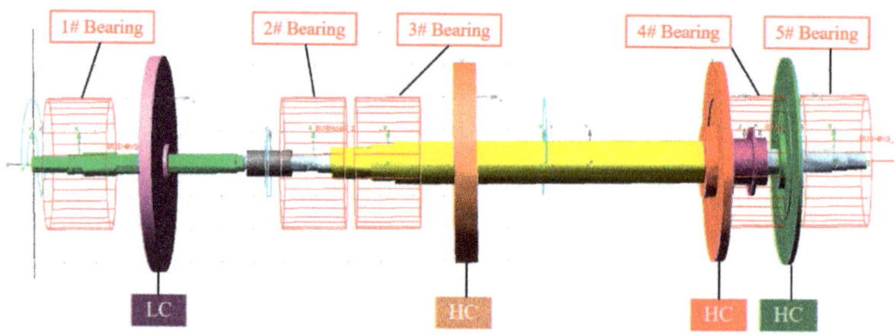

Figure 5. ADAMAS model of dual-rotor system.

Table 1. Equivalent disk parameters.

	Iner Diameter/mm	Outer Diameter/mm	Thickness/mm	Mass/Kg
LC	35	330	18	14.0
LT	48	316	10	11.3
HC	73	305	35	20
HT	62	304	14	10

The natural frequencies by order are shown in Table 2, where theoretical and simulation results are closely aligned, with a maximum error of 5%, validating the accuracy of the dynamic model.

Table 2. Comparison of natural frequency.

	First Order/Hz	Second Order/Hz	Third Order/Hz
Theory	55.9	75.5	122
Simulation	58	78.9	128.1
Error	4.1%	4.5%	5%

3. Analysis of the Influence of Rotor Dynamic Unbalance

3.1. Influence of Rotor Unbalance Value and Position

The rotor system features a completely axisymmetric design; however, factors such as uneven rotor material, unbalanced coupling, asymmetric keyways, and machining deviations introduce a degree of unbalance in the rotor system. It is assumed that the unbalanced mass exists in the y and z directions of the six degrees of freedom at the j-th node. In the OXYZ coordinate system, the unbalanced position fault model is established as

$$Q^{(j)} = [0 \quad \ldots \quad \underbrace{0 \quad i \quad 1 \quad 0 \quad 0 \quad 0}_{j\text{-}th \quad node} \quad \ldots \quad 0]^{\mathrm{T}} mr(j)\omega^2 e^{i\varphi_j} \tag{9}$$

where $Q^{(j)}$ represents the unbalanced force at the j-th node, $mr(j)$ denotes the unbalanced mass at the j-th node, and φ_j indicates its phase angle.

The unbalanced positions in this section are assigned to four equivalent disks: LC, LT, HC, and HT. To determine the appropriate measurement points for unbalanced vibration analysis, an investigation into the vibration modes of the dual-rotor system is conducted.

Based on the dual-rotor dynamics model established in the previous section, a modal analysis is conducted. The first three modes of the low-pressure rotor are illustrated in Figure 6, with corresponding resonance frequencies of 76.9 Hz, 122.6 Hz, and 150.6 Hz. The primary mode is characterized by bending of the turbine rotor, while the second mode involves both the bending of the fan rotor and turbine rotor. The third mode primarily represents the bending of the turbine. Figure 7 shows the first three vibration modes of the high-pressure rotor, with corresponding resonance frequencies of 60.4 Hz, 96.4 Hz, and 388.2 Hz. For the high-pressure rotor, the first-order mode is a translational mode, the second-order mode is a pitch mode, and the third-order mode is a bending mode. The first three vibration modes of the dual-rotor system are presented in Figure 8, with natural frequencies of 55.9 Hz, 75.5 Hz, and 122 Hz. In this system, the first-order mode combines the bending mode of the low-pressure turbine rotor and the translational mode of the high-pressure rotor. The second-order mode includes the bending mode of the low-pressure turbine rotor and the pitch mode of the high-pressure rotor. The third-order mode encompasses the bending mode of the fan rotor, the low-pressure turbine rotor, and the pitch mode of the high-pressure rotor.

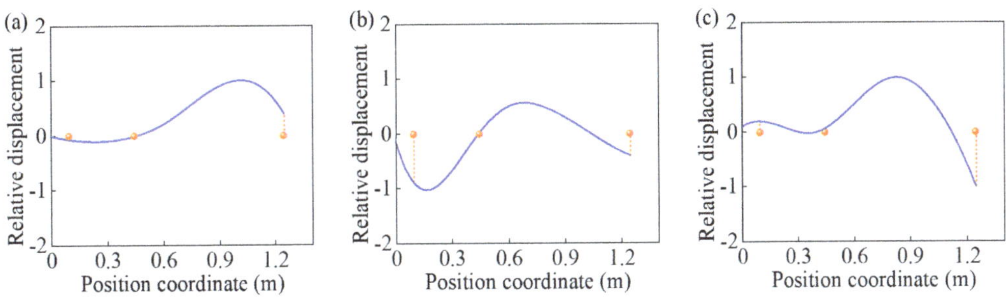

Figure 6. The first three order vibration modes of the low-pressure rotor: (**a**) first order, (**b**) second order, (**c**) third order.

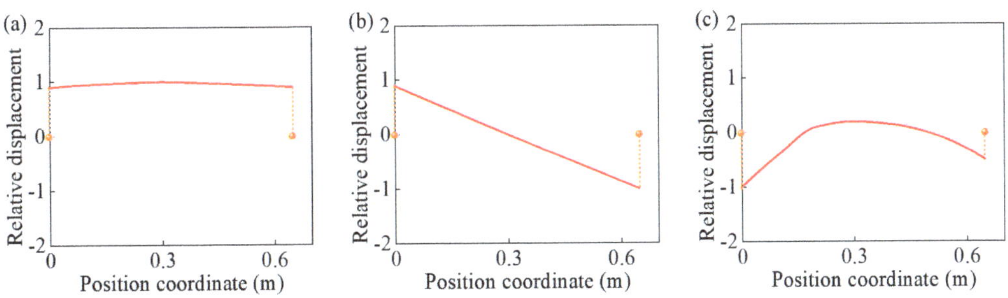

Figure 7. The first three order vibration modes of the high-pressure rotor: (**a**) first order, (**b**) second order, (**c**) third order.

Based on the first three modes of the low-pressure rotor, high-pressure rotor, and dual-rotor system, we have identified the main node and anti-node positions of vibration. Measuring points have been established in close proximity to these locations. The measuring points are strategically positioned at critical bearing nodes that exhibit significant vibration characteristics (including nodes 1, 3, 4, and 5), as well as at the four disk nodes. This results in a total of eight designated measuring points numbered from 1 to 8. Furthermore, it is worth noting that the operating speeds for the low-pressure rotor and high-pressure rotor are set at 1200 rpm and 1800 rpm, respectively.

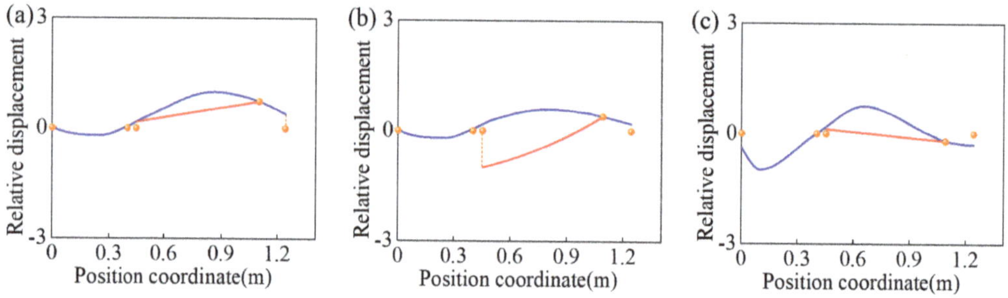

Figure 8. The first three order vibration modes of the dual-rotor system: (**a**) first order, (**b**) second order, (**c**) third order.

First, the influence of the unbalance position and magnitude on the vibration response of the dual-rotor system is examined. Unbalance is applied to the LC and LT positions of the low-pressure rotor, gradually increasing from 100 g·mm to 1000 g·mm. The resulting amplitudes at the eight measuring points are illustrated in Figure 9a,b. As the unbalance increases, the amplitude at all measuring points exhibits a gradual upward trend. Notably, in addition to the observation points directly on the two disks, the amplitudes at the measuring points of the 4# and 5# bearings are also significantly larger. This indicates that bearings 4 and 5 are highly sensitive to unbalance faults at the LC and LT positions.

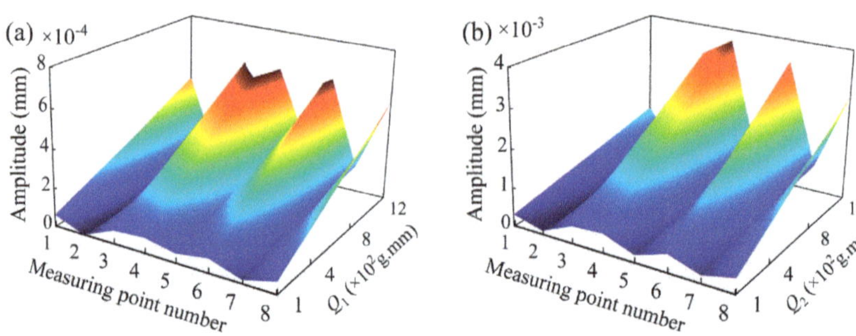

Figure 9. Single-disk unbalance of low-pressure rotor: (**a**) LC disk, (**b**) LT disk.

Similarly, when unbalance is applied to the HC and HT positions of the high-pressure rotor, the amplitudes at the corresponding measuring points are shown in Figure 10a and 10b, respectively. For the HC position, the amplitude at the 3# bearing shows significant variation, indicating high sensitivity to unbalance at this position. In contrast, when the unbalance is applied to the HT position, the amplitudes at the 3# and 4# bearings demonstrate considerable changes, suggesting that both bearings are also very sensitive to unbalance faults at the HT position. Therefore, these bearings can serve as effective measuring points for monitoring unbalance at the HT position.

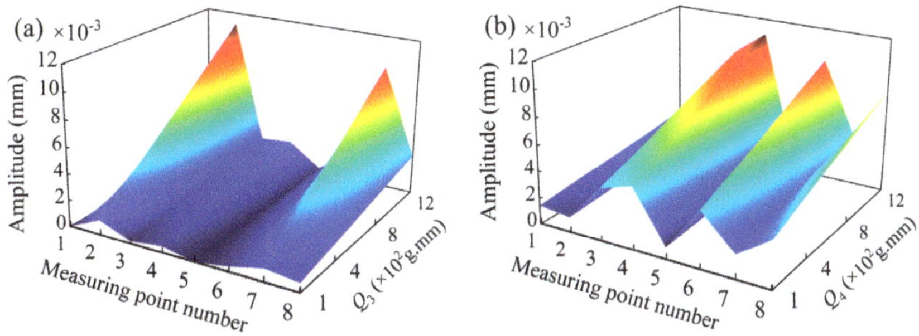

Figure 10. Single-disk unbalance of high-pressure rotor: (**a**) LC disk, (**b**) LT disk.

Next, the vibration response characteristics of the rotor system with two unbalanced faults are analyzed. As the unbalance of LC and HC gradually increases from 100 g·mm to 1000 g·mm, while maintaining the stable operating speeds of the low-pressure rotor at 1200 rpm and the high-pressure rotor at 1800 rpm, the amplitude changes at the eight key measuring points are recorded in detail. These changes are visually presented in Figure 11a,b.

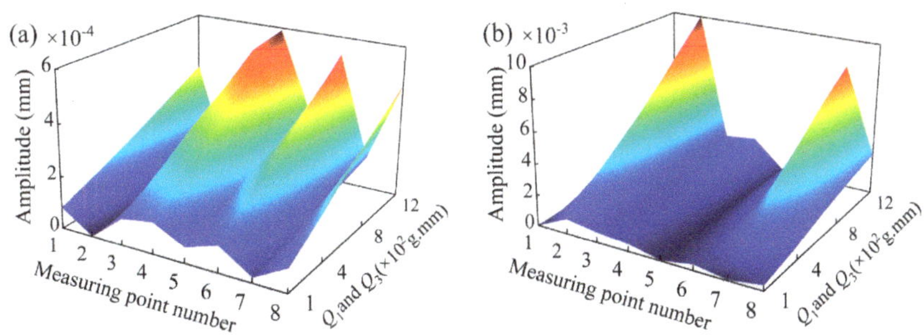

Figure 11. Single-disk unbalance of low-pressure rotor: (**a**) LC disk, (**b**) LT disk.

Figure 11 illustrates that the presence of two unbalances in the dual-rotor system significantly affects the amplitude of vibration. When the phase angle of the unbalanced positions remains constant, a synchronous increase in the unbalance of each turntable leads to a corresponding rise in vibration amplitude, demonstrating a multiplicative trend. Consequently, studying the relationship between the phase angles of unbalanced faults and their coupling effects is of considerable importance. By adjusting the relative phase between the rotors, it is anticipated that the unbalanced state of the rotor system can be reduced or even eliminated.

3.2. Probability Distribution of Assembly Phase Angle of Low-Pressure Rotor

The compressor part and the turbine part of the low-pressure rotor need to be accurately balanced before assembly, and then connected into a whole. However, when these two parts of unbalance are coupled, the vibration level often exceeds the specified standard, which has a serious impact on the performance and assembly efficiency of the engine. In the high-speed rotating state, the assembled compressor and turbine will produce a complex coupling imbalance effect. The result is that even if the rotor has been successfully balanced on the balancing machine, once assembled on the engine, the vibration value may still exceed the specified range. In view of this situation, it is particularly important to study the probability distribution characteristics of the assembly phase angle of the compressor and the turbine.

Prior to assembly, the compressor and turbine sections of the low-pressure rotor should undergo independent dynamic balancing in order to ensure that each rotor's residual unbalance meets the specified standard. In compliance with the international dynamic balancing standard ISO 1940 [30], rotors are classified according to their application scenarios and are assigned varying levels of balance accuracy. For aero-engine rotors, three levels of accuracy are specified: G2.5, G1, and G0.4, with G0.4 representing the highest precision in balance requirements [31]. The allowable residual unbalance can be calculated as

$$m_{per} = M \times G \times \frac{60}{2\pi r n} \times 10^3 \tag{10}$$

where m_{per} represents the allowable residual unbalance, M is the mass of the rotor, G denotes the rotor's balance accuracy grade, r is the radius of the rotor system, and n is the operating speed of the rotor.

When adopting the G2.5 balance accuracy standard, the allowable residual unbalance for the compressor disk and turbine disk is calculated to be 0.4074 g and 0.3244 g, respectively. It should be noted that while ensuring vibration levels remain within standard limits, both the compressor and turbine sections must adhere to a normal distribution for their residual unbalance. Within this framework, the minimum residual unbalance for both parts is set at 0.005 g, while the maximum is defined as the corresponding allowable unbalance value. Following conventional statistical principles, a mean of 0.2287 g with a standard deviation of 0.1039 g is assigned to the residual unbalance of the compressor section; whereas for the turbine section, a mean value of 0.1872 g accompanied by a standard deviation of 0.1098 g is established. The normal distribution characteristics of residual unbalance for both the compressor disk and the turbine disk are illustrated in Figure 12a and 12b, respectively.

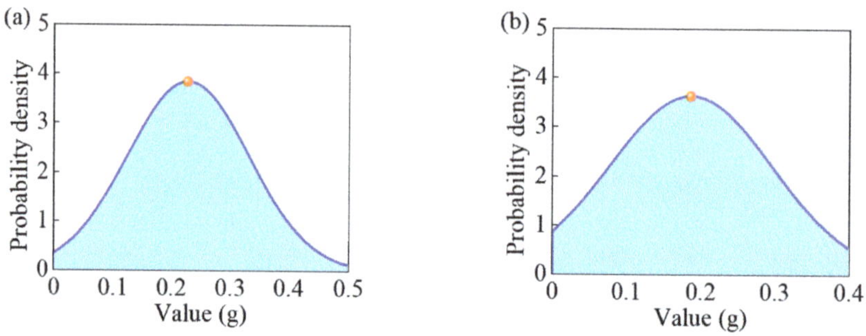

Figure 12. Residual unbalance distribution of low-pressure rotor system: (**a**) compressor part, (**b**) turbine part.

During the assembly of the low-pressure rotor's compressor and turbine sections, it is crucial to meticulously consider the individual unbalanced forces generated by each component. These forces are subsequently combined on the correction surface of intermediate bearing 4, adhering to the principle of force translation and synthesis. The unbalanced force produced by the compressor section is denoted as Q_{LC}, while that originating from the turbine section is represented as Q_{LT}. Both these forces encompass both magnitude and direction and can be mathematically expressed as follows:

$$Q_T = \sqrt{Q_{LC}^2 + Q_{LT}^2 + 2Q_{LC}Q_{LT}\cos\alpha} \tag{11}$$

where α is the intersection angle of the Q_{LC} and Q_{LC}.

Assuming a normal distribution for the residual unbalance of the compressor and turbine sections of the low-pressure rotor, we calculate the probability distribution of assembly phase angles for the compressor rotor and turbine rotor system. The acceptable balance standard is defined as a residual unbalance less than 0.3244 g once both components are assembled. Taking into account that assembly phase angles range from 0 to 360 degrees and that they can combine with each other, Equation (11) is employed to compute the corresponding residual unbalance for all possible combinations. Figure 13a illustrates the relationship between assembly phase angle combinations and calculated residual unbalances, while Figure 13b provides a detailed presentation of the probability density for combined residual unbalances. Statistical analysis reveals that out of a total of 129,600 angle combinations, there are 51,711 combinations resulting in residual unbalances exceeding 0.3244 g.

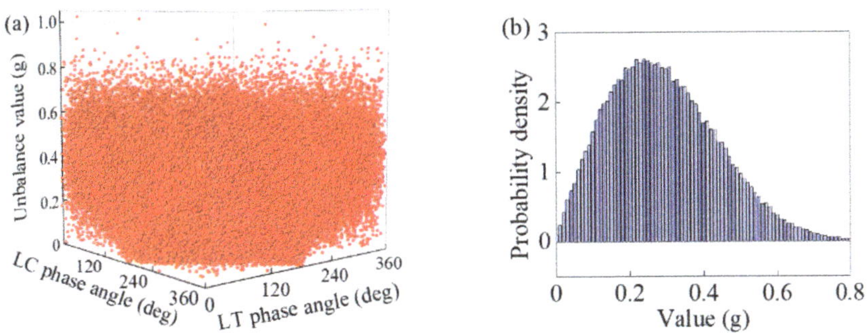

Figure 13. Combined residual unbalance: (**a**) phase angle change in LC and LT, (**b**) probability density.

Based on a field practice experience in the former Soviet Union, it was observed that reassembling the rotor compressor and turbine sections with a relative rotation of 180° resolved approximately 70% of excessive vibration issues encountered after engine assembly. Motivated by this observation, we investigated adjustments to the assembly phase angles for the 51,711 combinations exceeding the established threshold. In this study, we assumed a constant assembly phase angle for the compressor section while increasing the assembly phase angle of the turbine section by 180°. After implementing this adjustment, Figure 14a illustrates the distribution of residual unbalance in the low-pressure rotor system.

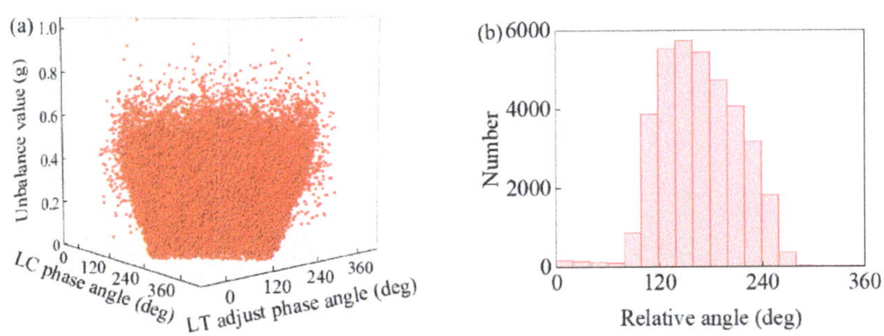

Figure 14. Combined residual unbalance assembly phase angle: (**a**) after adjustment, (**b**) relative phase angle distribution.

The results demonstrate that adjusting the turbine assembly phase angle by 180° reduced combinations with residual unbalances exceeding the set threshold to 15,439. Consequently, achieving a vibration qualification rate of 70.1435% for the low-pressure rotor system aligns with findings from field experience in former Soviet Union methodology. This study further validates that the selection of appropriate assembly phase angles significantly influences post-assembly unbalanced coupling effects between rotor compressor and turbine sections and offers potential reductions in such coupling effects. The assembly phase angle optimization method effectively suppresses the assembly imbalance of aero-engine rotors, ensuring that the vibration levels of the engine rotors remain within acceptable standards after assembly. This approach not only enhances engine performance but also improves assembly efficiency.

Additionally, the combinations of assembly phase angles for the low-pressure rotor system that exhibit qualified vibration after a 180° rotation are selected for statistical distribution analysis. The relative angle distribution between the compressor and turbine sections is presented in Figure 14b, revealing that the majority of qualified combinations cluster within the range of 100 to 240 degrees. This finding serves as a valuable reference for optimizing the assembly process of the low-pressure rotor system. When the relative angle between the compressor and turbine sections falls within this interval, it increases the likelihood of achieving a favorable balance state in the rotor system, thereby mitigating excessive vibration risks. Consequently, prioritizing these specific relative angle combinations during future assembly processes can significantly enhance both efficiency and performance.

4. Experimental Verifications

4.1. Construction of Dual-Rotor System Test Bench

To verify the influence of unbalanced phase difference combinations on rotor vibration response, a dynamic simulation test bench for the 31F aero-engine rotor system was designed and experimentally validated. This test bench closely resembles the prototype of the 31F aero-engine rotor system, adhering to similar design principles. The design is based on the actual rotor structure and follows the principles of similar vibration modes and equal critical speed ratios. The dynamics simulation test bench for the 31F aero-engine rotor system is illustrated in Figure 15. In addition, the parameters of the corresponding components of the 31F aero-engine rotor test bench are shown in Figure 16.

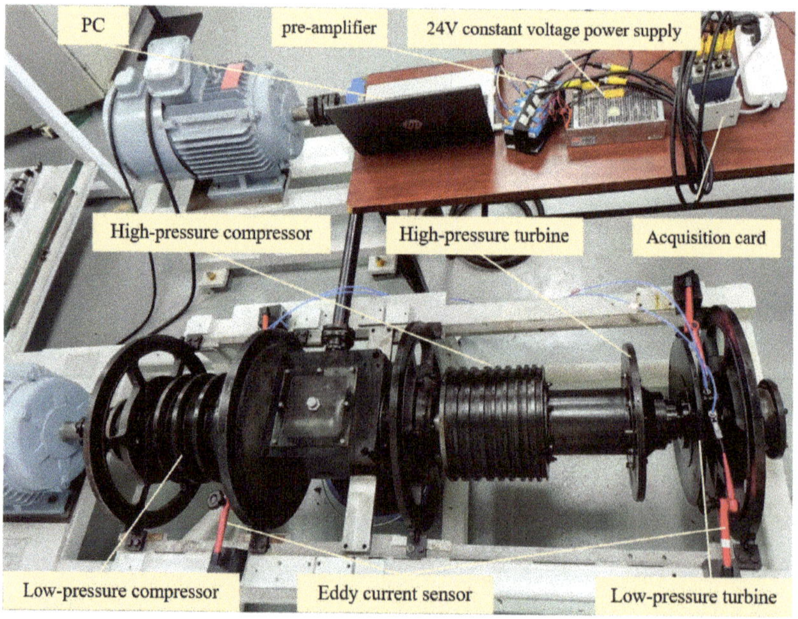

Figure 15. 31F aero-engine rotor system test bench.

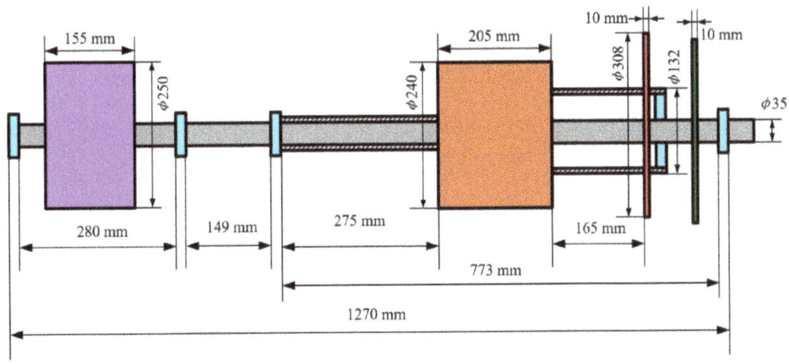

Figure 16. The parameters of the corresponding components of the rotor system.

Two pairs of eddy current displacement sensors are positioned between the compressor disk and the turbine disk. Each sensor is angled at 45° to the horizontal plane, with an angle of 90° between the two eddy current sensors. The signal acquisition system utilized is the Vib900 multi-channel field vibration signal analyzer from JiangSu YiYou Measurement and Control Technology Co., Ltd. (Wuxi, Jiangsu, China). For this study, the eddy current sensor located near the compressor disk of the low-pressure rotor serves as the primary measurement point. The critical speed of the low-pressure rotor is illustrated in Figure 17.

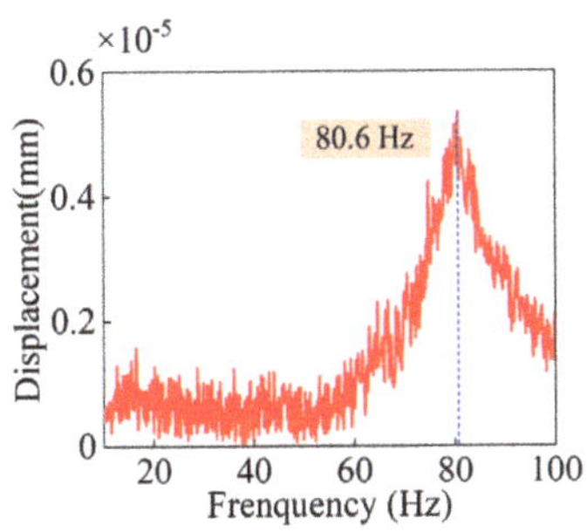

Figure 17. The first-order critical speed of the low-pressure rotor.

Figure 17 shows the first-order critical speed of the low-pressure rotor, where the maximum displacement is observed at 0.52 mm at a frequency of 80.6 Hz. This result closely aligns with the simulated first-order critical speed of 76.9 Hz (Figure 6), with an error of less than 5%. The slight discrepancy arises from modeling approximations and experimental conditions, confirming the reliability of the test bench in replicating the rotor system's dynamic behavior under real-world conditions.

4.2. Experimental Analysis

To investigate the assembly phase angle combinations of the compressor and turbine sections of the low-pressure rotor, a combined unbalance with phase difference combinations of 0 deg and 180 deg is applied to the low-pressure fan disk and the low-pressure turbine disk, respectively. The effect of these unbalanced phase difference combinations on the rotor's vibration response is studied experimentally. To account for influencing factors such as the weight of the simulation test bench and manufacturing assembly errors, a 20 g screw is added to the threaded holes of both the compressor disk and the turbine disk, serving as the initial unbalanced mass to induce the unbalanced vibration response of the rotor. Initially, the two unbalanced masses are positioned perpendicular to the axis from the top. Subsequently, the screw on the turbine disk is rotated by 180 deg, resulting in the unbalanced mass being perpendicular to the axis from the bottom. The vibration signal diagrams corresponding to these two assembly phase angles are illustrated in Figure 18. Clearly, the amplitude of the assembly phase angle of both LC and LT is 0 deg (black line), which is significantly larger than that of the assembly phase angle of LC and LT, which is 0 deg and 180 deg (red line), respectively. Adjusting the assembly phase angle of the turbine disk effectively reduces the vibration of the rotor system.

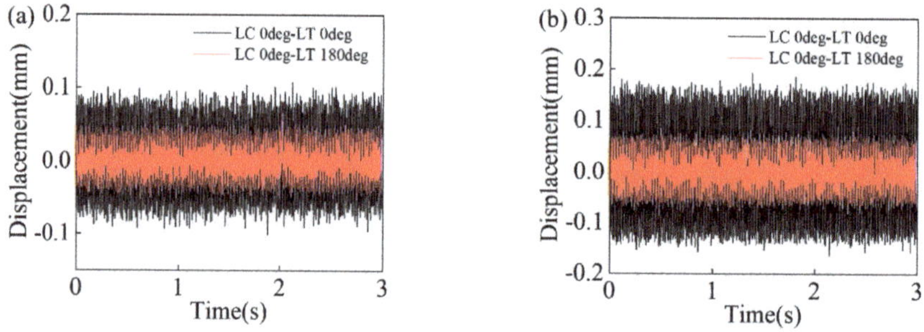

Figure 18. The vibration response of the combined assembly phase angles (**a**) at measuring point 1, (**b**) and at measuring point 2.

5. Conclusions

This paper investigates the probability distribution of assembly phase angle combinations for the low-pressure rotor. A four-disk, five-support double-rotor dynamic system is established, and the effects of unbalanced mass at varying positions and sizes on the vibration of the dual-rotor system are analyzed at different measuring points. The probability distribution of the assembly phase angles for the compressor and turbine sections of the low-pressure rotor is calculated. The conclusions are as follows:

(1) The position and size of the unbalanced mass significantly influence the vibration of the dual-rotor system. An increase in unbalanced mass correlates with an increase in the system's vibration response.
(2) For low-pressure rotors exhibiting excessive vibration, adjusting the assembly phase angle of either the compressor or the turbine section by 180° results in a vibration qualification rate of 70.1435%. This finding provides a theoretical explanation for the field experience method used in the former Soviet Union.
(3) The relative phase angles of the compressor and turbine sections that yield qualified rotating assembly phase angles are predominantly concentrated in the range of 100 to 240 degrees. This insight can serve as a valuable reference for optimizing the assembly process of the low-pressure rotor system.

The findings of this study offer practical implications for the design and assembly of dual-rotor systems in aircraft engines. By identifying the optimal assembly phase angle configurations to mitigate excessive vibration, this research provides valuable guidance for improving rotor balance and stability in aero-engines. These conclusions can be directly applied in the assembly processes of current and future aero-engine models to reduce maintenance requirements.

Author Contributions: Conceptualization, X.L.; methodology, H.L.; software, C.S.; validation, M.L.; investigation, J.B.; resources, C.S.; data curation, C.S.; writing—original draft preparation, H.L.; writing—review and editing, X.L.; visualization, M.L.; supervision, J.B.; project administration, X.L.; funding acquisition, X.L. All authors have read and agreed to the published version of the manuscript.

Funding: This research was funded by the National Natural Science Foundation (Grant no. 12072076).

Data Availability Statement: Data are contained within the article.

Conflicts of Interest: Author Mingfeng Li was employed by the company Aecc South Industry Company Limited. The remaining authors declare that the research was conducted in the absence of any commercial or financial relationships that could be construed as a potential conflict of interest.

References

1. Gao, T.; Yuan, S.-M.; Liu, Y.-Q.; Cao, S.-Q.; Yang, S.-P. Paroxysmal impulse vibration analysis of an aero-engine dual-rotor system with a defective inner ring for the inter-shaft bearing. *Commun. Nonlinear Sci. Numer. Simul.* **2024**, *131*, 107864. [CrossRef]
2. Yu, P.; Hou, L.; Jiang, K.; Jiang, Z.; Tao, X. Dynamic modeling and nonlinear analysis for lateral–torsional coupling vibration in an unbalanced rotor system. *Appl. Math. Model.* **2024**, *126*, 439–456. [CrossRef]
3. Dolev, A.; Tresser, S.; Bucher, I. Balancing high-speed rotors at low speed using optimized parametric excitation and tuned nonlinear feedback. *Int. J. Non-Linear Mech.* **2022**, *139*, 103873. [CrossRef]
4. Shuguo, L.; Yanhong, M.; Dayi, Z.; Jie, H. Studies on dynamic characteristics of the joint in the aero-engine rotor system. *Mech. Syst. Signal Process.* **2012**, *29*, 120–136. [CrossRef]
5. Lu, J.; Zhang, X.; Pan, X.; Zhang, M. Research on Unbalanced Vibration Suppression Method for Coupled Cantilever Dual-Rotor System. *Machines* **2022**, *10*, 758. [CrossRef]
6. Chen, J.; Yuan, K.; Zhu, L.; Zhao, X. Dynamic modelling and stability analysis of aero-engine rotor system considering aerodynamics. *Aerosp. Sci. Technol.* **2024**, *151*, 109294. [CrossRef]
7. Lu, Z.; Zhong, S.; Chen, H.; Wang, X. Nonlinear response analysis for a dual-rotor system supported by ball bearing. *Int. J. Non-Linear Mech.* **2021**, *128*, 103627. [CrossRef]
8. Yu, P.; Zhang, D.; Ma, Y.; Hong, J. Dynamic modeling and vibration characteristics analysis of the aero-engine dual-rotor system with Fan blade out. *Mech. Syst. Signal Process.* **2018**, *106*, 158–175. [CrossRef]
9. Guan, H.; Xiong, Q.; Ma, H.; Yang, Y.; Zeng, J.; Wang, P.; Bao, Y. Study on dynamic characteristics of the gear-dual-rotor system with multi-position rubbing. *Mech. Mach. Theory* **2024**, *191*, 105501. [CrossRef]
10. Chen, H.; Hou, L.; Chen, Y.; Yang, R. Dynamic characteristics of flexible rotor with squeeze film damper excited by two frequencies. *Nonlinear Dyn.* **2017**, *87*, 2463–2481. [CrossRef]
11. Wei, Z.; Zhang, S.; Jafari, S.; Nikolaidis, T. Gas turbine aero-engines real-time on-board modelling: A review, research challenges, and exploring the future. *Prog. Aerosp. Sci.* **2020**, *121*, 100693. [CrossRef]
12. Beaudoin, M.-A.; Behdinan, K. Analytical lump model for the nonlinear dynamic response of bolted flanges in aero-engine casings. *Mech. Syst. Signal Process.* **2019**, *115*, 14–28. [CrossRef]
13. Kang, Y.; Cao, S.; Hou, Y.; Chen, N. Analysis of backward whirling characteristics of a dual-rotor system caused by unbalance. *Measurement* **2022**, *203*, 111982. [CrossRef]
14. Zhang, Y.; Wang, W.; Wei, D.; Wang, G.; Xu, J.; Liu, K. Dynamic stability of unbalance-induced vibration in a turbocharger rotor-bearing system with the nonlinear effect of thermal turbulent lubricating fluid film. *J. Sound Vib.* **2022**, *528*, 116909. [CrossRef]
15. Yang, Y.; Cao, D.; Wang, D.; Jiang, G. Response analysis of a dual-disc rotor system with multi-unbalances–multi-fixed-point rubbing faults. *Nonlinear Dyn.* **2017**, *87*, 109–125. [CrossRef]
16. Wang, L.; Wang, A.; Yin, Y.; Heng, X.; Jin, M. Effects of unbalance orientation on the dynamic characteristics of a double overhung rotor system for high-speed turbochargers. *Nonlinear Dyn.* **2022**, *107*, 665–681. [CrossRef]
17. Song, L.; Chen, Y. A Prediction Model of Two-Sided Unbalance in the Multi-Stage Assembled Rotor of an Aero Engine. *Machines* **2024**, *12*, 503. [CrossRef]
18. Wang, L.; Bin, G.; Li, X.; Liu, D. Effects of unbalance location on dynamic characteristics of high-speed gasoline engine turbocharger with floating ring bearings. *Chin. J. Mech. Eng.* **2016**, *29*, 271–280. [CrossRef]
19. Li, T.; Wen, Z.; Zhao, B.; Sun, Q. A novel collaborative optimization assembly process method for multi-performance of aeroengine rotors. *Int. J. Adv. Manuf. Technol.* **2023**, *125*, 1827–1843. [CrossRef]
20. Zhang, M.; Liu, D.; Liu, Y. Recent progress in precision measurement and assembly optimization methods of the aero-engine multistage rotor: A comprehensive review. *Measurement* **2024**, *235*, 114990. [CrossRef]
21. Li, R.; Sun, C.; Liu, Y.; Mei, Y.; Tan, J. Prediction of the parallelism error and unbalance of aero-engine rotors based on intelligent algorithm. *IEEE Trans. Instrum. Meas.* **2023**, *72*, 3289542. [CrossRef]
22. Zhang, B.; Lu, H.; Liu, S.; Yang, Y.; Sang, D. Aero-Engine Rotor Assembly Process Optimization Based on Improved Harris Hawk Algorithm. *Aerospace* **2022**, *10*, 28. [CrossRef]
23. Yang, Z.; Hussain, T.; Popov, A.A.; McWilliam, S. Novel optimization technique for variation propagation control in an aero-engine assembly. *Proc. Inst. Mech. Eng. Part B J. Eng. Manuf.* **2011**, *225*, 100–111. [CrossRef]
24. Zhang, X.; Fu, X.; Fu, B.; Du, H.; Tong, H. Multi-objective optimization of aeroengine rotor assembly based on tensor coordinate transformation and NSGA-II. *CIRP J. Manuf. Sci. Technol.* **2024**, *51*, 190–200. [CrossRef]
25. Zhao, Y.; Mu, X.; Liu, J.; Sun, Q.; Zhou, P.; Fang, G. A decoupling algorithm-based technology for predicting and regulating the unbalance of aircraft rotor assembly considering manufacturing errors. *Machines* **2023**, *11*, 970. [CrossRef]
26. Lu, Z.; Wang, X.; Hou, L.; Chen, Y.; Liu, X. Nonlinear response analysis for an aero engine dual-rotor system coupled by the inter-shaft bearing. *Arch. Appl. Mech.* **2019**, *89*, 1275–1288. [CrossRef]

27. Wang, Q.-P.; Wang, F.; Zhang, W.-C.; Zhang, W.-F. Influences of unbalance phase combination on the dynamic characteristics for a turboprop engine. *Int. J. Turbo Jet-Engines* **2024**, *41*, 317–331. [CrossRef]
28. Zhao, S.; Zhang, L.; Zhu, R.; Han, Q.; Qin, Z.; Chu, F. Modeling approach for flexible shaft-disk-drum rotor systems with elastic connections and supports. *Appl. Math. Model.* **2022**, *106*, 402–425. [CrossRef]
29. Bauchau, O.A.; Craig, J.I. *Structural Analysis: With Applications to Aerospace Structures*; Springer: Dordrecht, Poland; New York, NY, USA, 2009.
30. *ISO 1940-1*; Mechanical Vibration—Balance Quality Requirements for Rotors in a Constant (Rigid) State—Part 1: Specification and Verification of Balance Tolerances. International Organization for Standardization: Geneva, Switzerland, 1986.
31. Zhang, T.; Jiang, F.; Yan, L.; Jiang, Z.; Xu, X. A novel ultrahigh-speed ball-on-disc tribometer. *Tribol. Int.* **2021**, *157*, 106901. [CrossRef]

machines

Article

Unbalanced Position Recognition of Rotor Systems Based on Long and Short-Term Memory Neural Networks

Yiming Cao [1], Changzhi Shi [1], Xuejun Li [1,*], Mingfeng Li [2] and Jie Bian [3]

[1] Key Laboratory of Rotor Vibration Monitoring and Diagnosis Technology for Machinery Industry, Foshan University, Foshan 528000, China; caoyiming@fosu.edu.cn (Y.C.); lhscz0110@163.com (C.S.)
[2] Aecc South Industry Company Limited, Zhuzhou 412000, China; lmf18007330425@163.com
[3] AVIC Hunan Power Machinery Research Institute, Zhuzhou 412002, China; hrbeu@163.com
* Correspondence: hnkjdxlxj@163.com

Abstract: Rotor unbalance stands as one of the primary causes of vibration and noise in rotating equipment. Accurate identification of unbalanced positions enables targeted measures for balance correction, thereby reducing vibration and noise levels and enhancing the operational efficiency and stability of the equipment. However, the complexity of rotor structures may lead to a diversity of vibration transmission paths, which complicates the identification of unbalanced positions. In this paper, an experimental platform for rotor systems is established to analyze the change patterns of vibration displacement in rotor systems at four unbalanced positions. Additionally, a rotor dynamics model is developed based on the finite element method and verified through experiments. Furthermore, an unbalanced rotor position identification method based on Long Short-Term Memory (LSTM) neural networks is proposed. This method utilizes multiple sets of measured response data and simulated data from unbalanced rotor positions to train the LSTM network, achieving precise identification of unbalanced positions at various rotational speeds. The research results indicate that under subcritical, critical, and supercritical speeds, the identification accuracy based on measured data reaches 95.5%, while the accuracy based on simulated data remains at a high level of 90.5%. These results fully validate the effectiveness and accuracy of the proposed model and identification method, providing new insights and technical means for identifying unbalanced rotor positions.

Keywords: rotor unbalance; dynamic modeling; aero-engine; neural network; vibration response

1. Introduction

The rotor system serves as a fundamental component of an aero-engine, primarily consisting of crucial parts such as the high-pressure compressor rotor and high-pressure turbine rotor, which are assembled into a unified whole through coupling devices. Before being reassembled and interconnected, both the compressor and turbine rotors undergo individual dynamic balancing. However, unbalanced faults in the rotors frequently lead to excessive vibration of the entire aero-engine [1]. Especially in extreme operating conditions, characterized by high temperatures, high pressures, and high speeds, unbalanced conditions are significantly amplified, exceeding controllable limits and potentially causing severe faults such as blade rubbing, which poses a severe threat to the longevity of the aero-engine [2]. Therefore, the accurate identification of unbalanced positions within the rotor system, followed by effective prediction, monitoring, and control of unbalance vibrations, has emerged as a critical technical challenge that demands immediate attention and innovative solutions.

Investigating the dynamic characteristics of rotor systems and constructing corresponding dynamic models can reveal the motion laws and vibration characteristics of rotors under complex operating conditions, thereby providing theoretical support for the identification and diagnosis of unbalance faults. In terms of dynamic modeling of rotor systems, the primary methods encompass the analytical method [3–5], the lumped parameter method [6–8], the transfer matrix method [9], and the finite element method [10–13]. Zhao et al. [3] established an analytical model for a flexible shaft-disk-drum rotor, examining the influence patterns of uncertain parameters on the vibration response of rotor systems with elastic supports. Miao et al. [4] built a rotor dynamic model considering the coupling of the coupling shaft based on the analytical method and subsequently explored the nonlinear vibration characteristics of the system on this basis, analyzing the characteristics of friction faults. Luo et al. [6] adopted the lumped mass method to establish a coupling model for rotor systems, taking into account angular misalignment and analyzed the influence of structural parameters of squeeze film dampers on the nonlinear vibration characteristics of the system. Li et al. [7] further established a lumped mass model for bolted-disk joints, deeply discussing the significant impact of tangential stiffness and bending stiffness at bolted connections on the vibration response of the system. Gupta et al. [9] utilized the extended transfer matrix method to establish a dynamic model for dual-rotor systems, studying the effects of rotational inertia and gyroscopic moments on the dynamic characteristics of the system and conducting a stability analysis of the system. Qin et al. [10] derived an analytical model for the bending stiffness of disk-drum connection structures and introduced it into the finite element model of rotor systems, solving the dynamic equations of rotor systems using the harmonic balance method. Lu et al. [11] established a finite element model for dual-rotor systems with breathing cracks and analyzed the dynamic response of these systems using the harmonic balance method. Chen et al. [12] considered spatial crack faults to establish a finite element model for gear-rotor systems, analyzing the impact of gear mesh stiffness and cracks on mesh stiffness and the vibration response of the system.

Rotor unbalance can induce severe vibrations in aero-engines, with the majority of mechanical failures in such engines being vibration-induced or vibration-related [14–19]. Therefore, an in-depth analysis of the influence patterns of rotor unbalance on system vibration characteristics is particularly significant. Numerous scholars have conducted extensive research on this issue. Yu et al. [14] performed dynamic modeling of a dual-rotor system in an aero-engine and found that, during the operation of turbofan engines, rotor blades not only introduce sudden unbalances and inertial asymmetries but also impose significant impact loads, leading to friction between the rotor and stator. Wang et al. [15] established a dynamic model that considered the nonlinear force coupling effects of squeeze film dampers and intermediate bearings in a dual-rotor system under multiple unbalanced forces. Through simulations, they observed that the rotor system exhibits quasi-periodic motion with four cycles near the critical speed. Yang et al. [16] addressed the issue of fixed-point friction in aero-engines by developing a mechanical vibration model for a dual-rotor system under the coupling effects of two unbalanced rotors and fixed-point friction. Chen et al. [20,21] proposed an assembly optimization method for multi-stage rotors in a specific type of aero-engine, using coaxiality and unbalance parameters as dual objective functions and employing genetic algorithms to solve for the optimal assembly angles of rotors at various stages. Song et al. [22] established a prediction model for unbalance in multi-stage assembled rotors, accurately forecasting bilateral unbalances and their acting phases in multi-stage rotors. In summary, the influence of rotor unbalance on the vibration characteristics of aero-engine is a complex and important issue. By conducting in-depth research on this issue, not only can the operational stability and reliability of aero-engines be improved, but valuable support can also be provided for the design and optimization of aero-engines.

Rotor unbalance can exacerbate vibrations in aero-engines, adversely impacting their operational stability and service life. In response to this challenge, numerous scholars have conducted in-depth research. Jiang et al. [23] proposed a method for dynamic unbalance compensation by generating control signals based on the real-time position of the rotor's unbalanced mass. They analyzed the control performance under different unbalance phases, optimization step sizes, and noise levels. Ma et al. [24] delved into the impact of unbalanced rotor vibration coupling, vibration transmission, and sensitivity on overall machine vibration and dynamic balance, revealing the intricate mechanisms through which unbalanced location, magnitude, and phase influence rotor vibration response. Wang et al. [25] further explored the influence of rotational speed and rotor unbalance on the deviation between static and dynamic balance positions of the rotor. They discovered that, under the influence of rotational unbalanced forces, there is a significant deviation between the rotor's dynamic and static balance positions, and this deviation approximates a linear relationship with rotational speed. Therefore, accurately identifying the unbalanced rotor position is crucial for precise fault diagnosis and rapid response. However, traditional methods for identifying unbalanced rotor positions often rely on empirical judgments and extensive test data, making it challenging to accurately capture subtle changes in unbalanced vibrations. With the advancement of signal processing technology and machine learning algorithms, in-depth analysis of vibration data can be conducted to achieve real-time monitoring and precise identification of unbalanced states, providing strong support for preventive maintenance and health management of engines. Han et al. [26,27] proposed a transfer learning method for unbalanced rotor fault localization that combines finite element models of rotor dynamics with experimental data. Although it achieves unbalanced rotor position identification to a certain extent, its accuracy and efficiency still require further enhancement.

In the research of rotor unbalance issues, Artificial Neural Networks (ANNs) have demonstrated immense potential in assisting the dynamic balancing process. Compared to traditional dynamic balancing methods, which often rely on experience and manual adjustments, ANNs can automatically adjust the rotor's balance state by deeply learning and simulating critical data such as vibration signals, thereby enhancing the precision and efficiency of rotor balancing. Santos et al. [28] established an analytical model for rotor systems supported by sliding bearings, utilizing plane separation techniques and ANNs to predict the position and magnitude of the balance correction masses for rotor systems. This method provides an effective pathway for continuous monitoring of dynamic balancing in rotating machinery. Walker et al. [29] further employed a data-driven approach, utilizing ANNs to achieve automatic localization of rotor unbalance and incorporating friction and misalignment faults into their research to validate the system's stability and reliability. Yan et al. [30] proposed an unbalanced rotor fault diagnosis method based on the Deep Belief Network (DBN), which can automatically learn typical features of rotor unbalance faults and accurately identify unbalance fault states. Zhong et al. [31,32] introduced a rotor balancing method based on unsupervised deep learning, capable of efficiently determining the magnitude and position of unbalance in two correction planes. Xing et al. [33] utilized Convolutional Neural Networks (CNNs) to detect the magnitude and position of dynamic unbalance in wind turbine rotors, with the natural modal functions of nacelle vibrations as the input variables, successfully discerning weak unbalance signals. Long Short-Term Memory (LSTM) networks, a variant of Recurrent Neural Networks (RNNs) designed for processing sequential data [34,35], have exhibited exceptional performance in many application scenarios due to their ability to learn and remember long-term dependencies. In the research of unbalanced rotor issues, LSTM networks also have broad application prospects and are anticipated to provide innovative solutions for the intellectualization and automation of rotor balancing processes.

This paper is organized as follows: In Section 2, the dynamic model of the rotor system is established using the finite element method, comprehensively considering the shaft, disks, support elements, and couplings. Following the establishment of the model, a rotor test tig is constructed, and preliminary modal validation is conducted to ensure the accuracy and reliability of the model. This validation step is crucial as it provides a solid foundation for subsequent experiments and analyses. In Section 3, vibration response experiments and simulation studies are carried out on the rotor system. These experiments are designed to assess the system's behavior at four different unbalanced positions and under various rotational speeds. The results obtained from these experiments further confirm the accuracy and reliability of the established dynamic model, reinforcing our confidence in its predictive capabilities. Additionally, an LSTM neural network structure is constructed, and deep training and intelligent analysis are performed on rotor vibration data. This advanced analysis enables us to effectively identify the rotor's unbalanced positions, which is a critical step in fault diagnosis and predictive maintenance. In Section 4, the identification results are analyzed and discussed. Section 5 summarizes the research conclusions of this paper. The innovation of this paper lies in the first application of the LSTM neural network to identify the unbalanced position of an aero-engine turbine rotor. LSTM networks are particularly suitable for processing time-varying vibration signal sequence data, which makes them ideal for dealing with unbalanced position identification problems. By training the LSTM model on a comprehensive dataset containing experimental and simulation data at different speeds, it is able to learn complex relationships between vibration patterns and unbalanced positions.

2. Rotor System Dynamics Modeling and Verification

During the operation of an aero-engine rotor system, its working rotational speed often exceeds the range of two critical speeds. Under such high-speed conditions, even minute unbalances on flexible shafts can trigger significant vibrations, thereby exacerbating the unbalanced vibration response of the rotor. Therefore, conducting in-depth research on the vibration characteristics of aero-engine rotor systems is particularly crucial, with the establishment of an accurate dynamic model crucial for accurately describing and predicting the system's vibration behavior. This section will mainly focus on the following three aspects: First, simplifying the model to eliminate unnecessary complexity while preserving the system's primary dynamic characteristics; Second, establishing differential equations of motion based on the simplified model to mathematically describe the system's vibration behavior; Finally, validating the model to ensure that it accurately reflects the vibration characteristics of the actual system.

2.1. Rotor Dynamics Modeling

The finite element method (FEM) is utilized for dynamic modeling of aero-engine rotor systems in this paper. The core principle of FEM involves dividing the entire system into multiple elements with specific shapes and degrees of freedom. These elements are interconnected through physical relationships at nodes to construct the dynamic equations of the entire system. The aero-engine rotor system constitutes a complex coupled system comprising components such as rotor shafts, disks, bearings, and couplings arranged in a specific quantity and sequence.

2.1.1. Rotating Shaft Model

The rotor shaft of an aero-engine is divided into several axial segments, and in this paper, Timoshenko beam elements are used for equivalence. Each beam element is comprised of two nodes, with each node possessing six degrees of freedom. A schematic diagram of the beam element is shown in Figure 1a. In the coordinate system $oxyz$, u, v, and w represent the displacements of the node in the x, y, and z axes, respectively, while θ_x, θ_y, and θ_z represent the angular displacements of the node in the x, y, and z axes, respectively. Therefore, the node displacement vector for the beam element is as follows:

$$\boldsymbol{q}^{\mathrm{e}} = [u_i, v_i, w_i, \theta_{xi}, \theta_{yi}, \theta_{zi}, u_j, v_j, w_j, \theta_{xj}, \theta_{yj}, \theta_{zj}]^{\mathrm{T}} \tag{1}$$

According to Hamilton's principle, the expressions of the beam element's mass matrix $\boldsymbol{M}^{\mathrm{e}}$, stiffness matrix $\boldsymbol{K}^{\mathrm{e}}$, and gyroscopic matrix $\boldsymbol{G}^{\mathrm{e}}$ are as follows:

$$\begin{cases} \boldsymbol{M}^{\mathrm{e}} = & \int_L \rho A(\boldsymbol{N}_u^{\mathrm{T}}\boldsymbol{N}_u + \boldsymbol{N}_v^{\mathrm{T}}\boldsymbol{N}_v + \boldsymbol{N}_w^{\mathrm{T}}\boldsymbol{N}_w) + \rho(I_x\boldsymbol{N}_u^{\mathrm{T}}\boldsymbol{N}_u + I_y\boldsymbol{N}_v^{\mathrm{T}}\boldsymbol{N}_v + I_z\boldsymbol{N}_w^{\mathrm{T}}\boldsymbol{N}_w)\mathrm{d}x \\ \boldsymbol{K}^{\mathrm{e}} = & \int_L EA\boldsymbol{N'}_u^{\mathrm{T}}\boldsymbol{N}'_u + \kappa_y GA\boldsymbol{N'}_v^{\mathrm{T}}(\boldsymbol{N}'_v - \boldsymbol{N}_\varphi) + \kappa_z GA\boldsymbol{N'}_w^{\mathrm{T}}(\boldsymbol{N}'_w + \boldsymbol{N}_\varphi) + GJ\boldsymbol{N'}_{\theta_x}^{\mathrm{T}}\boldsymbol{N}'_{\theta_x} \\ & +EI_y\boldsymbol{N'}_{\theta_y}^{\mathrm{T}}\boldsymbol{N}'_{\theta_y} + EI\boldsymbol{N'}_{\theta_z}^{\mathrm{T}}\boldsymbol{N}'_{\theta_z} + \kappa_z GA\boldsymbol{N'}_{\theta_y}^{\mathrm{T}}(\boldsymbol{N}'_w + \boldsymbol{N}_{\theta_y}) - \kappa_y GA\boldsymbol{N}_{\theta_z}^{\mathrm{T}}(\boldsymbol{N}'_v - \boldsymbol{N}_{\theta_z})\mathrm{d}x \\ \boldsymbol{G}^{\mathrm{e}} = & \int_L \Omega\rho I_x(\boldsymbol{N}_{\theta_y}^{\mathrm{T}}\boldsymbol{N}_{\theta_z} - \boldsymbol{N}_{\theta_z}^{\mathrm{T}}\boldsymbol{N}_{\theta_y})\mathrm{d}x \end{cases} \tag{2}$$

where $\boldsymbol{N}_n$ ($n = u, v, w, \theta_x, \theta_y, \theta_z$) represents the element-shaped function, ρ is the material density, and A is the cross-sectional area of the element. I_x, I_y, and I_z represent the moment of inertia of the section with respect to the x, y, and z axes, respectively. J is the torsion moment of inertia, Ω is the angular velocity of rotation, and κ_y and κ_z represent the shear coefficient with respect to the y and z axes, respectively. Since the cross-section is circular, the expressions for the shear coefficients κ_y, κ_z, and the torsional inertia J are given as follows:

$$\kappa_y = \kappa_z = \frac{6(1+v)}{(7+6v)} \tag{3}$$

$$J = \pi d_s^4/32 \tag{4}$$

where d_s is the diameter of the shaft section, and v is Poisson's ratio.

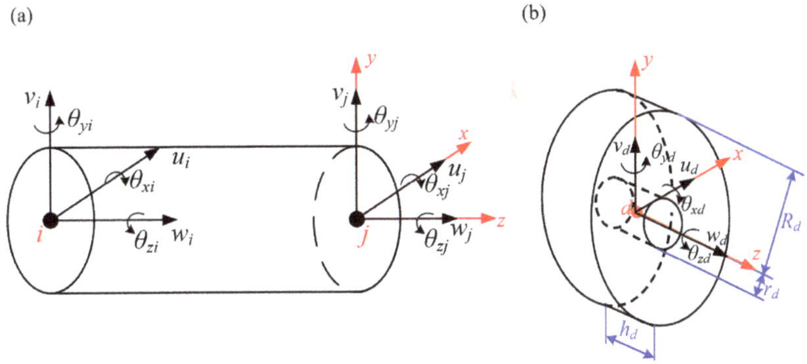

Figure 1. Schematic of element: (**a**) shaft element; (**b**) disk element.

2.1.2. Disk Model

In this paper, a lumped mass model is employed for the disk [36], equating them to mass points fixed at the nodes of the rotor shaft with specific mass and rotational inertia. It is assumed that the unbalanced forces of the rotor only act on the two nodes corresponding to the disks. The expressions for the element mass matrix $\boldsymbol{M}_{\mathrm{d}}^{\mathrm{e}}$ and the element gyroscopic moment matrix $\boldsymbol{G}_{\mathrm{d}}^{\mathrm{e}}$ of the disk are as follows:

$$\boldsymbol{M}_{\mathrm{d}}^{\mathrm{e}} = \mathrm{diag}(m_{\mathrm{d}}, m_{\mathrm{d}}, m_{\mathrm{d}}, J_{\mathrm{d}}, J_{\mathrm{d}}, J_{\mathrm{p}}) \tag{5}$$

$$\boldsymbol{G}_{\mathrm{d}}^{\mathrm{e}} = \begin{bmatrix} 0 & 0 & 0 & 0 & 0 & 0 \\ 0 & 0 & 0 & 0 & 0 & 0 \\ 0 & 0 & 0 & 0 & 0 & 0 \\ 0 & 0 & 0 & 0 & J_{\mathrm{p}} & 0 \\ 0 & 0 & 0 & -J_{\mathrm{p}} & 0 & 0 \\ 0 & 0 & 0 & 0 & 0 & 0 \end{bmatrix} \tag{6}$$

where the mass, diameter moment of inertia, and polar moment of inertia of the disk are m_{d}, J_{d}, and J_{p}, respectively, as shown in Figure 1b:

$$m_{\mathrm{d}} = \pi\rho h_{\mathrm{d}}(R_{\mathrm{d}} - r_{\mathrm{d}})^2,\ J_{\mathrm{d}} = \frac{\pi\rho h_{\mathrm{d}}(R_{\mathrm{d}} - r_{\mathrm{d}})^4}{4},\ J_{\mathrm{p}} = \frac{\pi\rho h_{\mathrm{d}}(R_{\mathrm{d}} - r_{\mathrm{d}})^4}{2} \tag{7}$$

where r_{d} is the inner radius of the disk, R_{d} is the outer radius of the disk, and h_{d} is the thickness of the disk.

2.1.3. Support Element Model

Bearings and supports are integrated to provide support for the rotor system. In this paper, a linear spring-damping model [37] is adopted for equivalent treatment of the support elements of the rotor system. Subsequently, the stiffness matrix $\boldsymbol{K}_{\mathrm{b}}^{\mathrm{e}}$ and damping matrix $\boldsymbol{C}_{\mathrm{b}}^{\mathrm{e}}$ of the support element can be expressed as follows:

$$\boldsymbol{K}_{\mathrm{b}}^{\mathrm{e}} = \mathrm{diag}(k_{\mathrm{b}u}, k_{\mathrm{b}v}, k_{\mathrm{b}w}, k_{\mathrm{b}\theta x}, k_{\mathrm{b}\theta y}, k_{\mathrm{b}\theta z}) \tag{8}$$

$$\boldsymbol{C}_{\mathrm{b}}^{\mathrm{e}} = \mathrm{diag}(c_{\mathrm{b}u}, c_{\mathrm{b}v}, c_{\mathrm{b}w}, c_{\mathrm{b}\theta x}, c_{\mathrm{b}\theta y}, c_{\mathrm{b}\theta z}) \tag{9}$$

where $k_{\mathrm{b}n}$ ($n = u, v, w, \theta_x, \theta_y, \theta_z$) represents the support stiffness in the n-direction, and $c_{\mathrm{b}n}$ ($n = u, v, w, \theta_x, \theta_y, \theta_z$) is the support damping in the n-direction.

2.1.4. Coupling Model

The bearing is equipped with a coupling to connect the a node of shaft 1 and the b node of shaft 2. The two halves of the coupling adopt a clearance fit, and the matching clearance is uniformly denoted as δ_{ab}, as shown in Figure 2. The nonlinear bearing forces of the coupling in all directions $F_{cn}(n = u, v, w, \theta_x, \theta_y, \theta_z)$ can be expressed as follows:

$$F_{cn} = \begin{cases} k_{cn}(n_a - n_b) + c_{cn}(\dot{n}_a - \dot{n}_b), |n_a - n_b| > \delta_{ab} \\ 0, |n_a - n_b| \leq \delta_{ab} \end{cases} \tag{10}$$

where k_{cn} and c_{cn} are the stiffness and damping of the coupling in each direction, respectively.

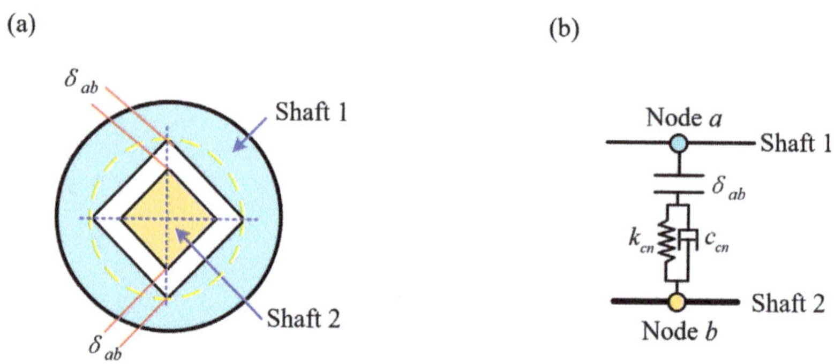

Figure 2. Diagram of a coupling with loose clearance: (**a**) structure diagram; (**b**) equivalent model.

Therefore, the dynamic equation of the rotor system under unbalanced excitation is as follows:

$$\boldsymbol{M}\ddot{\boldsymbol{q}} + (\boldsymbol{C} + \omega\boldsymbol{G})\dot{\boldsymbol{q}} + \boldsymbol{K}\boldsymbol{q} = \boldsymbol{F}_{\mathrm{b}} - \boldsymbol{G}_{\mathrm{s}} \tag{11}$$

where $\boldsymbol{q}$ represents the generalized displacement vector of the rotor system, and ω denotes the rotor speed. $\boldsymbol{M}$ is the mass matrix of the rotor system, which is obtained by the mass matrix of the shaft segment element $\boldsymbol{M}^{\mathrm{e}}$ and the mass matrix of the disk element $\boldsymbol{M}_{\mathrm{d}}^{\mathrm{e}}$. $\boldsymbol{K}$ is the stiffness matrix of the rotor system, which can be obtained by the combination of the axial segment element stiffness matrix $\boldsymbol{K}^{\mathrm{e}}$ and the supporting element stiffness matrix $\boldsymbol{K}_{\mathrm{b}}^{\mathrm{e}}$. $\boldsymbol{G}$ is the gyro matrix of the rotor system, which can be obtained by grouping the axis segment element gyro matrix $\boldsymbol{G}^{\mathrm{e}}$, the disk element gyro matrix $\boldsymbol{G}_{\mathrm{d}}^{\mathrm{e}}$, and the damping matrix of the supporting element $\boldsymbol{C}_{\mathrm{b}}^{\mathrm{e}}$. The rotor damping matrix $\boldsymbol{C}$ is simulated by proportional damping, which can be expressed as follows:

$$\boldsymbol{C} = \alpha_0\boldsymbol{M} + \beta_0\boldsymbol{K} \tag{12}$$

where the expressions of α_0 and β_0 are shown in Ref. [38].

G_s is the gravity vector, and F_b is the unbalanced rotor force vector. The unbalanced force vector and gravity vector at node i can be expressed in the following form:

$$F_b^i = \begin{bmatrix} F_{bui} & F_{bvi} & 0 & M_{bui} & M_{bvi} & 0 \end{bmatrix}^T \tag{13}$$

$$G_s^i = \begin{bmatrix} 0 & m_i g & 0 & 0 & 0 & 0 \end{bmatrix}^T \tag{14}$$

where g represents the acceleration of gravity, which is 9.8 m/s^2 in this paper. F_{bui}, F_{bvi}, M_{bui}, and M_{bvi} are the unbalanced rotor forces and moments located at node i, which can be expressed as follows:

$$\begin{Bmatrix} F_{bui} \\ F_{bvi} \\ M_{bui} \\ M_{bvi} \end{Bmatrix} = m_i e_{1i} \omega^2 \begin{Bmatrix} \cos(\omega t + \varphi_i) \\ \sin(\omega t + \varphi_i) \\ -e_{2i} \sin(\omega t + \varphi_i) \\ e_{2i} \cos(\omega t + \varphi_i) \end{Bmatrix} \tag{15}$$

where φ_i is the initial phase of the unbalance, m_i is the unbalanced mass, and e_{1i} and e_{2i} are the radial and axial eccentricity, respectively. Therefore, the vibration response of the rotor system is not only related to the unbalanced amplitude and position but also to the initial phase of the unbalance.

2.2. Model Verification

To validate the effectiveness of the established rotor dynamics model, a turbine rotor system test rig is designed and constructed in this paper, with its experimental setup illustrated in Figure 3. The test rig primarily comprises four core components: the high and low-pressure turbine rotor structures, the experimental platform, the drive motor, and the lubrication system. Among these, the low-pressure turbine section utilizes a hollow, flexible, slender shaft as the main transmission component, which is fixed and supported by bearings, elastic supports, and supporting structures. The power input end is connected to the drive motor through a short shaft, and the power is then transmitted to the hollow, flexible, slender shaft through a spline connection. In contrast, the high-pressure turbine structure is more complex, consisting of various components such as a power stub shaft, two-stage high-pressure turbines, end tooth coupling structures, bearings, damping devices, and supporting structures. Power is inputted through the hollow, flexible, slender shaft and transmitted via a spline connection.

During the modeling process, a linear spring-damper model is adopted to equivalently represent the bearings and support structures within the rotor system. The rotor shaft elements are connected through nodes, while the turbine disks are treated as rigid bodies, simplified as lumped masses with gyroscopic effects. Based on the above assumptions and simplifications, the geometric model of the turbine rotor system is established, as shown in Figure 4. With the known elastic modulus of the rotor shaft $E = 2.1 \times 10^{11}$ Pa, density $\rho = 7800\ kg/m^3$, and Poisson's ratio $v = 0.3$, the moment of inertia of the disk diameter is 0.12 $kg \cdot m^2$ and the moment of inertia of the disk is 0.13 $kg \cdot m^2$, the parameters of the shafts are shown in Table 1. The rotor dynamics model is constructed according to the generalized modeling method presented in Section 2.1.

Table 1. Comparisons of natural frequency between simulation and experiment.

Shaft Number	Length (mm)	Diameter (mm)
Shaft 1	95	32
Shaft 2	156.5	40
Shaft 3	104	70
Shaft 4	66	36
Shaft 5	162	34.8
Shaft 6	130	33.6

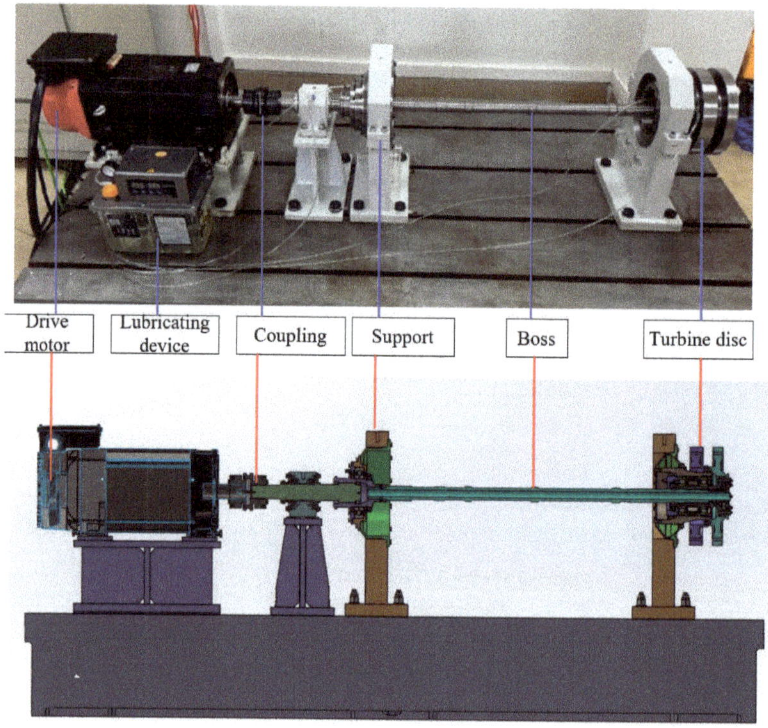

Figure 3. Turbine rotor system test bench and simplified structure diagram.

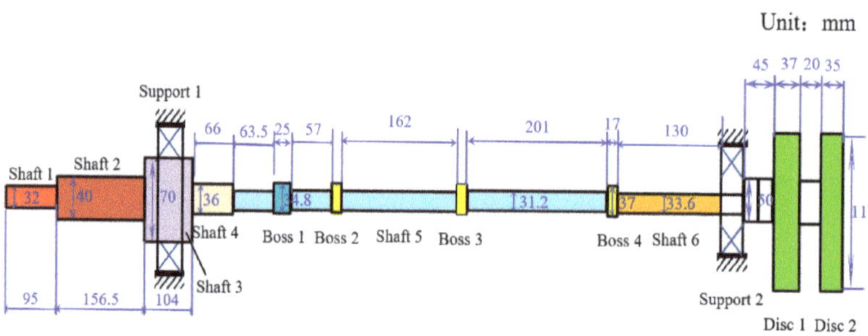

Figure 4. Geometric structure diagram of turbine rotor system.

The simulated rotor test rig designed in this paper employs a high-performance electric spindle motor as its driving component. This motor is securely connected to the transmission shaft via a coupling to ensure stable power transmission. The transmission shaft is further linked to the slender shaft through a precise spline connection, thereby effectively transmitting torque and driving the entire simulated rotor system to operate efficiently. To accurately measure the vibration displacement signals of the rotor system, four high-sensitivity eddy current sensors (Type: RP6600XL, Precision: <0.1%, Kaihang, Huanggang Hubei, China) and two acceleration sensors (Type: PCB-356A01, PCB, Buffalo, New York, NY, USA) are utilized to capture the vibration displacement signals at the measurement points, ensuring data accuracy and reliability. The test system selects an LMS acquisition instrument (LMS SCADAS Mobile) for data collection, and the experimental setup layout is shown in Figure 5. During the experiment, precise control of the driving motor is achieved through a frequency converter, enabling speed-up and speed-down tests on the rotor system to verify its inherent characteristics. The sampling frequency is 5120 Hz, and the sampling time is 10 s.

(a)
(b)
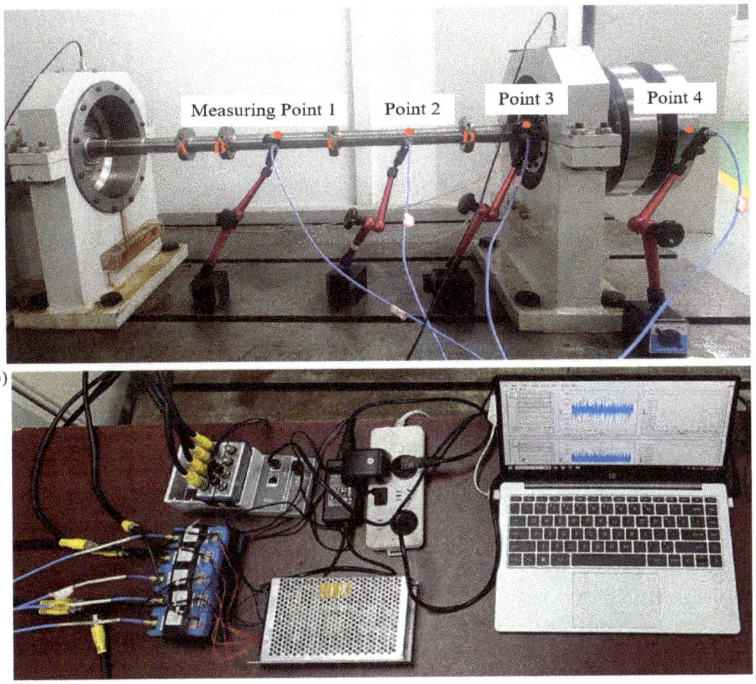

Figure 5. Rotor vibration test bench: (**a**) sensor layout; (**b**) data acquisition system.

A natural characteristic analysis is conducted on the rotor system dynamics model established using the finite element method. To validate the effectiveness of the model, detailed comparisons between the simulations and the experiments are performed, as presented in Table 2. In the table, the maximum error between the simulation results and the experimental results is only 4.3%, which fully demonstrates the high accuracy and validity of the constructed dynamics model. This finding not only provides a solid theoretical foundation for subsequent rotor dynamics research but also offers powerful technical support for rotor system design in practical engineering applications.

Table 2. Comparisons of natural frequency between simulation and experiment.

Frequency Order	f_{n1}	f_{n2}
Experiment (Hz)	121	236
Simulation (Hz)	115.8	227.3
Error (%)	4.3	3.7

3. Rotor Vibration Response and Unbalance Position Identification

In the previous section, the accuracy of the established dynamic model has been preliminarily verified by analyzing the inherent characteristics of the rotor system. To thoroughly validate the accuracy of the constructed dynamic model, a constant-speed test is conducted on the built test rig. During the testing process, vibration response data of the rotor system at various rotational speeds are meticulously recorded. A comprehensive comparison between these experimental data and the unbalanced vibration responses predicted by the finite element model is performed, further confirming the high accuracy and reliability of the established dynamic model. On this basis, through the construction of neural network architecture, the vibration signal data based on the measured data and the simulation data are deeply trained and intelligent analysis so as to identify the unbalanced position of the rotor, which provides a new method and method for improving the accuracy and reliability of the rotor balance.

3.1. Vibration Response Analysis of Rotor System at Different Speed

Based on Section 2.2, the first-order critical speed of the experimental system in this paper is determined to be 7260 rpm. To comprehensively evaluate the dynamic response of the rotor under different rotational speeds, constant-speed tests are conducted at subcritical (5600 rpm), critical (7260 rpm), and supercritical (8400 rpm) speeds. In the test, we set an unbalance of approximately 15 gmm millimeters (gmm) using a balance chuck to induce an observable unbalance response. This setting is designed to simulate the imbalances that may be encountered in actual operation and to assess their impact on system performance. Unbalance response analysis is an important part of rotating machinery fault diagnosis, through which we can understand the effects of unbalance on vibration, noise, and service life of the system. To verify the accuracy of the model, the signals collected by the eddy current sensor at measurement point 3 are selected for detailed analysis and comparison. The eddy current sensor is a non-contact measuring tool that accurately captures small vibration changes during the operation of rotating machinery. By comparing the measured signal with the simulation results, we can assess the predictive power of the model and identify potential sources of error.

It is noteworthy that the actual rotor test bench possesses a relatively complex structure, incorporating various friction coupling elements such as bearings and bolted connections, which may introduce nonlinear phenomena within the system. Consequently, the measured signals often exhibit greater complexity compared to simulated signals. Theoretically, unbalance solely induces rotational frequency components. However, in practical scenarios, other nonlinear frequency components may also influence the signals. To more accurately analyze the unbalanced response, band-pass filtering technology is used to eliminate the impact of other nonlinear frequencies, retaining only the rotational frequency components of the measured signals. This processing step is crucial for precisely identifying unbalanced responses.

In summary, vibration experiments are conducted at four different unbalanced positions under subcritical (5600 rpm), critical (7260 rpm), and supercritical (8400 rpm) conditions. At the same time, based on the established rotor finite element model, the unbalance is added to the corresponding position of the experiment to calculate the system vibration response. The experimental results are then compared with the simulation results. The comparison results are illustrated in Figures 6–8.

Through in-depth analysis of the experimental data, we can observe that, under the condition of exceeding the critical speed, the filtered experimental results of the rotor system exhibit a high degree of consistency with the simulation results. This comparison, to a certain extent, validates the rationality and accuracy of the established dynamic model. Specifically, the experimental waveform aligns well with the simulation waveform in terms of shape, amplitude, and phase. However, under subcritical and critical speed conditions, the time-domain waveform of the rotor system shows certain discrepancies. These discrepancies may primarily arise from the complexity of the test bench system. Specifically, the interfacial contact friction of components such as bearings and bolts within the test bench may more readily induce nonlinear behavior in the system. These nonlinear factors may include but are not limited to, variations in friction force, minute deformations at the contact interfaces, and the additional vibrations resulting from these. These factors collectively contribute to a certain deviation between the experimental waveform and the simulation waveform. To gain a deeper understanding of these discrepancies and further improve the accuracy of the model, future research should strengthen the study of nonlinear behavior in the test bench system, including the establishment of more precise nonlinear dynamic models.

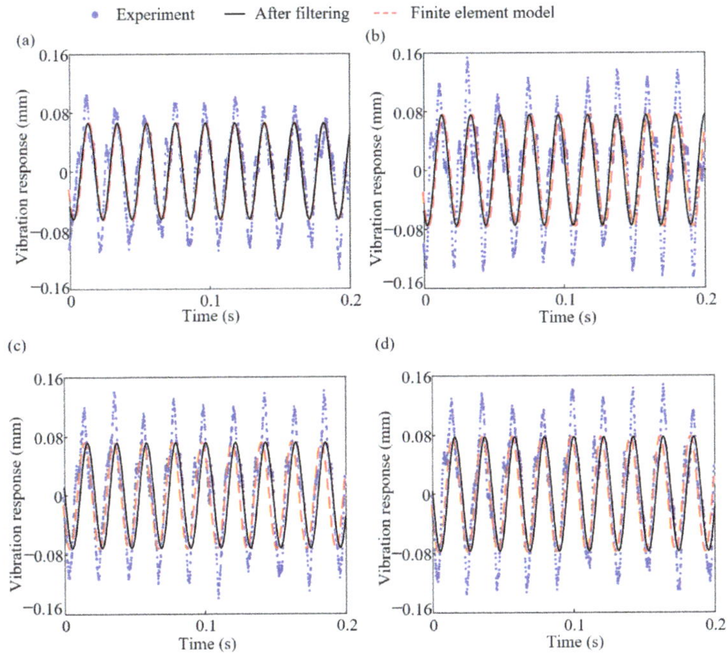

Figure 6. Comparisons between experimental results and simulation results of rotor system at a sub-critical speed of 5600 rpm: (**a**) boss 1 unbalanced; (**b**) boss 2 unbalanced; (**c**) boss 3 unbalanced; and (**d**) boss 4 unbalanced.

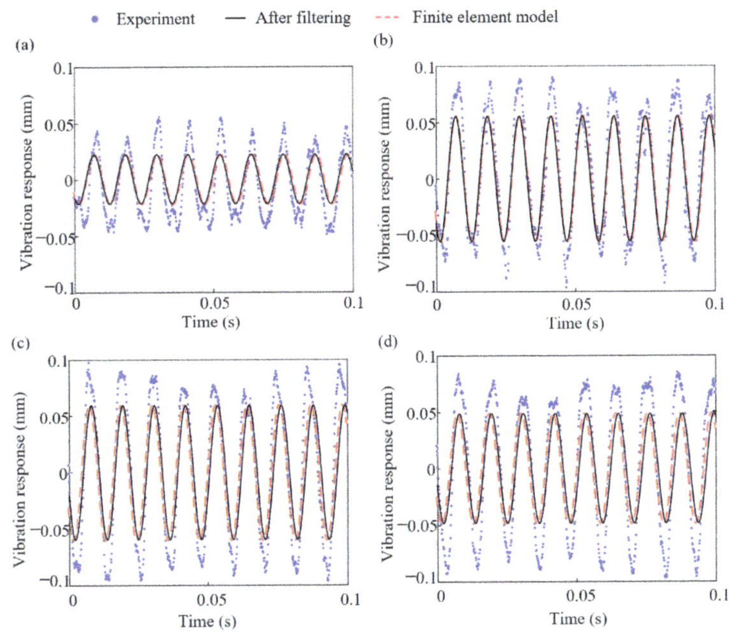

Figure 7. Comparisons between experimental results and simulation results of rotor system at a critical speed of 7260 rpm: (**a**) boss 1 unbalanced; (**b**) boss 2 unbalanced; (**c**) boss 3 unbalanced; and (**d**) boss 4 unbalanced.

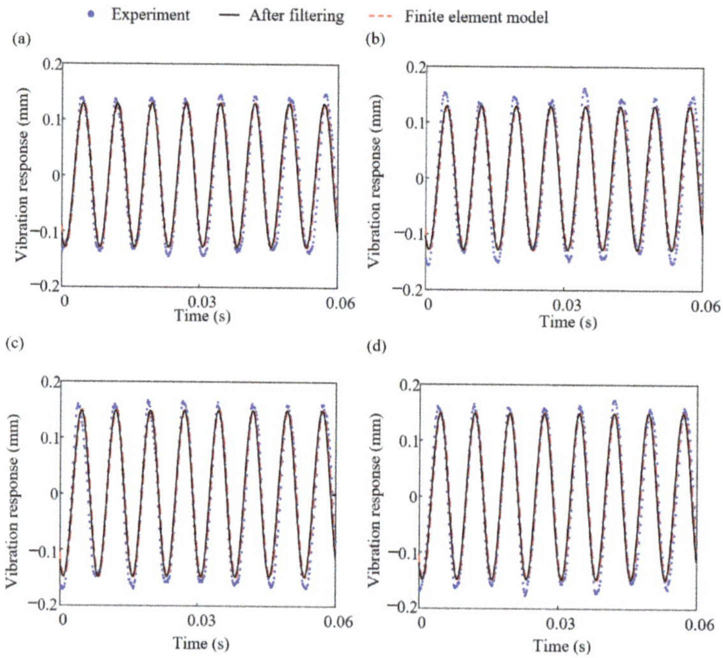

Figure 8. Comparisons between experimental results and simulation results of rotor system at an over-critical speed of 8400 rpm: (**a**) boss 1 unbalanced; (**b**) boss 2 unbalanced; (**c**) boss 3 unbalanced; and (**d**) boss 4 unbalanced.

3.2. Long Short-Term Memory Neural Network

In the research of high-speed dynamic balancing of flexible rotors, the identification of unbalanced rotor positions has always been a challenging problem. Especially for slender rotors, the axial location of unbalance has a significant impact on the high-speed dynamic balancing of the rotor. In recent years, deep learning, as a powerful machine learning technique, has achieved remarkable results in fields such as image recognition and speech recognition. In the area of unbalanced rotor position identification, deep learning techniques based on limited labeled data have also demonstrated great potential. By training deep neural networks, effective feature information can be extracted from the limited labeled data, thereby enabling accurate identification of unbalanced positions.

The LSTM (Long Short-Term Memory) network significantly enhances the neural network's capability in processing time-series data by introducing memory cells and control units such as input gates, forget gates, and output gates. The core of its working mechanism lies in the self-looping and feedback mechanism of the internal state of the memory cells. This mechanism, combined with sophisticated gating strategies, regulates the inflow and outflow of information streams, achieving effective truncation of error signals. This ensures the stability and integrity of errors during the backward propagation process, thereby avoiding common issues in traditional recurrent neural networks, such as gradient vanishing or gradient exploding. Figure 9 provides a detailed illustration of the structural composition of the LSTM unit.

The computational process of LSTM can be divided into three stages based on its three gate structures: the forget stage, the input stage, and the output stage. During the forget stage, the forget gate assesses the significance of the information in the cell state at time ($t-1$) and selects whether to retain or discard it. The important information that is retained is then passed on to the input stage:

$$f_{t1} = \sigma(\boldsymbol{W}_f[S_{t-1}, X_t] + \boldsymbol{b}_f) \tag{16}$$

where $\boldsymbol{W}_f$ and $\boldsymbol{b}_f$ are the weight matrix and bias vector of the forgetting gate, respectively, σ is the Sigmoid activation function, S_{t-1} represents the output value of the network hidden layer at time ($t-1$) and X_t is the input parameter of the structure at time t. In addition, when $f_{t1} = 0$, it represents the cell state of ($t-1$) at the moment of complete abandonment, while when $f_{t1} = 1$, it represents the cell state of complete retention of ($t-1$).

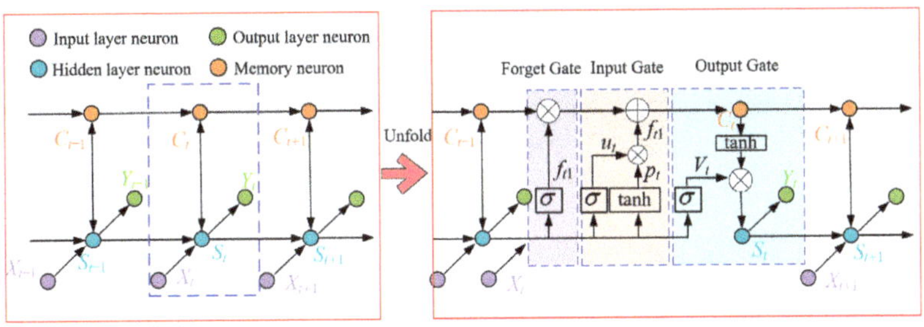

Figure 9. LSTM structure diagram.

The input stage mainly stores the input data at time ($t-1$) and the important information in the hidden state at time t in the cell state through the relevant calculation of the input gate:

$$U_t = \sigma(\boldsymbol{W}_u[S_{t-1}, X_t] + \boldsymbol{b}_u) \tag{17}$$

$$U_t = \sigma(\boldsymbol{W}_u[S_{t-1}, X_t] + \boldsymbol{b}_u) \tag{18}$$

$$f_{t2} = U_t \otimes P_t \tag{19}$$

where $\boldsymbol{W}_u$ and $\boldsymbol{W}_p$ are the weight matrices of input gate and loop cell states, respectively, $\boldsymbol{b}_u$ and $\boldsymbol{b}_p$ are the bias vectors of the input gate and the loop unit, respectively, and $\otimes$ represents the element-by-element product operation between vectors.

The output phase mainly determines the output of the LSTM unit at time t through the output gate:

$$o_t = sigmoid(\boldsymbol{W}_o[S_{t-1},\ X_t] + \boldsymbol{b}_o) \tag{20}$$

$$C_t = C_{t-1} \otimes f_{t1} \oplus f_{t2} \tag{21}$$

$$S_t = o_t \otimes tanh(C_t) \tag{22}$$

where C_t and C_{t-1} are the output of the memory unit at time t and ($t-1$), respectively. $\boldsymbol{W}_o$ and $\boldsymbol{b}_o$ are the weight matrix and bias vector of the output stage, respectively. For a single-output network, the output of the output layer is represented as follows:

$$Y_t = sigmoid(\boldsymbol{W}_Y S_t + \boldsymbol{b}_Y) \tag{23}$$

where $\boldsymbol{Y}_t$ is the corresponding matrix of X_t at time t, and $\boldsymbol{W}_Y$ and $\boldsymbol{b}_Y$ are the weight matrix and bias vector of the output gate.

3.3. Unbalanced Position Identification of Rotor Based on LSTM

In flexible rotor systems, unbalance is one of the primary causes of vibration and noise. For slender rotors, due to their long axial extension, the occurrence of unbalance at different locations will directly affect the dynamic characteristics and stability of the rotor. Therefore, accurately identifying the axial location of unbalance is of great significance for achieving high-speed dynamic balancing of the rotor and improving the overall performance of rotating machinery.

3.3.1. Data Preprocessing

Initially, four unbalanced positions, A, B, C, and D, are selected on the constructed rotor test bench, with their specific layout illustrated in Figure 5. At each of these positions, an unbalanced mass of 15 gmm with a phase of 30° is manually added to obtain the measured vibration response of the rotor at each unbalanced position. To ensure the accuracy and reliability of the data, preprocessing steps are taken to eliminate noise and outliers from the raw data. Furthermore, dynamic simulations are performed with consistent parameter settings to calculate the vibration displacement of the nodes. Based on this, both filtered measured vibration data and dynamic model simulation data are collected, covering vibration responses under different unbalanced positions (A, B, C, D). This provides a reliable data foundation for subsequent model training and validation. Given the significant advantages of LSTM models in processing sequential data, the vibration data are arranged in time series to form input sequences. To facilitate the recognition and classification of unbalanced positions by the LSTM model, the unbalanced positions (A, B, C, D) are converted into corresponding numerical labels for model recognition and classification.

3.3.2. Optimization Model Construction

In designing the LSTM neural network for rotor imbalance position recognition, careful consideration is given to the choice of network structure. Since an unbalanced distribution state corresponds to the vibration data composed of four measurement points of the rotor, the number of nodes in the input layer of the designed neural network is 4. Hidden Layer determines the ability of neural networks to learn features, and the number of hidden layers and nodes determines the complexity of the network to process data and the mapping ability of complex functions. In general, the formula for calculating the number of hidden layer nodes is as follows:

$$n_m = \sqrt{n_x + n_y} + a \tag{24}$$

where n_m indicates the number of hidden layer nodes, n_x is the number of nodes in the input layer, n_y is the number of nodes in the output layer, and a is the constant between the intervals [1, 10].

In addition, the choice of activation function is crucial for the performance of neural networks. In this case, the Sigmoid function is selected for the hidden layers compared to other activation functions like ReLU; the Sigmoid function has a smoother gradient that can help mitigate issues with gradient vanishing, especially in deeper networks. Additionally, its output range is well-suited for representing probabilities, although this is not directly leveraged in the output layer, which instead uses the Softmax function for classification. The Sigmoid function's properties made it a suitable choice for enhancing the network's nonlinear processing capabilities and improving its generalization performance on the rotor imbalance position recognition task. Following the LSTM layers, fully connected layers are added for feature extraction and classification. The output layer employed the Softmax function, which converted the outputs of the fully connected layer into a probability distribution, indicating the likelihood of each possible rotor imbalance position. Overall, the constructed multi-layer regression LSTM neural network model, as illustrated in Figure 10, is designed to strike an optimal balance between network complexity and convergence speed, making it well-suited for the task of rotor imbalance position recognition.

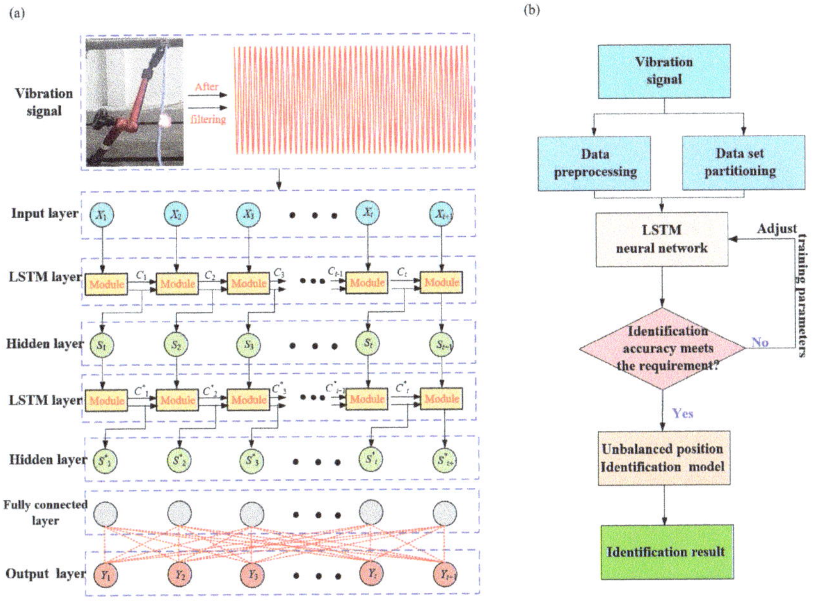

Figure 10. The LSTM neural network: (**a**) structure diagram; (**b**) identification process.

3.3.3. Model Training

The preprocessed data is divided into training, validation, and test sets, which are respectively utilized for model training, validation, and evaluation. For the purpose of model training and optimization, the cross-entropy loss function and Adam optimizer are selected. The training set data is input into the LSTM model for multiple iterations of training, continuing until the model's performance on the validation set stabilizes or begins to decline. It is noteworthy that the loss function serves as a crucial metric in neural network training, quantifying the error between the network's output and the desired output. A lower loss function value indicates better network performance. During the training process, the loss function value for all samples in the training set is calculated in each iteration, and the weights and biases of the neural network are updated according to the optimizer settings to minimize the loss function value. Simultaneously, the validation set is used to evaluate the network's performance and generalization ability, ensuring that the model does not suffer from overfitting. The loss curve during the training process at a subcritical rotational speed of 5600 rpm, as shown in Figure 11, demonstrates that the model exhibits good performance on both the training and validation data.

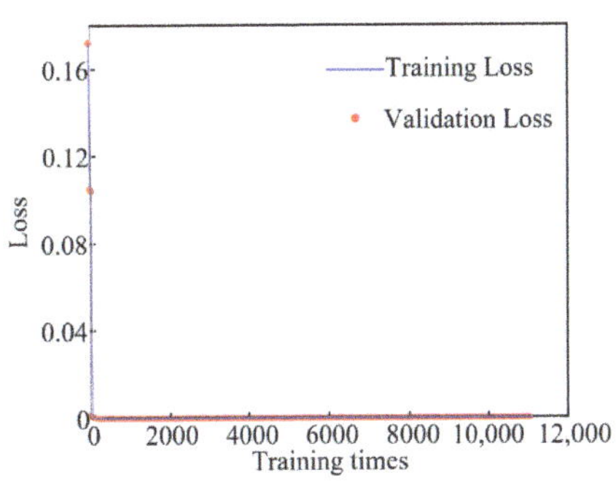

Figure 11. Loss curve of LSTM model.

4. Discussion

At present, researchers have explored various techniques for unbalanced identification in rotating machinery. Traditional methods often involve vibration analysis using Fast Fourier Transform (FFT) or other signal processing techniques to detect harmonic components indicative of unbalance. However, these methods can be limited in their ability to accurately pinpoint the exact location of the unbalance, especially in complex systems like aero engines. More advanced approaches, such as those utilizing machine learning algorithms, have emerged in recent years. While these methods have shown promise, they may struggle with the temporal dependencies inherent in vibration data, which can be crucial for accurate identification.

To solve the difficult problem of unbalanced position identification of an aero-engine rotor, a method of unbalanced position identification of an aero-engine power turbine rotor based on the LSTM is proposed. This method addresses this limitation by capturing the long-term dependencies in the vibration signals. LSTM networks are particularly well-suited for sequential data analysis, making them ideal for processing vibration signals recorded over time. By training the LSTM model on a comprehensive dataset that includes both experimental and simulation data from different rotational speeds, the method is able to learn the complex relationships between vibration patterns and unbalanced positions. The unbalanced position is identified by the neural network model trained by experimental data and simulation data at different rotational speeds. The results are as follows:

(1) At a subcritical rotational speed of 5600 rpm, the measured unbalanced rotor vibration response data from the test bench and the vibration response data obtained through simulation are input into the LSTM model for training, with the aim of identifying the unbalanced rotor position. The results of unbalanced rotor position identification based on the LSTM model are presented in Figure 12. Figure 12a shows that the accuracy of identification results based on measured data can be controlled above 97.5%, with an impressive 100% accuracy when unbalanced masses are present at positions B and D. Figure 12b demonstrates that the accuracy of identification results based on simulated data can be maintained above 91.2%, achieving a perfect 100% accuracy specifically when an unbalanced mass is at position B. These results validate the effectiveness of the proposed method in identifying unbalanced positions at subcritical speeds.

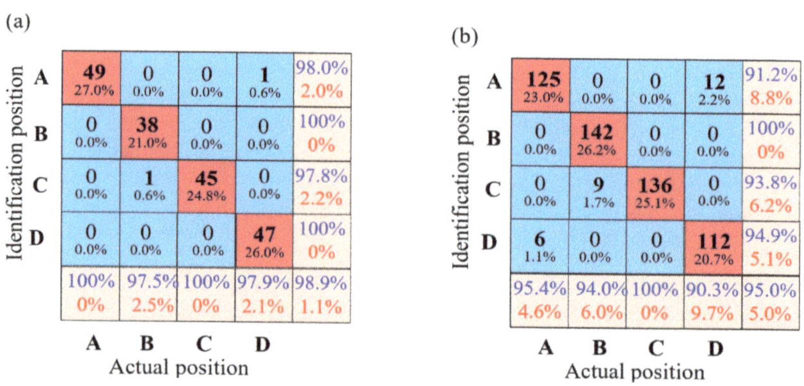

Figure 12. Identification results of unbalanced rotor position at subcritical speed based on LSTM: (**a**) based on experiment; (**b**) based on simulation.

(2) According to the experimental test results, the first-order critical speed of the rotor system is 7260 rpm. Therefore, by varying four unbalanced positions at the critical speed, the unbalanced vibration response data measured on the test bench and the vibration response data calculated through simulations are input into the LSTM model for training in order to identify the unbalanced position of the rotor at the critical speed, as shown in Figure 13. Figure 13a indicates that the accuracy of identification results based on measured

data can be maintained above 95.5%, with an impressive 100% accuracy, specifically when unbalanced masses are present at positions A and B. Figure 13b demonstrates that the accuracy of identification results based on simulated data can be controlled above 93.1%, achieving a perfect 100% accuracy when an unbalanced mass is at position A. These results validate the effectiveness of the proposed method in identifying unbalanced positions at the critical speed.

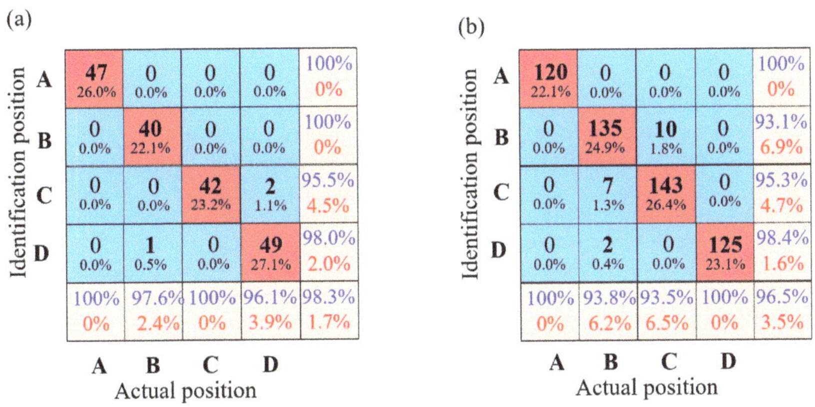

Figure 13. Identification results of unbalanced rotor position at a critical speed based on LSTM: (**a**) based on experiment; (**b**) based on simulation.

(3) When varying four unbalanced positions at supercritical speed (8400 rpm), the unbalanced vibration response data measured on the test bench and the vibration response data calculated through simulations are input into the LSTM model for training in order to identify the unbalanced position of the rotor at supercritical speed, as shown in Figure 14. Figure 14a indicates that the accuracy of identification results based on measured data can be maintained above 97.8%, with an impressive 100% accuracy, specifically when unbalanced masses are present at positions A and B. Meanwhile, Figure 14b demonstrates that the accuracy of identification results based on simulated data can be controlled above 90.5%, achieving a perfect 100% accuracy when an unbalanced mass is at position A. These results validate the effectiveness of the proposed method in identifying unbalanced positions at supercritical speed.

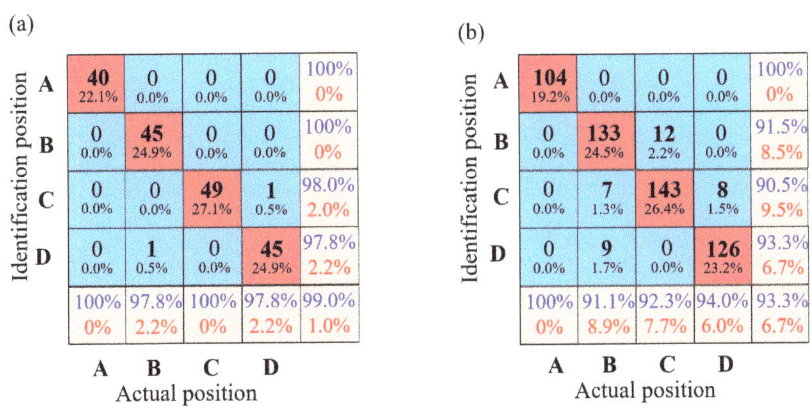

Figure 14. Identification results of unbalanced rotor position at supercritical speed based on the LSTM: (**a**) based on experiment; (**b**) based on simulation.

In summary, the LSTM-based approach offers superior performance in identifying unbalanced positions with high accuracy. Preliminary results, as presented earlier, indicate that the method can achieve accuracies above 90.5% for experimental data and above 95.5% for simulated data, with specific cases reaching perfect accuracy. By leveraging the strengths of LSTM networks and incorporating both experimental and simulation data, the method demonstrates the potential to overcome the limitations of traditional and other machine learning-based approaches, thereby enhancing the reliability and efficiency of aero engine maintenance and operation.

5. Conclusions

This paper focuses on the study of turbo-rotor systems in aero-engines, addressing critical issues such as excessive high-speed dynamic balance vibrations and the challenges associated with achieving dynamic balance. The main contributions of this paper include the following:

(1) A rotor system dynamics model is proposed based on the finite element method. In this paper, the finite element method is utilized to construct a dynamics model for the turbo-rotor system. This model can more accurately describe the dynamic behavior of the rotor system, providing a solid foundation for subsequent vibration analysis and research on unbalanced responses;

(2) A turbo-rotor vibration test bench is designed and validated. To verify the accuracy and effectiveness of the established model, this paper designed a turbo-rotor vibration test bench and, through comparison with measured data, validated that the prediction error of the model's natural frequencies is controlled within 5%, demonstrating the model's high precision and practicality;

(3) Unbalanced responses at different rotational speeds are studied. This paper conducted an in-depth analysis of unbalanced responses of the rotor at subcritical, critical, and supercritical speeds. By comparing experimental and simulation results, it verified the good consistency and minimal errors between them, providing important references for vibration control and optimal design of the rotor system.

(4) A method for identifying unbalanced positions based on LSTM neural networks is proposed. This paper innovatively proposed a method for identifying unbalanced positions in aero-engine turbo-rotors using Long Short-Term Memory (LSTM) neural networks. The recognition accuracy of this method reached 95.5% and 90.5% on experimental and simulation data, respectively, providing new ideas and technical means for intelligent detection and processing of rotor imbalance issues.

The innovation of this paper lies in the first application of LSTM neural networks to identify unbalanced positions in aero-engine turbo-rotors. Traditional methods for identifying unbalanced positions may rely on human experience and complex calculation processes. However, the method based on LSTM neural networks achieves intelligent recognition, is capable of automatically processing and analyzing data, reduces dependence on human experience, and improves the efficiency and accuracy of identification. By training and optimizing the LSTM neural networks, high accuracy in unbalanced position recognition is achieved on both experimental and simulation data, providing new ideas and technical means for vibration control and optimal design of aero-engine rotor systems.

Author Contributions: Conceptualization, X.L.; methodology, Y.C.; software, C.S.; validation, M.L.; investigation, J.B.; resources, C.S.; data curation, C.S.; writing—original draft preparation, Y.C.; writing—review and editing, X.L.; visualization, M.L.; supervision, J.B.; project administration, X.L.; funding acquisition, X.L. All authors have read and agreed to the published version of the manuscript.

Funding: This research was funded by the National Natural Science Foundation (Grant no. 12072076).

Data Availability Statement: Data are contained within the article.

Conflicts of Interest: The author Mingfeng Li is from Aecc South Industry Company Limited, Zhuzhou 412000, China. All authors declare that they have no conflicts of interest.

Nomenclature

A	Cross-sectional area of the element
$\boldsymbol{b}_f$	Bias vector of the forgetting gate
$\boldsymbol{b}_u$ and $\boldsymbol{b}_p$	Bias vectors of the input gate and the loop unit
c_{bn} $(n = u, v, w, \theta_x, \theta_y, \theta_z)$	Support damping of n-direction
c_{cn} $(n = u, v, w, \theta_x, \theta_y, \theta_z)$	Coupling damping of n-direction
C_t and C_{t-1}	Output of the memory unit at t and $(t-1)$
$\boldsymbol{C}$	Damping matrix of the rotor system
$\boldsymbol{C}_b^e\boldsymbol{K}_b^e$	Damping and stiffness matrix of the support element
d_s	Diameter of the shaft section
e_{1i} and e_{2i}	Radial and axial eccentricity
E	Young's modulus.
F_{cn} $(n = u, v, w, \theta_x, \theta_y, \theta_z)$	Nonlinear bearing forces of the coupling
$\boldsymbol{F}_b$	Unbalanced rotor force vector
F_{bui}, F_{bvi}, M_{bui}, and M_{bvi}	Unbalance forces and moments located at node i
m_d	Disk mass
$\boldsymbol{M}$	Mass matrix of the rotor system
g	Gravity acceleration
$\boldsymbol{G}$	Gyroscopic matrix of the rotor system
$\boldsymbol{G}_s$	Gravity vector
$\boldsymbol{G}_d^e\boldsymbol{M}_d^e$	Disk element gyroscopic, mass matrix
$\boldsymbol{G}^e, \boldsymbol{K}^e, \boldsymbol{M}^e$	Beam element gyroscopic, mass, and stiffness matrix
h_d	Thickness of the disk
I_x, I_y and I_z	Moment of inertia of the section
J	Moment of inertia of the torsion section
J_d and J_p	Diameter moment and polar moment of inertia of the disk
k_{bn} $(n = u, v, w, \theta_x, \theta_y, \theta_z)$	Support stiffness of n-direction
k_{cn} $(n = u, v, w, \theta_x, \theta_y, \theta_z)$	Coupling stiffness of n-direction
$\boldsymbol{K}$	Stiffness matrix of the rotor system
N_n $(n = u, v, w, \theta_x, \theta_y, \theta_z)$	Shape functions of elements in six directions.
$\boldsymbol{q}$	Generalized displacement vector
$\boldsymbol{q}^e$	Node element displacement vector
r_d, R_d	Inner and outer radius of the disk
S_{t-1}	Output value at $(t-1)$
$u_i, v_i, w_i, \theta_{xi}, \theta_{yi}, \theta_{zi}$	Displacements of node i in different directions
v	Poisson's ratio
$\boldsymbol{W}_u$ and $\boldsymbol{W}_p$	Weight matrices of input gate and loop cell states
$\boldsymbol{W}_o$ and $\boldsymbol{b}_o$	Weight matrix and bias vector of the output stage
$\boldsymbol{W}_f$	Weight matrix of the forgetting gate
X_t	Input parameter at t
α_0 and β_0	Rayleigh damping coefficients
ω	Rotor speed
κ_y and κ_z	Shear coefficient
Ω	Angular velocity of rotation
φ_i	Initial phase of the unbalance
σ	Sigmoid activation function
ρ	Material density
δ_{ab}	Matching clearance

References

1. Deepthikumar, M.B.; Sekhar, A.S.; Srikanthan, M.R. Modal balancing of flexible rotors with bow and distributed unbalance. *J. Sound Vib.* **2013**, *332*, 6216–6233. [CrossRef]
2. Wang, A.; Yao, W.; He, K. Analytical modelling and numerical experiment for simultaneous identification of unbalance and rolling-bearing coefficients of the continuous single-disc and single-span rotor-bearing system with Rayleigh beam model. *Mech. Syst. Signal Process.* **2019**, *116*, 322–346. [CrossRef]
3. Zhao, S.N.; Zhang, F.B.; Qin, Z.Y. Dynamic response analysis of a shaft-disk-drum rotor system with interval uncertainties. *Mech. Based Des. Struct. Mach.* **2024**, *52*, 3951–3973. [CrossRef]

4. Miao, J.; Wang, A.L.; Wang, Q.S. Rub-impact dynamic analysis of the central tie rod rotor-blade-casing coupling system with the Hirth couplings connection. *J. Vib. Eng. Technol.* **2024**, *12*, 1479–1503.
5. Xu, X.P.; Liu, Y.; Han, Q.K. A universal dynamic model and solution scheme for the electrical rotor system with wide range of eccentricity. *Int. J. Non-Linear Mech.* **2023**, *152*, 104402. [CrossRef]
6. Luo, Z.; Liang, B.B.; Sun, K. Dynamic characteristic analysis of the elastic ring squeeze film damper-rotor system considering coupling angular misalignment. *Nonlinear Dyn.* 2024. [CrossRef]
7. Li, Y.; Luo, Z.; Liu, J. Dynamic modeling and stability analysis of a rotor-bearing system with bolted-disk joint. *Mech. Syst. Signal Process.* **2021**, *158*, 107778. [CrossRef]
8. Li, X.B.; Xu, Y.J.; Huang, S.H. Dynamic analysis of a flexible rotor system considering the bearing slip. *Mech. Based Des. Struct. Mach.* **2024**. [CrossRef]
9. Gupta, K.; Kumar, R.; Tiwari, M. Effect of rotary inertia and gyroscopic moments on dynamics of two spool aeroengine Rotor. *Am. Soc. Mech. Eng.* **1993**. [CrossRef]
10. Qin, Z.Y.; Han, Q.K.; Chu, F.L. Analytical model of bolted disk-drum joints and its application to dynamic analysis of jointed rotor. *Proc. Inst. Mech. Eng. Part C J. Mech. Eng. Sci.* **2014**, *228*, 646–663. [CrossRef]
11. Lu, Z.Y.; Hou, L.; Chen, Y.S. Nonlinear response analysis for a dual-rotor system with a breathing transverse crack in the hollow shaft. *Nonlinear Dyn.* **2016**, *83*, 169–185. [CrossRef]
12. Chen, K.K.; Huangfu, Y.F.; Zhao, Z.F. Dynamic modeling of the gear-rotor systems with spatial propagation crack and complicated foundation structure. *Mech. Mach. Theory* **2022**, *172*, 104827. [CrossRef]
13. Ahmed, O.; El-Sayed, T.; Sayed, H. Finite element analyses of rotor/bearing system using second-order journal bearings stiffness and damping coefficients. *J. Vib. Control* **2023**, *30*, 3961–3984. [CrossRef]
14. Yu, P.; Zhang, D.; Ma, Y. Dynamic modeling and vibration characteristics analysis of the aero-engine dual-rotor system with Fan blade out. *Mech. Syst. Signal Process.* **2018**, *106*, 158–175. [CrossRef]
15. Wang, F.; Luo, G.H.; Yan, S. A comparison study on co- and counterrotating dual-rotor system with squeeze film dampers and intermediate bearing. *Shock Vib.* **2017**, *2017*, 1–25. [CrossRef]
16. Yang, Y.; Cao, D.Q.; Yu, T.H. Prediction of dynamic characteristics of a dual-rotor system with fixed point rubbing-Theoretical analysis and experimental study. *Int. J. Mech. Sci.* **2016**, *115*, 253–261. [CrossRef]
17. Li, R.; Liu, Y.; Sun, C. Measurement and adjustment of the centroid deviation of aero-engine rotors based on unbalance decoupling. *Measurement* **2023**, *212*, 112660. [CrossRef]
18. Yang, Y.J.; Yao, M.H.; Niu, Y. Model verification and vibration analysis of the four-disk hollow flexible shaft rotor system. *Int. J. Mech. Sci.* **2024**, *268*, 109051. [CrossRef]
19. Liu, Y.; Zhang, M.; Sun, C. A method to minimize stage-by-stage initial unbalance in the aero engine assembly of multistage rotors. *Aerosp. Sci. Technol.* **2019**, *85*, 270–276. [CrossRef]
20. Chen, Y.; Cui, J.; Sun, X. An assembly method for the multistage rotor of an aero-engine based on the dual objective synchronous optimization for the coaxality and unbalance. *Aerospace* **2021**, *8*, 94. [CrossRef]
21. Chen, Y.; Cui, J.; Sun, X. A vibration suppression method for the multistage rotor of an aero-engine based on assembly optimization. *Machines* **2021**, *9*, 189. [CrossRef]
22. Song, L.L.; Chen, Y. A prediction model of two-sided unbalance in the multi-stage assembled rotor of an aero engine. *Machines* **2024**, *12*, 503. [CrossRef]
23. Jiang, K.J.; Zhu, C.S.; Chen, L.L. Unbalance compensation by recursive seeking unbalance mass position in active magnetic bearing-rotor system. *IEEE Trans. Ind. Electron.* **2015**, *62*, 5655–5664.
24. Ma, P.P.; Zhai, J.Y.; Wang, Z.H.M. Unbalance vibration characteristics and sensitivity analysis of the dual-rotor system in aeroengines. *J. Aerosp. Eng.* **2021**, *34*, 04020094. [CrossRef]
25. Wang, W.B.; Song, P.Y.; Yu, H.C. Research on vibration amplitude of ultra-precision aerostatic motorized spindle under the combined action of rotor unbalance and hydrodynamic effect. *Sensors* **2023**, *23*, 496. [CrossRef] [PubMed]
26. Han, S.; Wang, Z.; Zhang, H. Flexible rotor unbalance fault location method based on transfer learning from simulation to experiment data. *Meas. Sci. Technol.* **2023**, *34*, 125053. [CrossRef]
27. Han, S.; Yang, T.G.; Zhu, J.Y. Unbalance position of aeroengine flexible rotor analysis and identification based on dynamic model and deep learning. *Proc. Inst. Mech. Eng. Part C J. Mech. Eng. Sci.* **2023**, *237*, 4410–4429. [CrossRef]
28. Santos, F.L.; Duarte, M.L.M.; de Faria, M.T.C. Balancing of a rigid rotor using artificial neural network to predict the correction masses. *Acta Sci. Technol.* **2009**, *31*, 151–157. [CrossRef]
29. Walker, R.; Vayanat, R.; Perinpanayagam, S. Unbalance localization through machine nonlinearities using an artificial neural network approach. *Mech. Mach. Theory* **2014**, *75*, 54–66. [CrossRef]
30. Yan, J.; Hu, Y.; Guo, C. Rotor unbalance fault diagnosis using DBN based on multi-source heterogeneous information fusion. *Procedia Manuf.* **2019**, *35*, 1184–1189. [CrossRef]
31. Zhong, S.; Han, H.X.; Hou, L. A two-stage deep-learning-based balancing method for rotating machinery. *Meas. Sci. Technol.* **2023**, *34*, 045903. [CrossRef]
32. Zhong, S.; Li, L.; Chen, H. A novel balancing method for rotor using unsupervised deep learning. *Shock Vib.* **2021**, *2021*, 1–9. [CrossRef]

33. Xing, Z.X.; Chen, M.Y.; Cui, J. Detection of magnitude and position of rotor aerodynamic unbalance of wind turbines using Convolutional Neural Network. *Renew. Energy* **2022**, *197*, 1020–1033. [CrossRef]
34. Hochreiter, S.; Schmidhuber, J. Long short-term memory. *Neural Comput.* **1997**, *9*, 1735–1780. [CrossRef]
35. Van Houdt, G.; Mosquera, C.; Nápoles, G. A review on the long short-term memory model. *Artif. Intell. Rev.* **2020**, *53*, 5929–5955. [CrossRef]
36. Sheu, H.C.; Chen, S.W. A lumped mass model for parametric instability analysis of cantilever shaft–disk systems. *J. Sound Vib.* **2000**, *234*, 331–348. [CrossRef]
37. Ma, H.; Li, H.; Niu, H. Numerical and experimental analysis of the first-and second-mode instability in a rotor-bearing system. *Arch. Appl. Mech.* **2014**, *84*, 519–541. [CrossRef]
38. Wang, P.; Zhao, X.; Xu, H. Nonlinear dynamic behaviors of a shaft-bearing-pedestal system with outer ring slip and damage. *Nonlinear Dyn.* **2024**. [CrossRef]

machines

MDPI

Article

Aero-Engine Fault Detection with an LSTM Auto-Encoder Combined with a Self-Attention Mechanism

Wenyou Du [1,*], Jingyi Zhang [1], Guanglei Meng [1] and Haoran Zhang [2]

1 College of Automation, Shenyang Aerospace University, Shenyang 110136, China; zhangjingyi2@stu.sau.edu.cn (J.Z.); mengguanglei@yeah.net (G.M.)
2 Tianjin Jepsen International Flight College Co., Ltd., Tianjin 300399, China; 15042840794@163.com
* Correspondence: wen-you.du@sau.edu.cn

Abstract: The safe operation of aero-engines is crucial for ensuring flight safety, and effective fault detection methods are fundamental to achieving this objective. In this paper, we propose a novel approach that integrates an auto-encoder with long short-term memory (LSTM) networks and a self-attention mechanism for the anomaly detection of aero-engine time-series data. The dataset utilized in this study was simulated from real data and injected with fault information. A fault detection model is developed utilizing normal data samples for training and faulty data samples for testing. The LSTM auto-encoder processes the time-series data through an encoder–decoder architecture, extracting latent representations and reconstructing the original inputs. Furthermore, the self-attention mechanism captures long-range dependencies and significant features within the sequences, thereby enhancing the detection accuracy of the model. Comparative analyses with the traditional LSTM auto-encoder, as well as one-class support vector machines (OC-SVM) and isolation forests (IF), reveal that the experimental results substantiate the feasibility and effectiveness of the proposed method, highlighting its potential value in engineering applications.

Keywords: fault detection; LSTM; auto-encoder; self-attention mechanism

1. Introduction

In recent years, the civil aviation industry has developed rapidly in China. With the ongoing expansion of domestic airlines and the rapid growth of the fleet, the safety and reliability of civil aircraft have garnered significant attention from both the government and the public [1–3]. The safe operation and dependable performance of civil aircraft are largely contingent upon the stability and reliability of the aero-engine. As the core component of the aircraft, the aero-engine is particularly susceptible to various failures due to its complex structure and harsh operational environment, especially as operating time increases [4,5]. This highlights the significance of fault diagnosis as a crucial area of research for guaranteeing aviation safety [6–8]. When an aviation engine failure occurs, it can not only disrupt flight operations, but also pose a serious threat to the safety of the passengers and crew onboard. Consequently, establishing an effective fault detection system to detect potential faults in aero-engines in a timely manner is crucial for facilitating prompt repairs and ensuring flight safety [9–11].

Fault diagnosis methods for aero-engines can be broadly categorized into three main types: physical model-based fault diagnosis methods, expert system-based fault diagnosis methods, and data-driven fault diagnosis methods. Model-based approaches involve constructing a mathematical model of the engine based on its operational mechanisms and conducting fault detection through sophisticated physical models employing various filtering algorithms [12–15]. However, the inherent complexity of aero-engines poses significant challenges in directly modeling physical faults. Expert system-based diagnostic methods leverage heuristic expertise and the knowledge of domain specialists to establish causal relationships between system failure modes and abnormal monitoring states [16–18].

Nonetheless, these approaches encounter difficulties in acquiring a priori knowledge and performing dynamic reasoning, which limits their accuracy. Conversely, data-driven diagnostic approaches diagnose failure modes by extracting failure features from monitoring data and establishing mapping relationships between these features and the corresponding failure modes [19–22]. These methods depend on monitoring data collected during system operation and do not necessitate physical models or extensive expertise. With the ongoing accumulation of aircraft maintenance data and advancements in deep learning techniques, the integration of data-driven and deep-learning-based fault diagnosis methods has gradually found application in aviation engine fault diagnosis tasks [23–26].

Current data-driven fault diagnosis methods for aero-engines primarily rely on time-series data collected from various engine components. Feature extraction and fault diagnosis are accomplished through the integration of deep learning and machine learning techniques. However, the collected aero-engine data are highly imbalanced, with normal data significantly outnumbering fault data. Additionally, labeling must be conducted manually by experts, complicating the acquisition of accurate and representative labeled samples [27]. In contrast, unsupervised fault detection methods do not require labeled data to identify anomalies or faults, making them more promising for development in the context of aero-engines. Liu et al. [28] employed a hyperplane-oriented over-sampling technique to address the issue of sparse fault samples in aero-engine training data. Kang et al. [29] proposed a fault detection model based on distillation learning to solve the problem of limited model generalization due to a lack of high-quality fault data samples for aero-engine rolling bearings. Zhong et al. [30] employed a deep learning approach based on stacked auto-encoder networks (SAE) to diagnose and classify various types and levels of bearing faults. Furthermore, Shi et al. [31] developed an innovative self-supervised domain adversarial generalization framework designed to address the fault diagnosis challenges in aviation high-speed bearings under complex operating conditions. Additionally, Li et al. [32] examined the use of LSTM neural networks along with Dempster–Shafer (D-S) evidence theory to develop a novel fault diagnosis method based on sensor information fusion. Guo et al. [33] presented a signal anomaly detection algorithm leveraging an attention mechanism, which first utilizes the attention mechanism to capture the long-range characteristics of spacecraft telemetry data and subsequently employs a stacked auto-encoder to reduce data dimensions and reconstruct the input signal, yielding an error reconstruction sequence. Anomaly detection of spacecraft telemetry signals is then performed by identifying anomaly indicators within the error reconstruction sequence using a window threshold method. Lastly, Zhao et al. [34] introduced a new model that integrates a variational auto-encoder (VAE) with a regression layer, referred to as a VAR. The VAE comprises a bidirectional LSTM network enhanced with an attentional mechanism. The VAR model was validated through simulations using the C-MAPSS dataset, demonstrating superior performance compared to existing methods.

While the above methods perform well in feature extraction, they often require a large amount of labelled data, limiting their ability to generalize, and have high computational complexity. Meanwhile, aero-engine operation process data are characterized by strong dynamic properties, multi-modal imbalance, and complex coupling. In the field of fault detection in aero-engines, the combination of LSTM and a self-attention mechanism provides a solution to the above problem. The self-attention mechanism further enhances the model's ability to recognise complex coupling relationships between features at different points in time by calculating correlations between different parts of the sequence, which is particularly important when dealing with multi-modal unbalanced data. This combination allows the model to not only understand the temporal dynamics of the data, but also to flexibly focus on the features that are most critical for fault detection, thus improving the accuracy and reliability of fault detection in complex and variable aero-engine operational data. In addition to this, self-attention mechanisms can be easily integrated with other model structures to improve the performance of these models, providing a promising solution to multi-modal dynamic system failure detection problems.

In summary, an unsupervised fault detection method using an LSTM auto-encoder combined with a self-attention mechanism is proposed in this paper, which is specifically applied to the fault detection of piston engines. The dataset is recorded by the ECU to record the flight status of the aircraft within two hours, resulting in a variety of working modes in the normal data, complex system coupling, and multiple linear relationships between multiple variables in the dynamic system. The attention mechanism can dynamically adjust its focus according to the different input data, so that the model can be more adapted to the dynamic system. It is compared with the LSTM auto-encoder, isolated forest (IF), and one-class support vector machine (OC-SVM). The results show that the proposed method is effective in detecting faults.

The structure of this paper is as follows: Section 2 provides an overview of related work, encompassing the foundational concepts of AE, LSTM, and self-attention mechanisms, along with the presentation of the SLAE fault detection method. Section 3 details the experiments conducted using aero-engine data and offers an analysis of the fault data. Finally, Section 4 presents the conclusions drawn from this study.

2. Self-Attention-Based LSTM Auto-Encoder Fault Detection Modeling

2.1. Auto-Encoder

The auto-encoder (AE) is an unsupervised learning model designed specifically for processing unlabeled data [35]. Given the significant imbalance in the collected aero-engine data, characterized by a predominance of normal data and a scarcity of fault data, as well as the requirement for manual labeling by experts, employing an auto-encoder structure for the fault detection task is especially appropriate.

Auto-encoders typically comprise two components: an encoder and a decoder, as illustrated in Figure 1. The encoder is tasked with mapping the input data to a low-dimensional latent space representation, while the decoder transforms the latent representation back to the input space, aiming to reconstruct the original data as accurately as possible.

The encoding and decoding processes are outlined as follows:

$$h = g_{\theta_1}(x) = \sigma(W_1 x + b_1) \tag{1}$$

$$\hat{x} = g_{\theta_2}(h) = \sigma(W_2 h + b_2) \tag{2}$$

Here, W represents the weights, b denotes the biases, x represents the input data, h is the output of the encoder, $\hat{x}$ is the output of the decoder, and an activation function is applied, with the sigmoid function being used in this context. The corresponding formula is presented below:

$$\sigma(\mathrm{x}) = \frac{1}{1+\mathrm{e}^{-\mathrm{x}}} \tag{3}$$

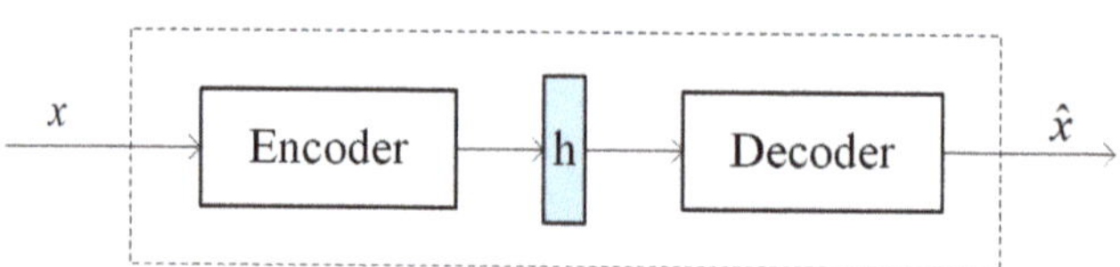

Figure 1. Auto-encoder.

The encoder receives the input X and maps it to the latent space representation h, while the decoder remaps h back to the input space to produce the reconstructed signal. The loss function is typically defined as the mean squared error, as shown below, where N represents the number of training samples. The objective of network training is to minimize this mean squared error.

$$\min \frac{1}{N} \sum_{i=1}^{N} \| x - \hat{x} \|_2^2 \tag{4}$$

The trained auto-encoder aims to learn the true probability distribution of the input data as accurately as possible. Its reconstruction error serves as an indicator of the likelihood of the input data and is straightforward to compute.

2.2. LSTM Network

LSTM networks are a specialized type of recurrent neural network (RNN) designed for processing and predicting time-series data. They address the challenges of learning long sequences, as well as mitigating issues related to gradient vanishing and gradient explosion [36].

LSTM is used in this paper, so the principles are briefly explained below. LSTM networks have a unique cellular structure, which manages the flow of information via an intricate gating mechanism. Unlike a traditional RNN, LSTM introduces input gates, forget gates, output gates, and a cell state. This gating mechanism can be viewed as a fully connected layer. LSTM facilitates information storage and updates through this gating system, specifically utilizing the Sigmoid function in conjunction with dot product operations. While RNNs only propagate hidden states, LSTM incorporates cell states as well. Figure 2 [37] illustrates the LSTM cell structure, where i_t, f_t, and o_t represent the input gate, forget gate, and output gate, respectively. Tanh is the hyperbolic tangent activation function that maps data to the range $[-1, 1]$. σ is the sigmoid activation function that maps the data to the range $[0, 1]$. W_i, W_f, W_c, W_o represent the weights associated with the input gate, forget gate, cell state, and output gate, respectively. b_f, b_i, b_c, b_o are the bias terms. x_t represents the input at the present time. h_{t-1} and h_t represent the output of the hidden layer at time $t-1$ and t, respectively. $\tilde{C}_t$ represents the temporary state of the input gate. C_{t-1} and C_t represent the state at time $t-1$ and t, respectively.

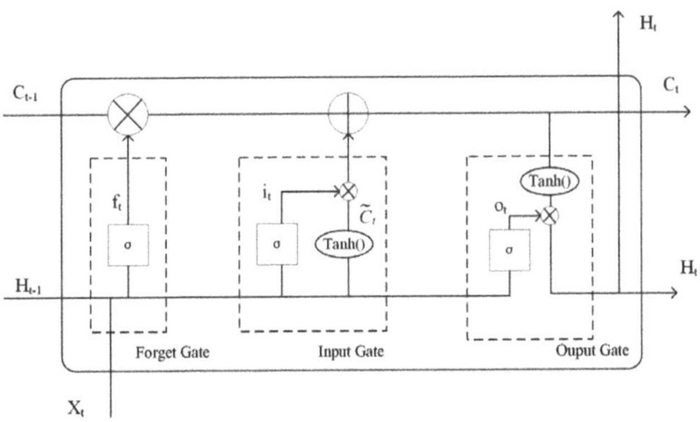

Figure 2. LSTM structure.

Forgetting Gate: This gate regulates whether past long-term memory information should be discarded.

$$f_t = \sigma(W_f \cdot [h_{t-1}, x_t] + b_f) \tag{5}$$

Input Gate: This gate determines the amount of information that is retrieved from the long-term memory unit for the output.

$$\tilde{C}_t = \tanh(W_c \cdot [h_{t-1}, x_t] + b_c) \tag{6}$$

$$i_t = \sigma(W_i \cdot [h_{t-1}, x_t] + b_i) \tag{7}$$

Output Gate: This gate regulates the amount of information that is extracted from the long-term memory unit for the output.

$$o_t = \sigma(W_o \cdot [h_{t-1}, x_t] + b_o) \tag{8}$$

$$h_t = o_t * \tanh(C_t) \tag{9}$$

Long-Term Memory Unit: Primarily utilized for storing and processing past information, as well as sifting and filtering data.

Short-Term Memory Unit: Designed to hold the most recent output and feed it back into the network.

$$C_t = f_t * C_{t-1} + i_t * \tilde{C}_t \tag{10}$$

The LSTM model inputs sequential data into an input gate, which writes the data to the long-term memory unit as required. The forget gate regulates which information should be discarded, while the output gate retrieves the necessary information from the long-term memory unit for output.

2.3. Self-Attention Mechanisms

Self-attention is a variant of the attention mechanism. Since its introduction in the 1990s, the Attention mechanism has evolved significantly, culminating in 2014 when Mnih et al. [38] first applied it to the visual domain. In the context of time-series data, the significance of the information carried by the time series signals varies across different windows [39]. The self-attention mechanism serves to extract the important information associated with each time window. Figure 3 [40] illustrates the structure of the self-attention mechanism in detail.

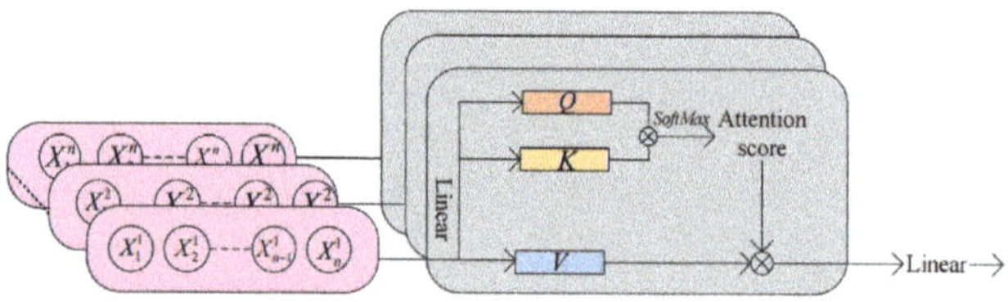

Figure 3. Self-attention structure.

Firstly, the n sequences of length m output from the encoder are passed through a linear layer to obtain the query matrix: $Q = [Q^1, Q^2, \cdots, Q^m]$, key matrix: $K = [K^1, K^2, \cdots, K^m]$, value matrix: $V = [V^1, V^2, \cdots, V^m]$. The calculation process is shown below:

$$\begin{cases} Q^i = W^q X^i \\ K^i = W^k X^i \\ V^i = W^v X^i \end{cases} \tag{11}$$

After that, the dot product between the query vector Q^i and the keyword vector K^i for each time window is calculated, and to prevent the dot product from being too large, the dot product is divided by $\sqrt{d_k}$, where d_k is the dimension of the keyword vector, and the result is normalized to a probability distribution using $SoftMax$. The attention value is multiplied by the value vector V_i , and the sum of the weights is calculated to be the attention score. The calculation process is shown below:

$$Attention(Q, K, V) = SoftMax(\frac{QK^T}{\sqrt{d_k}})V \tag{12}$$

2.4. Fault Detection Model

Building upon the analysis of the AE, LSTM, and self-attention mechanisms, we propose a self-attention-based LSTM auto-encoder fault detection method (SLAE) tailored to aero-engine fault detection. This method utilizes performance parameters from the time series, replacing the traditional AE neural network with an LSTM network. Additionally, a self-attention mechanism is integrated into the output of the encoding region to enhance the model's ability to learn and extract features from the data. The workflow of SLAE is illustrated in Figure 4.

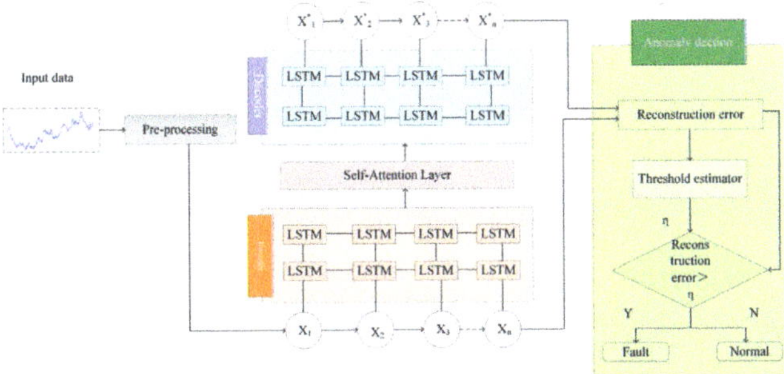

Figure 4. SLAE fault detection process.

First, the collected aero-engine data are divided into a training set and a testing set. The input data undergo pre-processing, primarily for standardization. Pre-processing is also a crucial factor influencing the outcomes of model training. The calculation process is detailed below, where μ represents the mean value of the original data and σ denotes the standard deviation of the original data. The fault detection flow chart is shown in Figure 5.

$$x_i' = \frac{x_i - \mu}{\sigma} \tag{13}$$

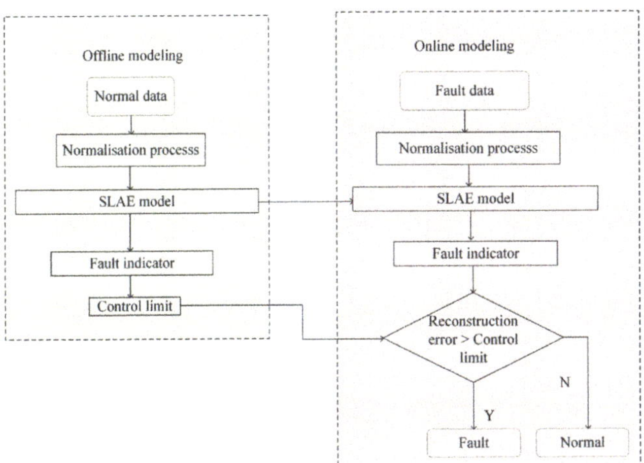

Figure 5. Fault detection flow chart.

The SLAE model utilizes preprocessed data as the input, with the encoder layer comprises a two-layer LSTM network. This LSTM captures dependencies within the input time series and generates an implicit representation. The output from the encoder passes through the self-attention layer, which efficiently extracts important features from the sequence by weighting different time points. The decoder then reconstructs the data into a time series format, allowing for the calculation of the reconstruction error.

$$reconstruction\_error = \| x - \hat{x} \|_2^2 \tag{14}$$

The network employs the Adam optimizer to minimize the loss function, which is defined using mean squared error (MSE). x and $\hat{x}$ represent the original data and the reconstructed data, respectively. The square of the L2 normal of the original data and the reconstructed data is calculated as the fault indicator. For the training data, fault indications are calculated for all samples, with a confidence level set at 0.95. The value at the 95th percentile of the reconstruction error serves as the control line. When calculating the reconstruction error for the testing data, any value exceeding this control line is deemed abnormal; otherwise, it is classified as normal.

The SLAE anomaly detection model is an unsupervised deep learning network designed for detecting anomalies. In the case of data bias in aero-engine fault diagnostic data, the SLAE model utilizes a large number of normal operation samples to perform high-precision fault detection on a limited number of anomaly samples.

3. Aero-Engine Fault Detection

3.1. Introduction to the Dataset

This dataset is generated from the simulation of a type of civil aircraft with a piston aircraft engine primarily used for flight training, and injected with fault information. The main components of the engine used in this aircraft include the air intake system, the exhaust system, the cooling system, the fuel system, the engine control system, and the propeller system. The main components of the intake system include the air filter, the intake manifold, and the pilot tube to feel the intake pressure; the main components of the exhaust system include the exhaust manifold, the turbocharger, the wastegate control solenoid valve, the wastegate opening control assembly, and the exhaust pipe; the cooling system mainly comprises the thermostat, the coolant radiator, the mid-cooling radiator, the heat exchanger, and the sliding oil radiator; the main components of the fuel system include the fuel filter, the low-pressure pump, the high-pressure pump, the common rail, and the fuel control system; and the fuel system mainly includes the fuel filter, the fuel low-pressure pump, the fuel high-pressure pump, the fuel common rail, the fuel injector nozzle, and the fuel common rail pressure control valve. The engine control system adopts digital full-authority electronic control technology to control the engine's work through the engine control unit. The main components of the propeller system include the propeller, the propeller hub, and the gear box.

The parameters that have a significant impact on the normal operation of this type of piston aero-engine include the crankshaft speed, power target value, intake pressure actual value, fuel pressure actual value, atmospheric pressure, DC generator voltage, power actual value, intake pressure target value, fuel pressure target value, and wastegate opening, and these ten parameters are used to make up the data file, as shown in Table 1:

Table 1. Aero-engine data composition.

Serial Number	Parameterisation	Abridge
1	Crankshaft Revolution Speed	CREV
2	Power Target Value	TP
3	Manifold Intake Air Pressure	MIAP
4	Measured Fuel Pressure	MFP
5	Atmospheric Pressure	AP
6	Direct Current Power Supply Voltage	DCPSV
7	Measured Power	MP
8	Target Intake Air Pressure	TIAP
9	Target Fuel Pressure	TFP
10	Exhaust Valve Opening Position	EVOP

In this type of engine, due to the use of digital full-authority electronic control technology, each key component is installed with sensors to collect engine operating parameters, which are recorded and saved by the ECU. During engine operation, the ECU collects and saves data once per second, and each data file contains about 12,000 pieces of data.The data file usually contains three complete flight data records, and the size of the data file is automatically determined by the ECU.

3.2. Fault Data Analysis

This civil aircraft aero-engine dataset contains 24 data files and 13 kinds of faults, among which the aero-engine of the No.1 aircraft is selected for fault diagnosis, containing three pieces of fault data, with the experimental samples containing 11,507, 11,458, and 11,421 entries and having 10 dimensions, containing 10 aero-engine parameters, respectively. The fault data are visualized and fault analysis is performed as follows:

Failure occurs due to sharp fluctuations in wastegate opening (EVOP) and the resulting sharp fluctuations in actual intake pressure. The EVOP ranges from 60–85 percent when flying at altitudes below a certain height and with the throttle stick in the full-throttle position. At a high altitude, an EVOP equaling 99 percent is normal. From Figure 6, it can be seen that at the sampling time from 6800 to 7500, the aero-engine wastegate opening fluctuates, resulting in a sharp fluctuation of the MIAP causing a malfunction. Under normal circumstances, there is a relatively fixed interpolation between the MIAP and the TIAP, instead of the fluctuation.

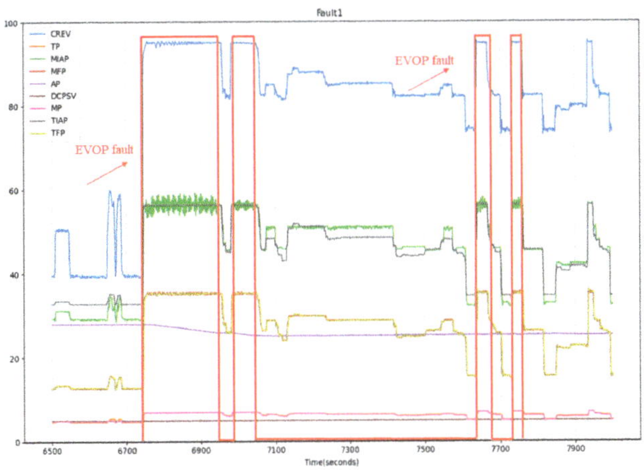

Figure 6. Fault 1 raw data.

The wastegate opening EVOP represents the opening of the solenoid valve, not the physical opening. The physical opening of the wastegate is controlled by the actuator cartridge, whose action is controlled by the air diaphragm cartridge. The solenoid valve controls the actuator cartridge, and thus the physical opening of the wastegate, by adjusting the air pressure to the air diaphragm cartridge.

From the results of the analysis in Figure 7, it is clear that the MIAP fluctuates dramatically during engine operation, especially at the sampling moments between 5500 and 6000. This fluctuation is abnormal, as it indicates a serious disturbance of the pressure stability within the intake system. At the same time, we observed an abnormally large difference between the TIAP and the MIAP, which was well outside the range allowed for normal engine operation. This increased difference usually means that the actual amount of air entering the engine does not match the target amount of air preset by the ECU and that there is an air leak in the aero-engine intake manifold.

From the results of the analysis in Figure 8, between the sampling moments 4000 and 4500, the value of the MIAP fluctuates dramatically. Such fluctuations are not normal and may be due to a problem in some part of the intake system. Further inspection revealed that there was an air leak in the intake manifold, which was the direct cause of the MIAP fluctuation. An air leak in the intake manifold causes a reduction in air flow, which affects the combustion efficiency and power output of the engine.

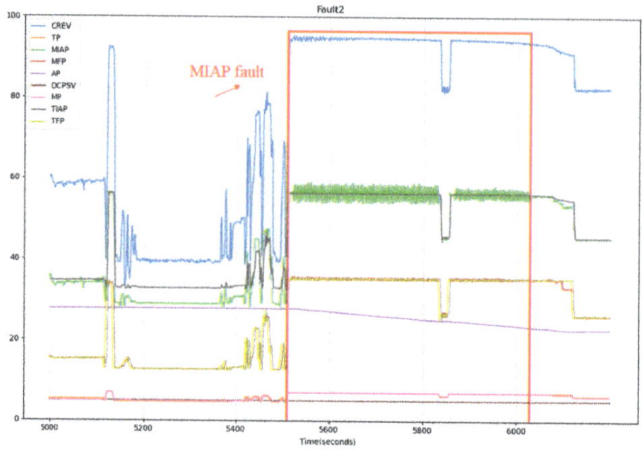

Figure 7. Fault 2 raw data.

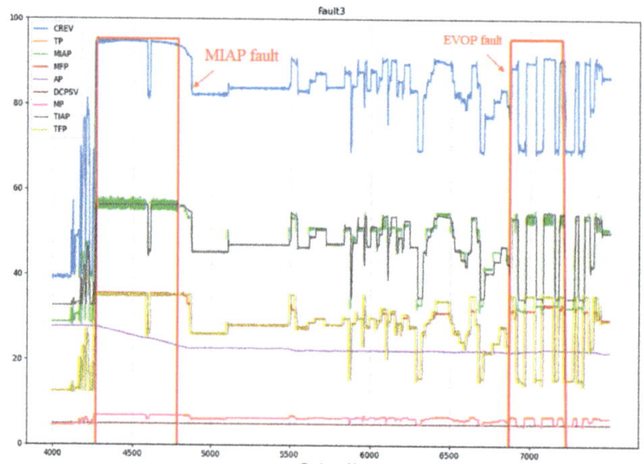

Figure 8. Fault 3 raw data.

Subsequently, between sampling moments 6900 and 7400, the EVOP fluctuated up and down even though the TIAP remained stable. This fluctuation indicates a problem with the wastegate opening control. The wastegate is an important component of the turbocharging system, and its opening directly affects the efficiency of the turbo and the air intake of the engine.

3.3. Experimental Indicators

To evaluate the performance of the model, the false positive rate (FPR) and false negative rate (FNR) are primarily selected as the experimental metrics. These two metrics provide a comprehensive reflection of the model's classification errors for both positive and negative class samples, thereby aiding in the assessment of the model's practicality and stability across various scenarios.

$$\mathrm{FPR} = \frac{\mathrm{FP}}{\mathrm{FP} + \mathrm{TN}} \tag{15}$$

$$\mathrm{FNR} = \frac{\mathrm{FN}}{\mathrm{FN} + \mathrm{TP}} \tag{16}$$

In the above equation, TP is the number of abnormal sample points that were correctly detected, TN is the number of correctly identified normal samples, FP is the number of normal sample points that were misjudged as abnormal, and FN is the number of abnormal sample points that were misjudged as normal.

A comparative analysis of the leakage rates across different methods was conducted while controlling the false alarm rate, aiming to verify the validity of the proposed model.

3.4. Analysis of Experimental Results

The trained SLAE model was compared with three methods, LSTM-AE, IF, and OC-SVM, and a 95 percent confidence level was taken as the control limit. The following are the detection effects of different methods for the three kinds of faults from Figures 9–11

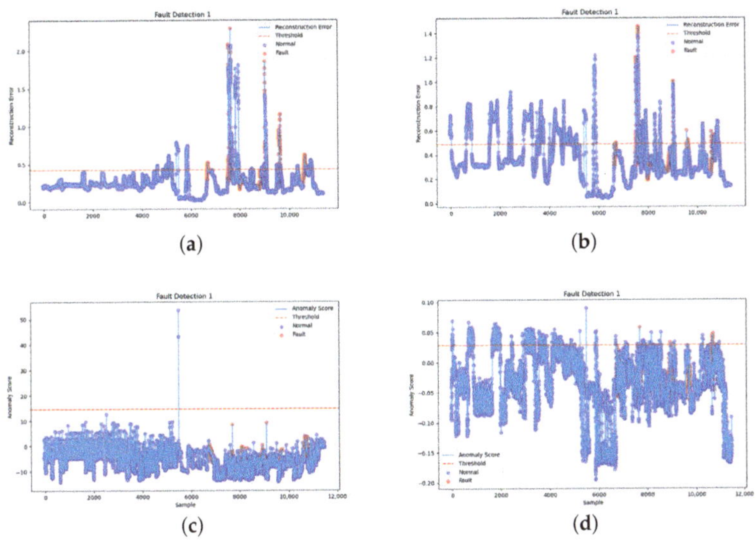

Figure 9. Fault 1 dection. (**a**) SLAE fault dection; (**b**) LSTM fault dection; (**c**) OC-SVM fault dection; (**d**) IF fault dection.

It can be seen from Table 2 that after the attention mechanism is introduced, the computational complexity is increased, and the training time is also increased. As the time step increases, the training time of the SLAE model is much longer than that of the AE-LSTM. When the time step is 100, the training time of SLAE is close to three times that of the AE-LSTM.

Table 2. Training time comparison.

Time Step	AE-LSTM (s)	SLAE (s)
10	300	443
20	408	686
30	512	911
40	608	1178
50	705	1502
60	793	1768
70	873	2082
80	980	2372
90	1079	2833
100	1175	3262

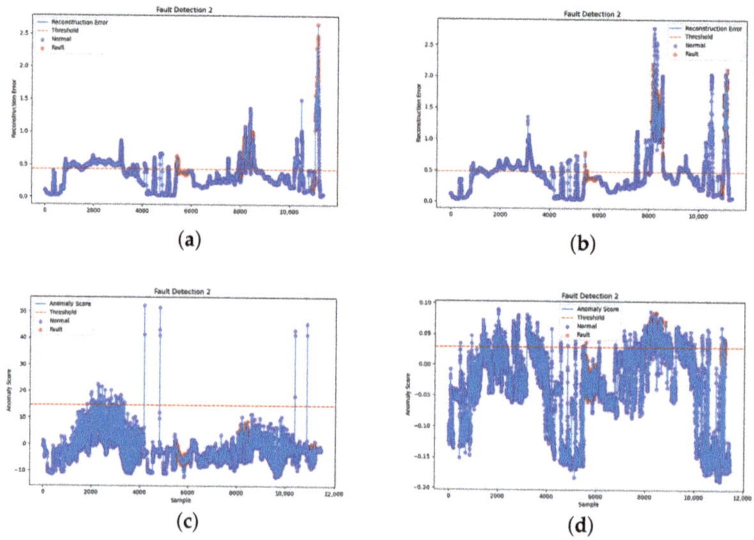

Figure 10. Fault 2 dection. (**a**) SLAE fault dection; (**b**) LSTM fault dection; (**c**) OC-SVM fault dection; (**d**) IF fault dection.

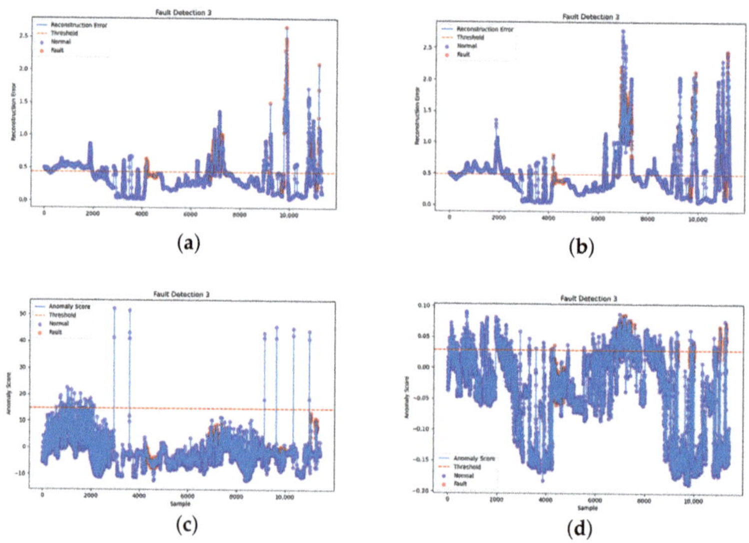

Figure 11. Fault 3 dection. (**a**) SLAE fault dection; (**b**) LSTM fault dection; (**c**) OC-SVM fault dection; (**d**) IF fault dection.

Analyzing the experimental results from Table 3, the LSTM auto-encoder combined with the self-attention fault detection method proposed in this paper is compared to the traditional LSTM auto-encoder method, as well as the OC-SVM and IF. When the control limit is a 95 percent confidence level, the SLAE model significantly outperforms the other three methods in detecting Fault 1. Additionally, with the introduction of the attention mechanism in the SLAE model, the false alarm rate decreased compared to the traditional LSTM model.

Table 3. Evaluation indicators.

Fault Number	SLAE		AE-LSTM		OC-SVM		IF	
	FPR	FNR	FPR	FNR	FPR	FNR	FPR	FNR
1	0.0732	0.5974	0.2463	0.8033	0.0003	1	0.0725	0.9781
2	0.3244	0.5691	0.3462	0.6289	0.0062	1	0.1320	0.6600
3	0.3188	0.5339	0.3749	0.5818	0.0067	1	0.1336	0.6073

4. Conclusions

To address the fault detection problem in aero-engines, this paper proposes a fault detection method that combines an LSTM auto-encoder with a self-attention mechanism. This approach effectively captures complex patterns and long-term dependencies in time-series data. The LSTM auto-encoder excels in handling time-series data, while the self-attention mechanism captures global features. Faults are detected based on the reconstruction error between the input data and the model output, and comparative experiments are conducted to validate the feasibility and advantages of the proposed method.

The detection method, while showing superior capabilities compared to traditional auto-encoders in detecting engine faults, is not without its limitations. This unsatisfactory detection could be attributed to factors such as strong correlations between variables and the complexity introduced by multiple operating modes in the normal data, compounded by unbalanced samples between these modes—a challenge that warrants further investigation in future studies. Moreover, this method's reliance on normal data from each specific engine to train the model presents a practical constraint, as it necessitates the detection of abnormal data associated with these particular engines. The individual training for each engine component is not only time-consuming and labor-intensive but also highlights the need for a more efficient approach. In response to this challenge, further work could use transfer learning for engine fault detection, aiming to improve the efficiency of fault detection and reduce the resources required for model training.

Author Contributions: W.D.: methodology, experiments, data analysis, writing-review and editing, funding acquisition, supervision. J.Z.: coding, validation, writing—original. H.Z.: data curation. G.M.: supervision. All authors have read and agreed to the published version of the manuscript.

Funding: This work is supported by National Natural Science Foundation of China (61903262), Natural Science Foundation of Liao Ning province (2024-MS-133), and the Fundamental Research Funds for the Universities of Liaoning province (20240206, LJ212410143047).

Data Availability Statement: The data presented in this study are available on request from the corresponding author because the data are not publicly available due to privacy.

Conflicts of Interest: The author Haoran Zhang was employed by the company Tianjin Jepsen International Flight College Co., Ltd. The remaining authors declare that the research was conducted in the absence of any commercial or financial relationships that could be construed as a potential conflict of interest.

Nomenclature

AE	Auto-encoder
Rnn	Recurrent neural network
$LSTM$	Long short-term memory
b	Bias term
W	Weight
h	Encoder output
f_t	Forget gate
o_t	output gate
i_t	Input gate
$CREV$	Crankshaft revolution speed
TP	Power target value

MIAP	Input gate
MFP	Manifold intake air pressure
AP	Atmospheric pressure
DCPSV	Direct current power supply voltage
MP	Measured power
TIAP	Target intake air pressure
TFP	Target fuel pressure
EVOP	Exhaust valve opening position
ECU	Engine control unit

References

1. Xu, J.; Wang, Y.; Wang, Z.; Wang, X.; Zhao, Y. Transient gas path fault diagnosis of aero-engine based on domain adaptive offline reinforcement learning. *Aerosp. Sci. Technol.* **2024**, *155*, 109701. [CrossRef]
2. Huang, Y.; Tao, J.; Zhao, J.; Sun, G.; Yin, K.; Zhai, J. Graph structure embedded with physical constraints-based information fusion network for interpretable fault diagnosis of aero-engine. *Energy* **2023**, *283*, 129120. [CrossRef]
3. Liao, Z.; Zhan, K.; Zhao, H.; Deng, Y.; Geng, J.; Chen, X.; Song, Z. Addressing class-imbalanced learning in real-time aero-engine gas-path fault diagnosis via feature filtering and mapping. *Reliab. Eng. Syst. Saf.* **2024**, *249*, 110189. [CrossRef]
4. Chen, J.; Yan, Z.; Lin, C.; Yao, B.; Ge, H. Aero-engine high speed bearing fault diagnosis for data imbalance: A sample enhanced diagnostic method based on pre-training WGAN-GP. *Measurement* **2023**, *213*, 112709. [CrossRef]
5. Li, B.; Xue, S.K.; Fu, Y.H.; Tang, Y.D.; Zhao, Y.P. Kernel adapted extreme learning machine for cross-domain fault diagnosis of aero-engines. *Aerosp. Sci. Technol.* **2024**, *146*, 108970. [CrossRef]
6. Chi, J.; Liu, Y.; Luo, D.; Cao, L. Fault Diagnosis of Aero-engine Gas Path Base on PSO-SVM. In Proceedings of the 2019 CAA Symposium on Fault Detection, Supervision and Safety for Technical Processes (SAFEPROCESS), Xiamen, China, 5–7 July 2019; IEEE: Piscataway, NJ, USA, 2019; pp. 82–86.
7. Liao, Z.; Zhang, R.; Zhao, H.; Gao, F.; Geng, J.; Chen, X.; Song, Z. Towards domain shifts: Stream fine-tuning via feed-forward fault data generation for on-board aero-engine gas-path diagnosis. *Measurement* **2024**, *237*, 115207. [CrossRef]
8. Lu, Z.; Li, B.; Fu, C.; Wu, J.; Xu, L.; Jia, S.; Zhang, H. Remaining Useful Life Prediction Method Based on Dual-Path Interaction Network with Multiscale Feature Fusion and Dynamic Weight Adaptation. *Actuators* **2024**, *13*, 413. [CrossRef]
9. Yang, T.; Han, S.; Han, Q.; Zhu, Q.; Yu, X. A Fault Diagnosis Study of Aero-engine Hydraulic Piping Clamp Based on Deep Learning. In Proceedings of the 2021 Global Reliability and Prognostics and Health Management (PHM-Nanjing), Nanjing, China, 15–17 October 2021; IEEE: Piscataway, NJ, USA, 2021; pp. 1–5.
10. Yang, T.; Li, G.; Yuan, S.; Qi, Y.; Yu, X.; Han, Q. The LST-SATM-net: A new deep feature learning framework for aero-engine hydraulic pipeline systems intelligent faults diagnosis. *Appl. Acoust.* **2023**, *210*, 109436. [CrossRef]
11. Nie, L.; Xu, S.; Zhang, L.; Yin, Y.; Dong, Z.; Zhou, X. Remaining Useful Life Prediction of Aeroengines Based on Multi-Head Attention Mechanism. *Machines* **2022**, *10*, 552. [CrossRef]
12. Zhao, Z.; Zhou, R.; Dong, Z. Aero-engine faults diagnosis based on K-means improved Wasserstein GAN and relevant vector machine. In Proceedings of the 2019 Chinese Control Conference (CCC), Guangzhou, China, 27–30 July 2019; IEEE: Piscataway, NJ, USA, 2019; pp. 4795–4800.
13. Liu, F.Z.; Shi, Y.; Sun, C.Y.; Sun, X.M. Improved real-time H∞ control for aero-engines based on the equilibrium manifold expansion model. *J. Frankl. Inst.* **2024**, *361*, 107342. [CrossRef]
14. Gan, C.; Ding, S.; Qiu, T.; Liu, P.; Ma, Q. Model-based safety analysis with time resolution (MBSA-TR) method for complex aerothermal–mechanical systems of aero-engines. *Reliab. Eng. Syst. Saf.* **2024**, *243*, 109864. [CrossRef]
15. Wang, B.; Xuan, Y. Heuristic deepening of aero engine performance analysis model based on thermodynamic principle of variable mass system. *Energy* **2024**, *306*, 132481. [CrossRef]
16. Xiang, F.; Zhang, Y.; Zhang, S.; Wang, Z.; Qiu, L.; Choi, J.H. Bayesian gated-transformer model for risk-aware prediction of aero-engine remaining useful life. *Expert Syst. Appl.* **2024**, *238*, 121859. [CrossRef]
17. Sun, A.; Guo, D.; Wang, R. A data-based expert system for aero-engine gas path fault diagnosis. In Proceedings of the 2021 33rd Chinese Control and Decision Conference (CCDC), Kunming, China, 22–24 May 2021; IEEE: Piscataway, NJ, USA, 2021; pp. 2917–2922.
18. Chen, M.; Qu, R.; Fang, W. Case-based reasoning system for fault diagnosis of aero-engines. *Expert Syst. Appl.* **2022**, *202*, 117350. [CrossRef]
19. Ma, S.; Yu, X.M.; Sun, T.; Zhao, G.; Guo, M. Research on Data-Driven Fault Diagnosis of Aero-Engine Transmission Systems. In Proceedings of the 2023 42nd Chinese Control Conference (CCC), Tianjin, China, 24–26 July 2023; IEEE: Piscataway, NJ, USA, 2023; pp. 4922–4927.
20. Wang, Y.Q.; Zhao, Y.P. A novel inter-domain attention-based adversarial network for aero-engine partial unsupervised cross-domain fault diagnosis. *Eng. Appl. Artif. Intell.* **2023**, *123*, 106486. [CrossRef]
21. Liu, K.; Wang, P.; Li, Q. The research of aero-engine digital twin model based on model-driven and data-driven fusion method. In Proceedings of the 2021 IEEE 10th Data Driven Control and Learning Systems Conference (DDCLS), Suzhou, China, 14–16 May 2021; IEEE: Piscataway, NJ, USA, 2021; pp. 322–327.

22. Kiakojouri, A.; Lu, Z.; Mirring, P.; Powrie, H.; Wang, L. A Generalised Intelligent Bearing Fault Diagnosis Model Based on a Two-Stage Approach. *Machines* **2024**, *12*, 77. [CrossRef]
23. Huang, Y.; Tao, J.; Sun, G.; Wu, T.; Yu, L.; Zhao, X. A novel digital twin approach based on deep multimodal information fusion for aero-engine fault diagnosis. *Energy* **2023**, *270*, 126894. [CrossRef]
24. Song, Y.; Gou, L.; Zhang, M.; Huang, Y.; Chen, Y. Data-Driven Equivalent Space Aero-Engine Sensor Robust Fault Diagnosis. In Proceedings of the 2023 14th International Conference on Mechanical and Aerospace Engineering (ICMAE), Porto, Portugal, 18–21 July 2023; IEEE: Piscataway, NJ, USA, 2023; pp. 237–242.
25. Chen, Y.; Wang, X.; Wang, Y.; Gao, Y.; Qu, J.; Dai, H.; Xu, C. Multi-combination fault data augmentation method of aero-engine gas path based on Extraction TimeGAN. *Measurement* **2025**, *242*, 115778. [CrossRef]
26. Wang, C.; Peng, Z.; Liu, R. Prediction of Aero-Engine Remaining Useful Life Combined with Fault Information. *Machines* **2022**, *10*, 927. [CrossRef]
27. Hu, J.; Li, W.; Zhang, Y.; Tian, Z. Cross-domain few-shot fault diagnosis based on meta-learning and domain adversarial graph convolutional network. *Eng. Appl. Artif. Intell.* **2024**, *136*, 108970. [CrossRef]
28. Liu, D.; Zhong, S.; Lin, L.; Zhao, M.; Fu, X.; Liu, X. HOOST: A novel hyperplane-oriented over-sampling technique for imbalanced fault detection of aero-engines. *Knowl.-Based Syst.* **2024**, *300*, 112142. [CrossRef]
29. Kang, Y.; Chen, G.; Wang, H.; Sheng, J.; Wei, X. Fault anomaly detection method of aero-engine rolling bearing based on distillation learning. *ISA Trans.* **2024**, *145*, 387–398. [CrossRef] [PubMed]
30. Lin, X.; Li, B.; Yang, X.; Wang, J. Fault diagnosis of aero-engine bearing using a stacked auto-encoder network. In Proceedings of the 2018 IEEE 4th Information Technology and Mechatronics Engineering Conference (ITOEC), Chongqing, China, 14–16 December 2018; IEEE: Piscataway, NJ, USA, 2018; pp. 545–548.
31. Shi, H.; Cao, S.; Zuo, H.; Ma, J.; Lin, C. Deep subdomain adversarial network with self-supervised learning for aero-engine high speed bearing fault diagnosis with unknown working conditions. *Measurement* **2025**, *241*, 115668. [CrossRef]
32. Pengyu, L.; Wenkui, H.; Kun, Y.; Qiuying, Y.; Yong, Y. Fault diagnosis based on multi-sensor information fusion using LSTM and DS evidence theory. In Proceedings of the 2020 7th International Conference on Information Science and Control Engineering (ICISCE), Changsha, China, 18–20 December 2020; IEEE: Piscataway, NJ, USA, 2020; pp. 617–621.
33. Guo, P.; Wei, C.; Yin, Z. Anomaly detection of spacecraft reconstructed signals based on attention mechanism. In Proceedings of the 2023 6th International Symposium on Autonomous Systems (ISAS), Nanjing, China, 23–25 June 2023; IEEE: Piscataway, NJ, USA, 2023; pp. 1–5.
34. Zhao, Y.; Suo, C.; Wang, Y. Remaining Useful Life Estimation Using a Variational Autoencoder-Regression Framework. In Proceedings of the 2023 IEEE 6th International Conference on Pattern Recognition and Artificial Intelligence (PRAI), Haikou, China, 18–20 August 2023; IEEE: Piscataway, NJ, USA, 2023; pp. 1267–1272.
35. Zhai, J.; Zhang, S.; Chen, J.; He, Q. Autoencoder and its various variants. In Proceedings of the 2018 IEEE International Conference on Systems, Man, and Cybernetics (SMC), Miyazaki, Japan, 7–10 October 2018; IEEE: Piscataway, NJ, USA, 2018; pp. 415–419.
36. Tan, Z.; Pan, P. Network fault prediction based on CNN-LSTM hybrid neural network. In Proceedings of the 2019 International Conference on Communications, Information System and Computer Engineering (CISCE), Haikou, China, 5–7 July 2019; IEEE: Piscataway, NJ, USA, 2019; pp. 486–490.
37. Xu, C.; Wang, Z. Fault Detection for Satellite Gyroscope Using LSTM Networks. In Proceedings of the 2023 IEEE 12th Data Driven Control and Learning Systems Conference (DDCLS), Xiangtan, China, 12–14 May 2023; pp. 633–638.
38. Mnih, V.; Heess, N.; Graves, A.; Kavukcuoglu, K. Recurrent models of visual attention. In Proceedings of the 27th International Conference on Neural Information Processing Systems, NIPS'14—Volume 2, Cambridge, MA, USA, 8–13December 2014; pp. 2204–2212.
39. Shi, X.; Lu, R. Attention-based bidirectional hierarchical LSTM networks for text semantic classification. In Proceedings of the 2019 10th International Conference on Information Technology in Medicine and Education (ITME), Qingdao, China, 23–25 August 2019; IEEE: Piscataway, NJ, USA, 2019; pp. 618–622.
40. Yang, Q.; Li, H.; Hua, C.; Wu, D. Fault Diagnosis of Multimodal Chemical Process Based on Attention Mechanism and Deep Subdomain Countermeasure Adaptive Network. In Proceedings of the 2022 2nd International Conference on Algorithms, High Performance Computing and Artificial Intelligence (AHPCAI), Guangzhou, China, 21–23 October 2022; pp. 686–690.

machines

MDPI

Article

Enhanced Non-Destructive Testing of Small Wind Turbine Blades Using Infrared Thermography

Majid Memari [1,*], Mohammad Shekaramiz [2], Mohammad A. S. Masoum [2] and Abdennour C. Seibi [3]

1 Computer Science Department, Utah Valley University, Orem, UT 84058, USA
2 Machine Learning and Drone Laboratory, Electrical and Computer Engineering Department, Utah Valley University, Orem, UT 84058, USA; mshekaramiz@uvu.edu (M.S.); m.masoum@ieee.org (M.A.S.M.)
3 Mechanical and Civil Engineering Department, Utah Valley University, Orem, UT 84058, USA; aseibi@uvu.edu
* Correspondence: mmemari@uvu.edu

Abstract: This study presents a foundational step in a broader initiative aimed at leveraging thermal imaging technology to enhance wind turbine maintenance, particularly focusing on the challenges of detecting defects and object localization in small wind turbine blades. Serving as a preliminary experiment, this research project tested methodologies and technologies on a smaller scale before advancing to more complex applications involving large, operational wind turbines using drone-mounted cameras. Utilizing thermal cameras suitable for both handheld and drone use, alongside advanced image processing applications, we navigated the significant challenge of acquiring high-quality thermal images to detect small defects. This required a concentrated analysis of a select subset of data and a methodological shift towards object detection and localization using the You Only Look Once (YOLO) model versions 8 and 9. This effort not only paves the way for applying these techniques to larger-scale turbines but also contributes to the ongoing development of an integrated maintenance strategy in the wind energy sector. Highlighting the critical impact of environmental conditions on thermal imaging, our research underscores the importance of continued exploration in this field, especially in enhancing object localization techniques for the future drone-based maintenance of operational wind turbine blades (WTBs).

Keywords: infrared thermography; wind turbine blade; deep learning; defect detection; object localization; YOLO

1. Introduction

The shift toward renewable energy sources is essential for mitigating global carbon emissions, with wind energy playing a crucial role in this transition. Wind turbines, a fundamental component of this shift, require meticulous maintenance to maintain efficiency and reliability. This underscores the vital importance of continuous innovation in maintenance and inspection practices. Wind turbine blades (WTBs) are particularly susceptible to various damages that can drastically affect their performance. Among the technologies used in maintenance, thermal imaging stands out for its ability to detect subsurface defects that are otherwise invisible, enhancing the effectiveness of maintenance protocols. Our comprehensive review [1] delves into the latest advancements and explores the transformative potential of these technologies in detecting defects in WTBs. Recent advancements in thermal imaging have increasingly facilitated the inspection of WTBs, highlighting its potential as a non-invasive diagnostic tool. Despite these technological strides, several critical shortcomings persist in the application of thermal imaging within the context of wind turbine maintenance. The primary limitation lies in the generic nature of the datasets used in most studies. While thermal imaging has been proven to detect subsurface defects such as delaminations, voids, and moisture ingress effectively, the datasets typically employed do not adequately reflect the specific and varied nature of defects that WTBs exhibit [2]. These datasets often lack the resolution or specific environmental conditions under which real-world turbine blades operate, limiting the practical applicability of the findings [3]. Furthermore, existing research often does not address the full spectrum of defect types that can affect WTBs. For example, while thermal imaging is effective at identifying larger thermal anomalies, smaller yet critical defects might not be detected due to the limitations in sensor resolution or the methodologies employed [4]. This gap is crucial because small defects can evolve into significant damage if not identified and addressed early. Moreover, the integration of Unmanned Aerial Vehicles (UAVs) with thermal imaging technologies, although innovative, faces challenges related to the stability and precision of the UAV platforms. The accuracy of thermal imaging is highly dependent on the stability of the camera during flight, which can be affected by wind and other environmental factors. This introduces variability in the data quality, which can compromise the defect detection process [5]. In terms of data processing, most existing approaches utilize conventional image processing techniques that may not adequately handle the complexities of thermal data, which require advanced algorithms to be interpreted effectively. The need for more sophisticated machine learning models that can learn from and adapt to the unique patterns of thermal images is evident but underexplored in the current literature [6]. To address these deficiencies, there is a need for research that develops high-fidelity, high-resolution thermal datasets that are reflective of the actual operating conditions of wind turbines. Additionally, enhancing UAV stability and employing advanced machine learning techniques for data analysis are critical in overcoming the current limitations of thermal imaging applications in this field.

1.1. *Inferared Thermography (IRT) Non-Destructive Testing (NDT)*

Wind turbine blade infrared thermography (IRT) non-destructive testing (NDT) is an advanced diagnostic technique used to detect defects and assess the integrity of wind turbine blades. This method involves using infrared cameras to capture thermal images of the blades, which reveal temperature variations that indicate the presence of internal flaws such as delaminations, cracks, or moisture ingress. IRT NDT is valued for its ability to provide rapid, accurate, and comprehensive inspections without causing damage to the blades. As a result, it enhances the reliability and efficiency of wind turbines by ensuring the early detection and remediation of potential issues, thereby extending the lifespan of the blades and optimizing energy production.

Galleguillos et al. [7] discussed a novel NDT technique using IRT combined with UAVs for inspecting WTBs. The study validated the feasibility of using passive IRT, which leverages natural thermal effects rather than external heat sources, to identify common defects such as delaminations, cracks, and impact damage in wind blades. The authors conducted both ground-based tests with artificially introduced defects and flight tests using a unmanned aerial systems equipped with an IR camera. The results demonstrated the method's effectiveness in detecting significant defects rapidly and safely, suggesting that this approach could significantly reduce the time and risk associated with traditional blade inspection methods.

Doroshtnasir et al. [8] introduced an NDT method for inspecting WTBs using IRT. This method is particularly useful for offshore turbines where accessibility is limited. The authors described a novel procedure that involves creating difference thermograms to minimize misinterpretations caused by environmental reflections, identifying signals unrelated to structural or dynamic features, and comparing these signals to processed photographs to rule out false positives from surface effects. The study confirmed that the approach can effectively detect subsurface defects from a distance, offering a significant advancement over traditional IRT methods which often only consider a subset of these steps. This technique not only increases the reliability of inspections but also enhances the ability to conduct them under operational conditions, thereby improving maintenance efficiency and safety in wind energy operations.

Traphan et al. [9] explored the application of IRT for identifying surface defects in WTBs while in operation. This non-invasive method provides a promising alternative to traditional inspections, which often require turbine downtime and physical access to blades. The study compared IRT data with stereoscopic particle image velocimetry to validate the method's efficacy. It further investigated the aerodynamic effects of detected surface defects using both experimental setups and computational fluid dynamics simulations. Results indicated that IRT could effectively visualize surface anomalies by detecting thermal patterns associated with defective areas, thereby enabling the early detection of potential damages that could lead to significant maintenance issues if left unaddressed.

Chen et al. [10] introduced an innovative method called AQUADA for the automated detection and quantification of structural damages in WTBs using thermography and computer vision. The study highlighted the application of this technology to reduce the Levelized Cost of Energy (LCOE) by simplifying and economizing the inspection process. By automating damage detection and integrating analysis with image acquisition, AQUADA minimizes human error and operational downtime. The method was tested on WTBs under laboratory conditions to validate its effectiveness. Results showed that AQUADA could potentially cut inspection costs by half and reduce the LCOE by 1–2%, proving its efficacy in identifying both superficial and subsurface blade damages without interfering with turbine operation.

Xu et al. [11] introduced an innovative approach to defect depth recognition using Gated Recurrent Unit (GRU) networks in conjunction with active IRT . This method leverages Principal Component Analysis (PCA) to process raw thermal sequences, reducing dimensionality and correlation before training the GRU model. The methodology was validated through experiments on polymethyl methacrylate specimens with flat-bottom hole defects. The results demonstrated that the PCA-enhanced GRU model achieved superior accuracy in recognizing defect depths compared to traditional approaches, showing its potential for enhancing non-destructive testing by efficiently processing thermal data and accurately determining defect characteristics.

Wang et al. [12] discussed an advanced method for detecting thermal defects in a substation equipment using a Convolutional Neural Network (CNN) integrated with a Support Vector Machine (SVM). The method enhances image processing by using improved pre-processing techniques to isolate the region of interest (RoI) and refine the thermal image data. Temperature values were segmented and categorized into 11 distinct labels, forming the T-IR11 dataset, which was essential for training the CNN model. The approach was evaluated using precision, recall, and F1 score metrics, demonstrating a high accuracy of 99.50%. This method significantly automates and improves the efficiency and accuracy of thermal defect detection in substation equipment, which is crucial for maintaining the reliability and stability of power systems.

Jiang et al. [13] evaluated the use of You Only Look Once (YOLO) [14] models for object detection in UAV-based thermal infrared images and videos. The study emphasized overcoming challenges like low resolution, complex backgrounds, and a lack of labeled datasets in UAV thermal infrared imagery by adapting YOLO models for thermal infrared application. These models were tested on ground-based and UAV datasets, demonstrating high effectiveness in various scenarios, including day and night operations. The results were promising, with a mean average precision (mAP) of 88.69% for detecting specific objects like cars and people, and the models achieved up to 50 frames per second in the processing speed. This research underlined the YOLO models' adaptability to thermal infrared data, enhancing UAV capabilities for real-time surveillance and monitoring applications.

Wang et al. [15] investigated the effectiveness of using passive IRT for detecting various types of damage in small WTBs under natural outdoor conditions. This method optimized the light conditions required for effective detection by analyzing the interaction between light intensity and the surface temperatures of damaged blades. The experimental findings indicated that the optimal detection of different damage types—such as foreign body attachment, surface wear, and cracks—occurred at specific light intensities, with light around 1000 W/m^2 being most effective. Additionally, the study utilized numerical simulations to further validate the experimental results and establish a thermodynamic coupling model, demonstrating a linear relationship between light intensity and the thermal responses of the blade, such as temperature, stress, and strain, thereby enhancing the practical application of this technology in field conditions.

Yu et al. [16] introduced an innovative method for stitching infrared images of WTBs using data from UAVs and a U-Net-based neural network. This method was designed to overcome challenges associated with the weak texture and low contrast of infrared images. By employing U-Net for semantic segmentation to isolate the blades from complex backgrounds, the method calculated necessary geometric transformations to align and stitch the images accurately. This process enhanced the quality of the stitched images, as evidenced by comparative analysis against traditional stitching algorithms. The approach not only ensures accurate blade orientation and alignment but also significantly improves the detection and quantification of defects, contributing to more effective maintenance strategies for wind turbines.

Chen et al. [17] introduced AQUADA Plus, a novel method leveraging thermography and computer vision to detect damage in large-scale composite structures under cyclic loads. The method automates the detection, localization, and evaluation of damage through thermal image analysis, addressing challenges posed by complex thermal backgrounds and environmental conditions. A significant aspect of this study was the integration of drone-based thermography, demonstrating the method's application in real-world settings without stopping wind turbine operations. The approach not only enhances damage detectability and accuracy but also promises considerable reductions in inspection times and costs. The potential of this method to revolutionize structural health monitoring in wind turbines and similar large structures is underscored by its successful application in both laboratory settings and field tests.

Oliveira et al. [18] proposed a comprehensive framework for detecting faults in large-scale photovoltaic plants using aerial IRT combined with orthomosaicking techniques. Orthomosaicking, often referred to simply as mosaicking, is a process used in image processing and remote sensing; it involves stitching together multiple overlapping images into a single, seamless composite image, known as an orthomosaic. Utilizing the Mask R-CNN algorithm for instance segmentation, the study addressed the challenges of analyzing extensive photovoltaic arrays by automating fault detection and localization. This framework effectively reduces the time and potential errors associated with manual inspections through its integration of deep learning and image processing technologies. Initial tests demonstrated high accuracy in detecting and pinpointing hotspots and other defects, offering significant improvements in operational efficiency for maintenance operations in photovoltaic plants.

Attallah et al. [19] presented an advanced method for diagnosing inter-turn faults in induction generators used in offshore wind turbines. This method uses multiple CNNs in combination with IRT and a sophisticated feature selection mechanism to improve fault detection efficiency and accuracy. The approach leverages the strengths of different CNN architectures by merging their features, followed by a selective reduction in these features to enhance diagnostic effectiveness. It eliminates the need for contact-based measurement, making it ideal for the challenging offshore environment. The proposed method not only increases the accuracy of fault diagnosis in wind turbine generators but also significantly reduces the computational load and diagnosis time. This was demonstrated through a detailed comparison to traditional and other deep learning methods, highlighting its superiority in accuracy and efficiency.

Haidong et al. [20] detailed an innovative fault diagnosis method for rotating machinery that operates under varying speed conditions. This approach introduced a new Generative Adversarial Network (GAN) [21], called the Dual-Threshold Attention-Guided GAN (DTAGAN), which is tailored to generate high-quality infrared thermal images for fault diagnosis. The proposed DTAGAN addresses several challenges typical of traditional GANs such as gradient vanishing and poor feature extraction. By combining Wasserstein distance and gradient penalty within the loss function and incorporating an attention mechanism to highlight significant global features from thermal images, DTAGAN significantly enhances the quality and efficiency of image generation. The study demonstrated that DTAGAN outperformed existing methods in diagnosing faults in rotor-bearing systems through comparative experiments, particularly under conditions of limited data and speed variations.

Tanda et al. [22] investigated the efficacy of using UAVs and manned aircraft equipped with IRT for monitoring solar photovoltaic (PV) systems. It provided a detailed comparison of these two remote sensing platforms in detecting thermal anomalies in solar panels, which are indicative of faults. Both methods demonstrated high accuracy in identifying defects, with UAVs offering detailed imaging at lower operational costs and manned aircraft covering larger areas more quickly, thus potentially reducing overall inspection time for large-scale installations. The economic analysis highlighted that while UAVs are generally more cost-effective for smaller or medium-sized PV plants, aircraft can be more economically viable for larger installations exceeding 40 MW due to their quicker survey capabilities and the ability to cover extensive areas in a single day.

Petersen et al. [23] detailed a study on the use of Mid-Infrared Optical Coherence Tomography (MIR OCT) for inspecting leading edge coatings of WTBs. The study addressed the critical issue of erosion, which affects the aerodynamic efficiency and longevity of turbine blades. Through detailed experiments, the authors demonstrated that MIR OCT could effectively identify subsurface defects such as bubbles and cracks that were not detectable by other non-destructive methods like ultrasonic testing. This technology allowed for a deeper inspection of the protective coatings applied to turbine blades, offering a significant improvement over traditional methods by providing detailed, high-resolution images of internal structural defects without damaging the blades. The research suggests that implementing MIR OCT in the manufacturing process could enhance the quality control of turbine blade coatings, potentially reducing maintenance costs and downtime due to blade erosion.

Zhao et al. [24] explored advancements in active IRT technology for identifying defects in the renewable energy and electronics sectors. It provided a detailed examination of different active IRT methods like pulsed, lock-in, vibration-induced, and eddy current thermography, highlighting their applications and effectiveness in pinpointing defects in various materials and components. Specifically, the article discusses the integration of

IRT with deep learning to enhance defect detection in photovoltaic systems and electronic components, noting significant improvements in both accuracy and efficiency. It addresses the practical challenges and benefits of these technologies, such as non-contact and large-area imaging capabilities, and suggests future research directions to further optimize defect detection and system monitoring.

Wang et al. [25] focused on enhancing defect detection in WTBs through advanced image segmentation techniques using UAV-carried infrared imaging. The study introduced a novel segmentation method that incorporates both global and local image information to address the common challenges of image inhomogeneity in infrared images, which often impedes accurate segmentation. This method significantly improves segmentation accuracy and speed by utilizing a detail map for the edge stopping function and applying a variable threshold based on local statistics as the initial contour for the level-set method. The approach was demonstrated to be more efficient than traditional methods, achieving high segmentation accuracy while also correcting for bias in the image, thus ensuring precise defect localization in wind turbine maintenance applications.

Li et al. [26] presented a comprehensive study on utilizing solar radiation as a passive excitation source for WTB IRT NDT. By leveraging natural solar conditions, the research explored the detectability of subsurface defects like delaminations in turbine blades, employing numerical simulations and experimental validations. The findings suggested that the time of day significantly affected the thermal visibility of defects due to the solar angle and intensity, providing critical insights into the optimal conditions for conducting thermal inspections. This method represents a cost-effective alternative to active heating techniques, potentially enhancing routine maintenance strategies for wind turbines.

Zhou [27], who published in IEEE Transactions on Instrumentation and Measurement, explored a novel defect detection method for wind turbines using combined RGB and IRT. The study introduced a Regression Crop data-processing method and an adaptive feature fusion module that integrates RGB and IR data, significantly enhancing the detection accuracy and precision of actual defects in WTBs. The approach employs advanced deep learning techniques and was tested using a dataset of wind turbine images, showing an increase in detection precision to 99%. This method addresses the challenges of distinguishing between actual defects and false positives like dust or debris on the blades, presenting a substantial improvement over traditional methods.

Zheng et al. [28], who published in the IEEE Sensors Journal, introduced a novel method for detecting temperature anomalies in blower components using infrared video analysis. This method utilizes advanced segmentation algorithms and a hierarchical multi-scene anomaly detection strategy to improve the accuracy and robustness of temperature anomaly detection. Key techniques include improved point rendering for better image segmentation, K-means++ clustering for initial data classification, and k-nearest neighbor analysis for finer anomaly detection. The approach is enhanced with dynamic time warping to handle different data lengths and sliding window techniques to estimate anomaly durations, offering significant improvements in industrial monitoring applications.

Jia et al. [29] proposed an advanced method, AQUADA-Seg, for segmenting WTBs using fused optical and thermal video data. This technique utilizes both modalities to provide comprehensive information about the blades, enhancing the segmentation process especially in complex background conditions that typically hinder accurate segmentation. By incorporating both modalities, the method significantly outperforms traditional single-modal approaches, achieving near real-time processing capabilities and high accuracy, thus facilitating more efficient and timely inspections of wind turbines while in operation.

Sheiati [30] described a novel deep learning method for segmenting fatigue damage in WTBs using passive IRT. This method efficiently segments the blade from complex dynamic backgrounds in thermal images, ensuring accurate damage detection under operational conditions. The proposed approach utilizes a two-step process, starting with an innovative background segmentation using pre-trained deep learning models, followed by damage segmentation that isolates and analyzes fatigue damage. This technique significantly improves the accuracy and efficiency of thermal imaging analysis for structural health monitoring in wind turbines, achieving impressive recall and precision rates in both background and damage segmentation tasks.

While some valuable thermal imaging datasets exist for WTB inspection, such as those presented by Duan et al. [31] and Aminzadeh et al. [32], there remains a need for more comprehensive, high-quality datasets that specifically address the challenges of small-scale wind turbines and early-stage defect detection. Our work complements existing datasets by focusing on minute defects and their thermal signatures in small-scale turbines, contributing to the development of more precise, automated defect detection systems capable of early-stage damage identification, which is crucial for proactive maintenance.

1.2. Object Detection and Localization

Object detection and localization have been significantly improved by the advent of CNNs, making it possible to identify and locate objects within images with high precision. The demand for real-time object detection has grown with applications in autonomous driving and surveillance. Methods like CNNs have evolved to meet real-time requirements by reducing computational overhead and increasing processing speed. The YOLO (You Only Look Once) series has significantly influenced the domain of object detection, integrating rapid processing speeds with the precision required for real-time applications. This evolution began in 2016 with YOLOv1, developed by Joseph Redmon et al. [14], and has since seen several iterations, each building on the last in terms of detection accuracy, processing speed, and adaptability across various objects and scenarios. The successive versions have continually addressed specific challenges in object detection, leading to advancements in feature extraction, bounding box predictions, class predictions, and overall algorithmic and architectural enhancements.

YOLOv1, introduced by Redmon et al. in 2016 [14], revolutionized object detection by treating it as a single regression problem, directly converting image pixels to bounding box coordinates and class probabilities. This approach eliminated the need for separate region proposal generation, significantly boosting processing speed and enabling real-time performance.

YOLOv2, or YOLO9000, introduced by Redmon et al. in 2017 [33], enhanced the original YOLOv1 by integrating batch normalization, utilizing high-resolution classifiers from the start of training, and adopting a new convolutional architecture. These improvements not only retained the high speed of the original but also increased the accuracy of the detections.

YOLOv3, detailed by Redmon et al. in 2018 [34], advanced its predecessors by incorporating multi-scale predictions and leveraging a deeper feature extractor network, known as Darknet-53 [34]. This version improved detection capabilities across various object sizes, further balancing speed and accuracy enhancements.

YOLOv4, developed by Bochkovskiy et al. in 2020 [35], incorporated additional features such as Weighted-Residual-Connections (WRC) [36], Cross-Stage-Partial connections (CSP) [37], and Cross Mini-Batch Normalization (CmBN) [38]. These features significantly improved the model's robustness and generalization over different datasets and conditions, making it more effective in diverse scenarios.

YOLOv5, Developed and released by Ultralytics in 2020 [39], a technology company known for its contributions to machine learning and computer vision. Ultralytics' release of YOLOv5 represents an unofficial continuation of the YOLO project, which they have enhanced with proprietary modifications and improvements. This version introduces adaptive anchor box calculations, employs a new compound scaling method [40], and extensively uses data augmentations. YOLOv5's architecture is designed to be flexible, offering various model sizes that balance computational demands with detection performance, making it highly adaptable for different applications. The release by Ultralytics has been widely adopted in the community due to its performance improvements and ease of use, setting a new benchmark for real-time object detection systems.

YOLOv6, presented by Li et al. in 2023 [41], aimed at refining the architecture and training methods of YOLO, focusing on optimizing the network structure to enhance performance in real-time applications. It features advanced training techniques and structural optimizations that further improve the speed-accuracy trade-off.

YOLOv7, developed by Peng et al. in 2023 [42], introduces improvements in feature extraction and fusion techniques, which significantly enhance detection precision, particularly in complex and challenging environments [43].

In this study, we focus on the latest and most advanced versions of the YOLO architecture, YOLOv8 and YOLOv9, which offer significant improvements in detection accuracy and computational efficiency specifically relevant to our application.

YOLOv8 [42] emphasizes improving detection across various object sizes, especially small objects. It employs a novel network structure and methods to maintain contextual feature information, which boosts detection accuracy.

YOLOv9 was proposed by Wang et al. in Feb 2024 [44] and implemented by Ultralytics in April 2024 [45], innovated with architectural features such as Programmable Gradient Information (PGI) and a Generalized Efficient Layer Aggregation Network (GELAN), aiming to preserve semantic information more effectively and improve computational efficiency.

In Table 1, we present a summary and comparison of different YOLO object detection model iterations, highlighting their main features, advantages, and limitations. The table demonstrates that the latest models, YOLOv8 and YOLOv9, significantly enhance detection accuracy, processing speed, and performance in complex environments. These improvements make them especially well suited for real-time applications like defect detection in wind turbine blades. Therefore, we employed these latest YOLO models in our research, leveraging their superior performance to achieve more accurate and efficient defect localization in thermal images. The use of YOLOv8 and YOLOv9 in our study underscores our commitment to utilizing cutting-edge technology to enhance wind turbine maintenance processes. Section 3 provides an in-depth look at the implementation, training (with our dataset), optimization, analysis, and comparison of these two models.

Table 1. Comparison of YOLO model versions.

Model	Key Features	Advantages	Limitations
YOLOv1	Single regression problem for object detection [14]	Fast processing speed, real-time detection	Lower accuracy compared to later versions
YOLOv2	Batch normalization, high-resolution classifiers, new convolutional architecture [33]	Improved accuracy over YOLOv1, retains high speed	Still limited in handling small objects
YOLOv3	Multi-scale predictions, deeper feature extractor (Darknet-53) [34]	Better detection across various object sizes, improved balance of speed and accuracy	Increased computational complexity

Table 1. *Cont.*

Model	Key Features	Advantages	Limitations
YOLOv4	Weighted Residual Connections (WRCs), Cross-Stage Partial (CSP) connections, Cross Mini-Batch Normalization (CmBN) [35]	Enhanced robustness and generalization, effective across diverse scenarios	Larger model size
YOLOv5	Adaptive anchor box calculation, compound scaling, extensive data augmentation [39]	Flexible architecture, ease of use, widely adopted	Proprietary modifications, unofficial continuation
YOLOv6	Optimized network structure for real-time performance [41]	Improved speed–accuracy trade-off, efficient for real-time applications	Primarily focused on optimizing existing architectures
YOLOv7	Improved feature extraction and fusion techniques [43]	Higher detection precision, better performance in complex environments	Incremental improvements over YOLOv6
YOLOv8 *	FPN and PAN architectures, transfer learning [42]	High accuracy, better handling of small objects, improved contextual feature information	Higher training complexity, requires fine-tuning
YOLOv9 *	Programmable Gradient Information (PGI), Generalized Efficient Layer Aggregation Network (GELAN) [44]	Advanced gradient handling, optimal balance of accuracy and speed, lightweight	Newer model, requires more testing and validation

* Model versions used in this paper for object detection and localization.

1.3. Contributions

Addressing these gaps, our research makes several critical contributions to the field. The creation of a new thermal dataset: we developed a bespoke thermal dataset that captures a wide range of small, critical defects specific to WTBs, greatly enhancing defect detectability and analysis accuracy. Advancement in defect localization: our work extended beyond simple defect identification to sophisticated localization, utilizing state-of-the-art object detection models for more detailed and actionable insights. Drone-compatible camera testing: initially testing our thermal camera in a handheld setup, we ensured its capability to produce high-quality images, a prerequisite for the future deployment on drones for comprehensive blade inspections. Comparative analysis of advanced object detection models: we provide an analytical comparison of the latest YOLO models, assessing their strengths and adaptability to the nuanced requirements of thermal imaging for wind turbine inspection.

The novelty of this research lies in its targeted approach to dataset creation and the application of advanced deep learning techniques tailored for wind turbine maintenance. This approach not only fills a significant void in current maintenance practices but also sets a benchmark for future technological advancements in the sector. The importance of this work is underscored by its potential to significantly reduce maintenance costs, enhance turbine safety and longevity, and support the broader adoption of wind energy globally.

1.4. Research Roadmap

Building on the foundational work presented in this paper, future research is expected to progress through several distinct phases to further develop and apply the techniques introduced. The first phase may involve expanding the dataset to include thermal images from large, operational wind turbines under various environmental conditions, allowing for more comprehensive training and testing of object detection models. The second phase could focus on integrating these models with drone-mounted thermal cameras, enabling automated, real-time defect detection in operational environments. Finally, the third phase should involve rigorous field testing to validate the effectiveness and reliability of the proposed system in real-world applications. These phases are anticipated to lead to the development of an advanced, scalable maintenance strategy applicable across the renewable energy sector.

2. Methodology

This study outlines our methodical approach, which used advanced thermal imaging cameras and analytical models to inspect wind turbines. We detail the deployment of the FLIR FLIR (Forward Looking InfraRed) Vue TZ20 camera, capable of capturing high-resolution images to detect defects in turbine blades. Our methodology aimed to bridge theoretical research and practical application, providing a framework for robust and adaptable inspections. This introduction sets the stage for a deeper discussion of the specific equipment and procedures used in our research.

2.1. The Studied Wind Turbine Specification

For our experiments, we used the Primus Air Max, a small-scale wind turbine frequently employed in maritime settings. This turbine is capable of generating approximately 30 kWh per month under average wind speeds of 5.5 m/s. It features a swept area of 1.07 m^2, a rotor diameter of 1.17 m, and a weight of 5.9 kg. The construction of the turbine includes a cast aluminum body with corrosion-resistant paint and three carbon fiber blades. In addition, it is equipped with a microprocessor-based control system, a brushless alternator, and electronic torque control, designed to withstand wind speeds of up to 40.2 m/s.

To simulate typical operational damage, intentional defects, such as cracks, holes, and erosions, were artificially created on the blades. These defects mimicked the real-world damage observed in wind turbine blades (WTBs), enhancing the relevance of our study for practical applications.

Figure 1 shows the Primus Air Max turbine, highlighting the overall configuration and the positioning of its three carbon fiber blades, including visible defects.

(**a**) A front view of the small-scale wind turbine.

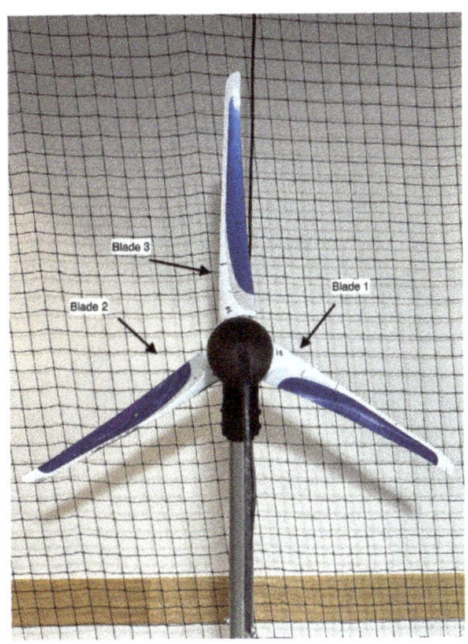

(**b**) A back view of the small-scale wind turbine.

Figure 1. The small-scale wind turbine with intentionally created blade defects. The turbine has three carbon fiber blades (labeled as 1, 2, and 3 clockwise from the top) with the following dimensions: rotor diameter = 1.17 m; blade length = 0.585 m; and swept area = 1.07 m^2 .

Figure 2a–c display different perspectives of the front blades. These images are particularly valuable for examining the blade profiles and their surface condition, highlighting the artificially created defects for detailed assessment. Similarly, Figure 2d–f show the back sides of the blades. These images help in assessing the wear and the specific artificially induced damage that could affect the performance of the turbine over time.

These detailed views of each blade are crucial for assessing the structural integrity and surface conditions of the blades, with particular focus on the artificially induced damages to simulate real-world scenarios.

2.2. Camera Deployment and Imaging Setup

This research utilized the FLIR Vue TZ20 camera, shown in Figure 3 specifically engineered for drone operations. The camera features a spectral range of 7.5–13.5 μ m and thermal sensitivity (NETD) < 50 mK at f/1.0. It features a dual thermal camera system with a 20× digital thermal zoom, allowing for both wide and focused field-of-view options. This versatility is crucial for capturing detailed thermal data across various inspection scenarios, from public safety to industrial inspection. The camera operated in conjunction with FLIR Boson GUI 3.0 on a Windows platform, designed to enhance the development and deployment of thermal imaging solutions. The suite offers a user-friendly interface for the efficient management and processing of thermal images. The camera was connected to the drone using the gimbal depicted in Figure 3b, which ensured stable mounting and precise control during flight. The interface and connection setup are illustrated in Figure 3c. For further technical details, the FLIR website provides extensive information on the Vue TZ20-R's specifications and integration capabilities with DJI Matrice M200 v2- and M300-series drones via the Skyport V2.0 gimbal. The deployment of this camera in our project focused on capturing high-quality thermal images that revealed minute defects in the WTBs, a task compounded by the complexities of outdoor environmental conditions.

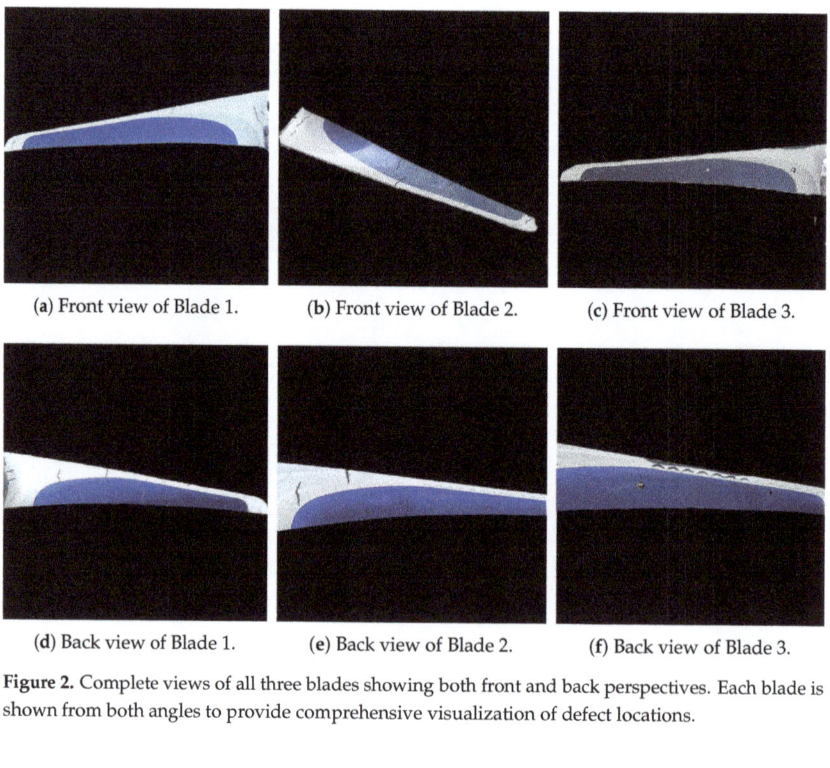

(a) Front view of Blade 1. (b) Front view of Blade 2. (c) Front view of Blade 3.

(d) Back view of Blade 1. (e) Back view of Blade 2. (f) Back view of Blade 3.

Figure 2. Complete views of all three blades showing both front and back perspectives. Each blade is shown from both angles to provide comprehensive visualization of defect locations.

Limitations and Biases: Despite the advanced capabilities of the FLIR Vue TZ20 camera, several limitations and potential biases were identified during its deployment. The precision of defect detection was highly dependent on environmental conditions, such as temperature fluctuations and varying sunlight levels, which could introduce biases in the thermal data captured. Furthermore, the current setup's dependence on manual camera operation may have impacted the repeatability and consistency of the results. This underscores the necessity for drone integration in future phases to enhance data capture accuracy and reliability.

(**a**) FLIR Vue TZ20 camera.

(**b**) Camera connector gimbal.

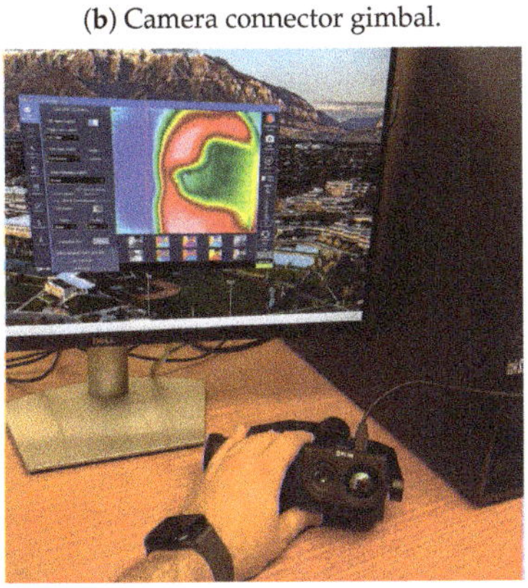

(**c**) FLIR Boson GUI 3.0.

Figure 3. (**a**) FLIR Vue TZ20 camera, (**b**) camera connector gimbal, and (**c**) camera connected to FLIR Boson GUI 3.0 application.

Practical Implementation of Drone Deployment: Future practical approaches to drone deployment are expected to involve addressing several technical challenges such as ensuring the stability of the drone platform during flight, particularly in adverse weather conditions. Additionally, the integration of a thermal camera with the drone's navigation and control systems will require careful calibration to maintain image quality while in motion. Future drone deployments should focus on automating the inspection process, minimizing manual intervention, and allowing for real-time analysis of wind turbine blade conditions.

Temperature and Sunlight Effects: Temperature and direct sunlight significantly affect the quality of defect detection in thermal imaging. Our study, conducted during midday at noon in Utah, benefited from intense sunlight, which significantly improved the clarity of the thermal images. The high ambient temperature and direct sunlight increased the visibility of defects. The increased thermal contrast between defective and non-defective areas made defects more visible and detection more accurate. These findings indicate that conducting inspections under similar environmental conditions can enhance thermal camera performance. However, the effectiveness of this approach may vary under different weather conditions, requiring careful consideration in scheduling and executing inspections.

2.3. Dataset Creation Setup

Creating the dataset for WTB inspection presented numerous challenges, particularly in capturing thermal images that accurately reflected small defects. The imaging setup involved connecting the FLIR Vue TZ20 camera to a laptop, a necessary configuration that, while restrictive, was vital for initial data capture phases. Our experiments spanned both laboratory and field settings to emulate the operational environment of wind turbines as closely as possible. However, drone deployment in the future could potentially mitigate some of the current limitations by offering more flexible and comprehensive data capture capabilities.

Dataset Construction and Differentiation: The specialized dataset constructed in this study differs from existing datasets by focusing specifically on small-scale wind turbines and capturing high-resolution thermal images that emphasize minute defects. Unlike other datasets, which may use generic or low-resolution images, our dataset was tailored to reflect the specific conditions and defect types encountered in wind turbine blades. This focus on high fidelity and relevance to real-world conditions enhances the applicability of our models to operational wind turbines.

Operational Feasibility: The method developed in this study can be implemented with operational wind turbines without the need for the turbines to be stopped. This capability makes it suitable for online monitoring, where the inspection can be conducted periodically without interrupting the turbine's operation. However, the effectiveness of the inspection during operation may be influenced by the turbine's rotational speed and the resulting motion blur, which must be accounted for in future model training and calibration.

2.4. Dataset Structure

The dataset was structured into directories based on the component condition and the imaging technique used. It included an Indoor Directory with 1200 thermal images, equally divided between faulty and healthy conditions, and an Outdoor Directory that mirrored the Indoor Directory with another 1200 images, also equally divided between faulty and healthy conditions. Additionally, the dataset featured 6 high-resolution RGB images, offering extra visual context for the inspections, with an equal number of faulty and healthy images.

The comparative analysis between laboratory and field images underscored the superior accuracy of data captured under outdoor conditions, closely aligning with our operational objectives. This led to a strategic focus on enhancing the dataset with images that accurately represent the diverse and challenging conditions faced by wind turbines.

The annotated thermal images depicted below serve as a visual representation of the defects encountered. These images underscore the model's ability to recognize and demarcate defects across diverse conditions and blade positions.

Figure 4 demonstrates the defects, including cracks, holes, and erosion, annotated in the thermal images of the WTBs. These annotations are essential for training machine learning models to detect and classify types of damage in similar real-world scenarios.

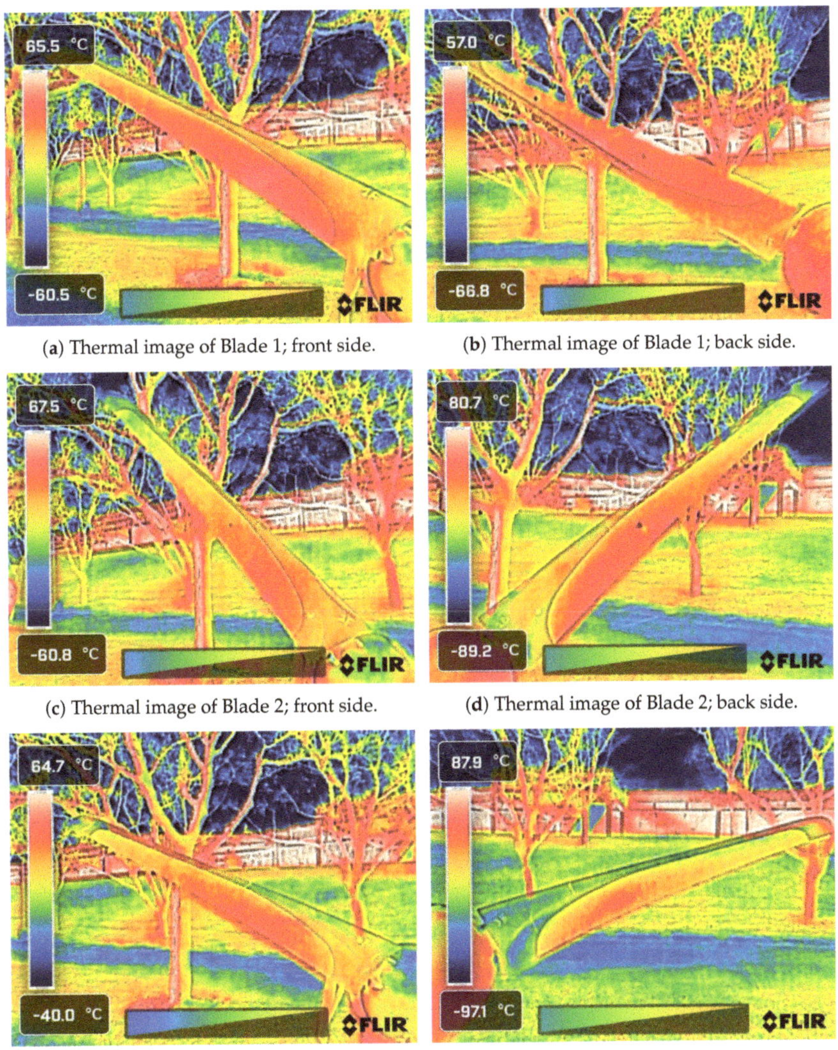

(**a**) Thermal image of Blade 1; front side. (**b**) Thermal image of Blade 1; back side.

(**c**) Thermal image of Blade 2; front side. (**d**) Thermal image of Blade 2; back side.

(**e**) Thermal image of Blade 3; front side. (**f**) Thermal image of Blade 3; back side.

Figure 4. Thermal images of the small-scale wind turbine blades, showing the front and back views of each blade. The images were processed in FLIR Thermal Studio using Digital Detail Enhancement, enabling clearer visualization of the defects and making them more discernible for human interpretation.

2.5. Models

In this section, we delve into the latest advancements within the YOLO series: YOLOv8 [42] and YOLOv9 [44]. These versions represent the zenith of the series's development, introducing novel features and optimizations that redefine benchmarks for performance and efficiency in the realm of real-time object detection.

The decision to include both YOLOv8 and YOLOv9 in our analysis served multiple important purposes. First, while YOLOv9 is the newer version, YOLOv8 has been extensively validated in various industrial applications and serves as a robust baseline for performance comparison. Second, the inclusion of both models allowed us to quantitatively assess the specific improvements that YOLOv9's architectural innovations (such as PGI and GELAN) bring to defect detection in thermal images. Third, from a practical implementation perspective, YOLOv8's established stability and broader hardware compatibility make it a valuable alternative in scenarios where the additional computational requirements of YOLOv9 may not be justified by the marginal performance gains. This comparative approach provides valuable insights for practitioners in choosing the most appropriate model for their specific wind turbine maintenance needs.

YOLOv8 marks a significant leap forward in real-time object detection. The architecture of YOLOv8, shown in Figure 5, introduces several enhancements that bolster detection accuracy and operational efficiency. Notably, YOLOv8 excels in processing environments with complex object spatial configurations and varied aspect ratios. It achieves this through a sophisticated architecture that amalgamates the Feature Pyramid Network (FPN) and the Path Aggregation Network (PAN), facilitating the extraction of comprehensive, multi-scale feature representations.

A pivotal innovation in YOLOv8 is its strategic employment of transfer learning, which substantially augments its detection capabilities in specific real-world settings. This model initially undergoes training on a wide-ranging dataset and is then finely tuned on datasets that closely reflect targeted application environments. This approach ensures an exemplary balance between processing velocity and detection precision, essential for scenarios necessitating swift decision-making.

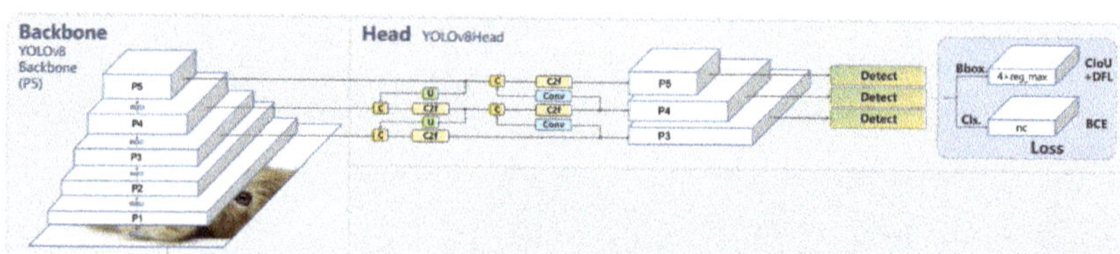

Figure 5. Architectural overview of YOLOv8, illustrating its key components and operational flow [42].

The YOLOv8 loss function, $L(\theta)$, is a combination of several components, the box loss L_{box}, the classification loss L_{cls}, and the distribution focal loss L_{dfl}, each normalized by the number of positive examples, N_{pos}. And, it includes a regularization term with the weight decay ϕ:

$$L(\theta) = \frac{\lambda_{\text{box}}}{N_{\text{pos}}} L_{\text{box}}(\theta) + \frac{\lambda_{\text{cls}}}{N_{\text{pos}}} L_{\text{cls}}(\theta) + \frac{\lambda_{\text{dfl}}}{N_{\text{pos}}} L_{\text{dfl}}(\theta) + \phi\|\theta\|_2^2 \tag{1}$$

Specifically, the box loss component is defined as

$$L_{\text{box}} = \sum_{x,y} 1_{c^*_{x,y}} \left[(1 - q_{x,y}) + \frac{\|b_{x,y} - \hat{b}_{x,y}\|^2}{\rho^2} + \alpha_{x,y} v_{x,y} \right] \tag{2}$$

where we have the following:

- $b_{x,y}$ and $\hat{b}_{x,y}$ represent the predicted and ground truth bounding box coordinates at the position (x, y), respectively.
- $q_{x,y}$ is the intersection over union (IoU) between the predicted and ground truth boxes.
- $\alpha_{x,y}$ is a dynamic weighting factor that adjusts the contribution of each spatial location.
- $\nu_{x,y}$ is a modulation factor that helps balance the learning of different sized objects.
- ρ is a normalization parameter for the coordinate differences.
- $1_{c^*_{x,y}}$ is an indicator function that equals 1 when a defect is present at the position (x, y).

YOLOv9 introduces a paradigmatic shift with its Programmable Gradient Information (PGI), shown in Figure 6, and the Generalized Efficient Layer Aggregation Network (GELAN), shown in Figure 7, targeting the pervasive challenge of information loss in deep neural networks. The PGI mechanism employs an auxiliary branch that preserves gradient information during backpropagation, ensuring that important features for defect detection are not lost during training. This is particularly crucial for thermal images where subtle temperature variations can indicate defects. The GELAN architecture optimizes the network's computational efficiency while maintaining detection accuracy, making it suitable for real-time defect detection applications.

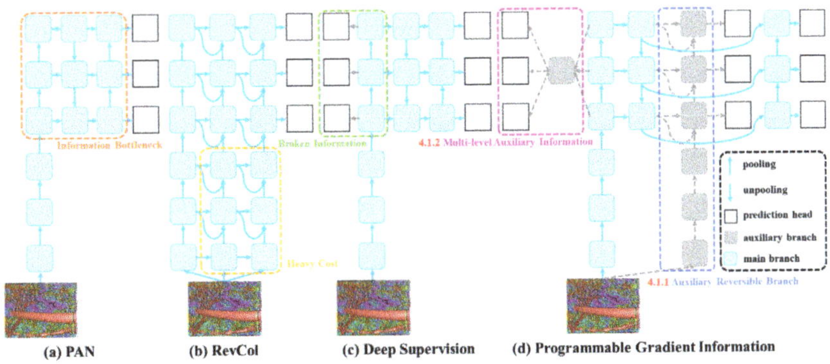

Figure 6. Depiction of PGI integration within YOLOv9, showcasing its innovative approach to gradient information programming [44].

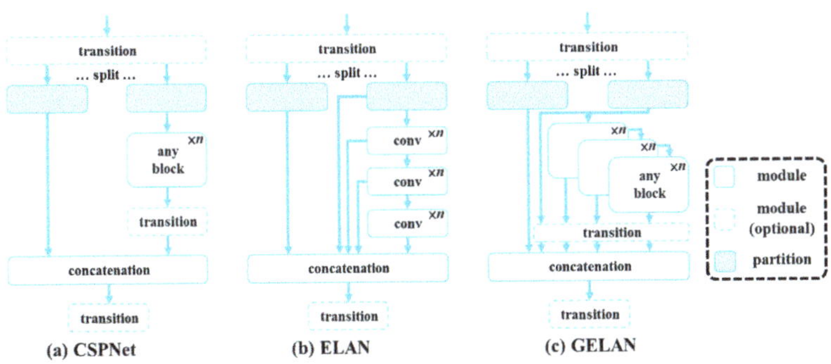

Figure 7. The GELAN architecture in YOLOv9, highlighting its adaptability to support various computational blocks [44].

For YOLOv9, which employs the Programmable Gradient Information (PGI) approach, the exact mathematical formulation of the loss function is conceptually represented as

$$L_{\text{PGI}} = \text{Function}(L_{\text{main}}, L_{\text{aux}}, \text{Gradients from PGI})$$

Here, L_{main} and L_{aux} represent losses from the main and auxiliary branches of the network, respectively. The loss dynamically integrates gradients managed by the PGI system to optimize the learning process, ensuring a minimal loss of information and maximizing the reliability of the gradient updates.

3. Results and Analysis

Following the detailed exposition of the methodology, which outlined the technical setup, camera deployment, and specifics of the wind turbine used in this study, we now transition into the Results and Analysis Section. This portion of the paper evaluates the effectiveness of the deployed technologies and methods in detecting and analyzing defects in WTBs.

3.1. Parameter Tuning and Optimization

While we utilized the established YOLO architecture as our foundation, our implementation involved significant customization and optimization for the specific challenges of wind turbine blade defect detection. Our approach included the following:

1. **Custom Dataset Preparation and Annotation**: We developed a specialized dataset comprising thermal images of wind turbine blade defects, meticulously annotated to capture various defect types and severities. This dataset ensured that the model was trained on relevant and high-quality data specific to the application domain.
2. **Extensive Hyperparameter Optimization**: Leveraging Bayesian optimization techniques, we systematically explored the hyperparameter space to identify the optimal configurations that enhanced model performance for our specific application. This included tuning parameters such as learning rates, batch sizes, and momentum coefficients.
3. **Modified Loss Function Weights**: The loss functions were adjusted to prioritize accurate localization of defects, ensuring that the model emphasized critical areas during training. By assigning higher weights to certain loss components, the model became more sensitive to the nuances of defect detection.
4. **Specialized Data Augmentation Strategies**: We implemented data augmentation methods tailored to the characteristics of thermal images, such as thermal noise addition, brightness variations, and geometric transformations. These strategies improved the model's robustness and ability to generalize across different thermal imaging conditions.
5. **Fine-Tuned Confidence Thresholds**: Confidence thresholds were meticulously calibrated to balance precision and recall, optimizing the model's defect detection accuracy in real-world scenarios. This calibration helped in reducing false positives and enhancing the reliability of detections.

These customizations transformed the base YOLO architecture into a specialized tool for wind turbine maintenance, moving beyond default configurations to address the unique challenges of thermal defect detection. By tailoring the models to our specific needs, we enhanced their potential effectiveness and reliability for real-world applications.

Both YOLOv8 and YOLOv9 feature significant architectural enhancements that demanded comprehensive parameter tuning for optimal performance. Both models use a batch size of 64 and are trained for 100 epochs. The initial learning rate is set to 0.001 and is applied with the Adam optimizer, which is typically configured with a momentum of 0.9. Regularization techniques include a dropout rate of 0.5 and a weight decay of 0.0005 to prevent overfitting. Both models utilize transfer learning, as they are pretrained on the COCO (Common Objects in Context) dataset [46]. Bayesian optimization is employed, starting from these default values, to determine the optimal model configurations. For YOLOv8, specific parameter tuning and optimization include configuring the loss function. YOLOv8 utilizes standard cross-entropy for classification and smooth L1 for bounding box prediction, effectively training the detection model. YOLOv9, on the other hand, focuses on optimizing gradient information programming to ensure maximum information flow through the gradient. The default settings mirror those of YOLOv8. Additionally, YOLOv9 incorporates CSP-ELAN blocks with a default depth of three layers and filter sizes starting from 256, balancing efficiency and performance. Customized loss functions are developed to leverage the unique features of PGI and GELAN within YOLOv9. The parameter tuning and optimization for both YOLOv8 and YOLOv9 are critical, iterative processes that are heavily data-driven. These processes involve extensive experimentation and validation on benchmark datasets, ensuring that the models not only meet but exceed current standards in object detection while being practical for real-world high-speed and accurate detection tasks.

3.2. YOLOv8 Results

This section presents the experimental results obtained by applying YOLOv8 to defect detection in thermal images of WTBs, where the model was tasked with identifying three classes of defects: holes, erosion, and cracks.

The confusion matrix for normalized data, as seen in Figure 8, indicates the model's performance in terms of relative predictions per class. The matrix shows moderate to high diagonal values, suggesting an acceptable level of accuracy in predictions, with the highest normalized value for erosion detection (0.85). However, some confusion is noticeable between the classes, particularly between holes and background, where a normalized value of 0.39 is observed.

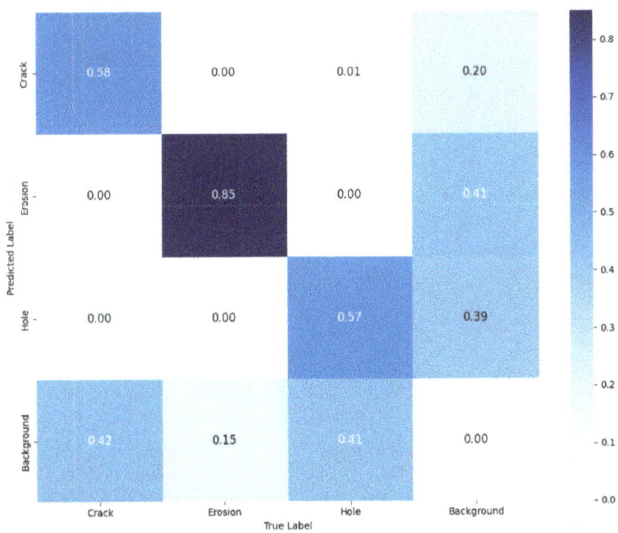

Figure 8. Confusion matrix normalized for YOLOv8.

The model's performance across different defect types revealed a particularly strong capability in detecting erosion, achieving an impressive detection rate of 85%. This high accuracy was likely due to the distinct thermal signature of erosion defects, which allowed them to be more easily identified. In contrast, the model faced greater challenges when distinguishing holes from the background, with a lower detection rate of 57% for holes and a notable confusion rate of 39% with the background. This difficulty may have stemmed from the similarity in thermal profiles between the holes and certain background elements, leading to misclassification. Cracks, although detected with a 58% accuracy, also exhibited some confusion with the background, indicated by a 20% misclassification rate. These findings suggest that while the model excels in identifying defects with clear thermal distinctions, it requires further refinement to enhance its accuracy in detecting defects with subtler thermal characteristics.

To evaluate the statistical significance of these results, we conducted a detailed analysis of the confidence intervals for each class's detection rate. The high accuracy observed in erosion detection, accompanied by a confidence interval indicating minimal variance, demonstrates that the model's performance in this category is statistically robust. However, the broader confidence intervals found for holes and cracks suggest that the reliability of the results in these categories may be lower, potentially due to the limited dataset size or the inherent complexity of detecting these types of defects. These insights underscore the necessity for additional model training and dataset expansion to bolster the reliability and statistical significance of the detection rates across all defect types.

The Precision-Recall curve, illustrated in Figure 9, were employed to evaluate the model's precision and recall against various confidence thresholds. Erosion defects were detected with a high precision of 0.866, while the model achieved a precision of 0.592 for cracks and 0.608 for holes. The average precision across all classes at an intersection over union (IoU) of 0.5 was 0.689, indicating a balanced detection capability across all defect types.

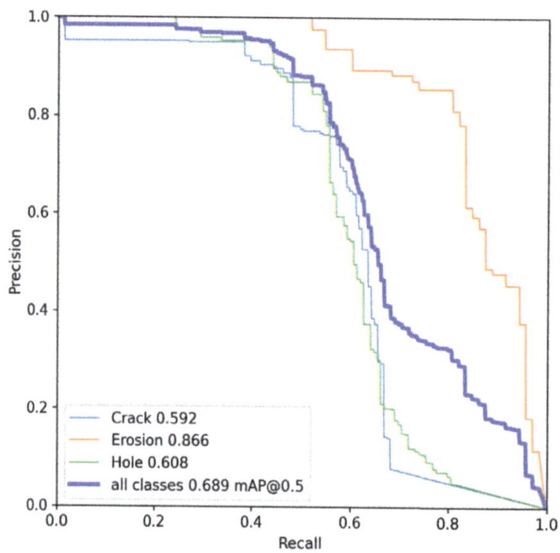

Figure 9. Precision–recall curve for YOLOv8. Curves are displayed with improved spacing and non-overlapping labels for better readability.

The F1–confidence curve, shown in Figure 10, plots the harmonic mean of precision and recall for all the three classes against different confidence levels. The curve peaks at an F1 score of 0.69 at a confidence threshold of 0.416, suggesting an optimal balance between precision and recall at this threshold for all classes.

The loss and performance metrics, as visualized in Figure 11, indicate a steady decline in both training and validation loss over time, suggesting that the model learned effectively without overfitting. The precision and recall metrics gradually increase, with the model's mean average precision (mAP) for the box predictions reaching a peak, demonstrating the efficacy of YOLOv8 in localizing the defects.

Figures 12 and 13 provide comprehensive insights into the distribution of dataset labels and the quality of bounding box annotations. These visual aids are instrumental in analyzing the spatial distribution of defect types, their frequency, and the precision of bounding box placements, which are crucial for the YOLOv8 model's training effectiveness and its ability to generalize across varied defect presentations in WTBs.

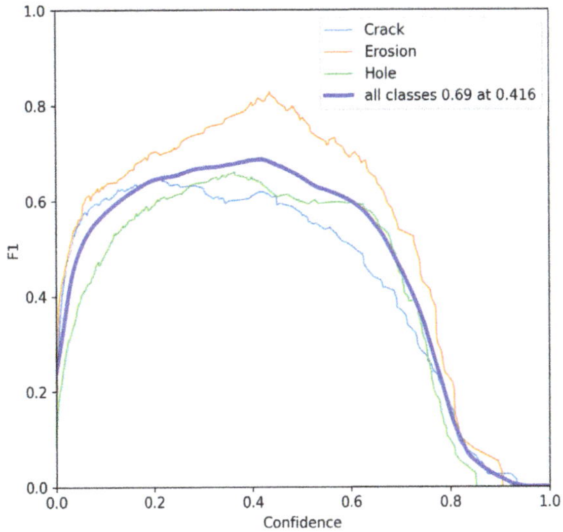

Figure 10. F1–confidence curve for YOLOv8.

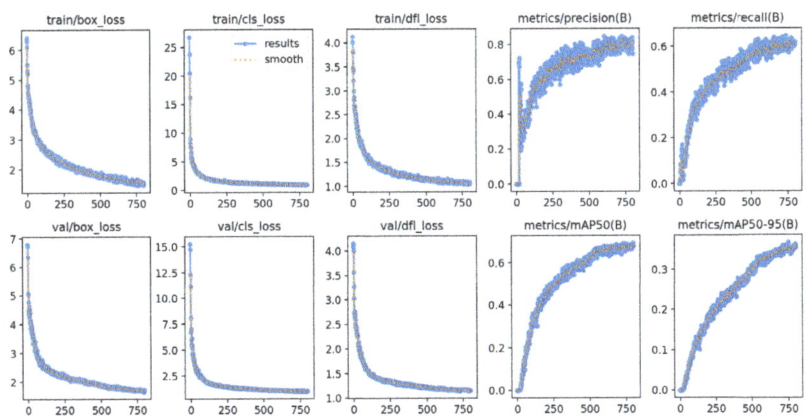

Figure 11. Loss and performance metrics for YOLOv8.

Figure 12 provides insights into the relationships between different bounding box attributes used in the YOLOv8 model. The x- and y-axes represent the coordinates (x, y) and dimensions (width, height) of the bounding boxes used to annotate defects in the dataset. Each cell in the matrix shows the relationship between the corresponding bounding box attributes. The color intensity indicates the density of occurrences, with darker colors representing higher densities. This visualization helps in understanding the distribution and correlation of bounding box attributes, which is important for improving the model's detection and localization capabilities.

Figure 13 provides insights into the distribution of defect labels and the dimensions of the bounding boxes used to annotate defects in the dataset for the YOLOv8 model.

The top left bar chart shows the frequency of different defect types (cracks, erosion, holes) on the x-axis, with the number of instances of each defect type on the y-axis. This chart indicates that cracks were the most common defect type, followed by holes and erosion.

The bottom left scatter plot illustrates the spatial distribution of defects on the wind turbine blade (WTB) surface, with x- and y-coordinates representing the position on the WTB surface. This plot shows where defects were commonly located on the WTB.

The bottom right scatter plot displays the relationship between bounding box width and height, with width on the x-axis and height on the y-axis. This plot illustrates the range of defect sizes and highlights the variability in defect dimensions that the model needed to handle.

The top right part of the figure depicts the anchor boxes used by the YOLOv8 model. These anchor boxes are preset dimensions applied during model training to efficiently identify objects of varying shapes and sizes. This visualization emphasizes the model's capacity to adjust to diverse defect dimensions, which is critical for accurate localization and classification in practical applications.

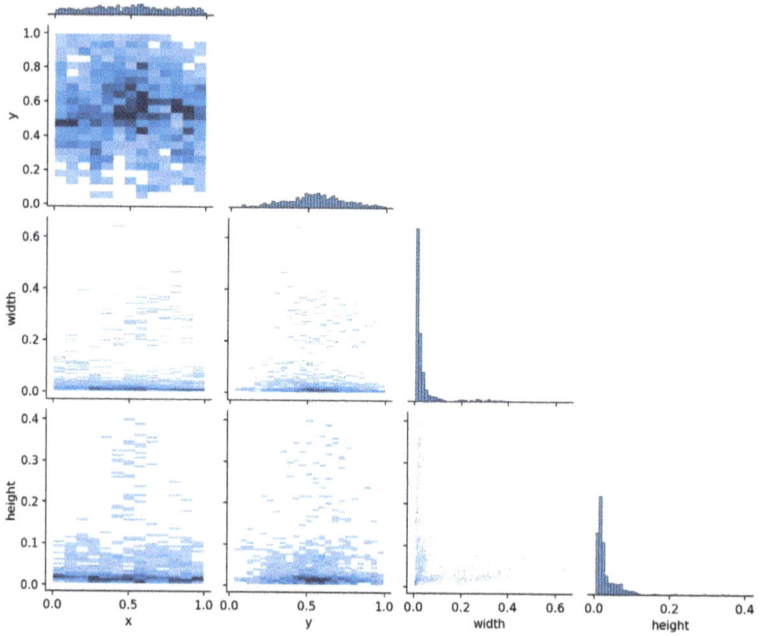

Figure 12. This figure illustrates the relationships between different bounding box attributes (x, y, width, height) used in the YOLOv8 model. The x- and y-axes represent these attributes. The color intensity indicates the density of occurrences, with darker colors representing higher densities. This visualization helps in understanding the distribution and correlation of bounding box attributes, which is important for improving the model's detection and localization capabilities.

Additionally, Figure 14 presents examples of defect detection results using YOLOv8. The figure shows the model's ability to localize and classify defects accurately in the images of WTBs, highlighting specific instances of holes, erosion, and cracks.

YOLOv9's normalized confusion matrix is depicted in Figure 15, showing each class's predictive accuracy. A high degree of precision can be observed in the matrix, particularly for the erosion class, similarly to the YOLOv8 model. However, the distribution of misclassifications will be a point of comparison to assess YOLOv9's improvement over its predecessor.

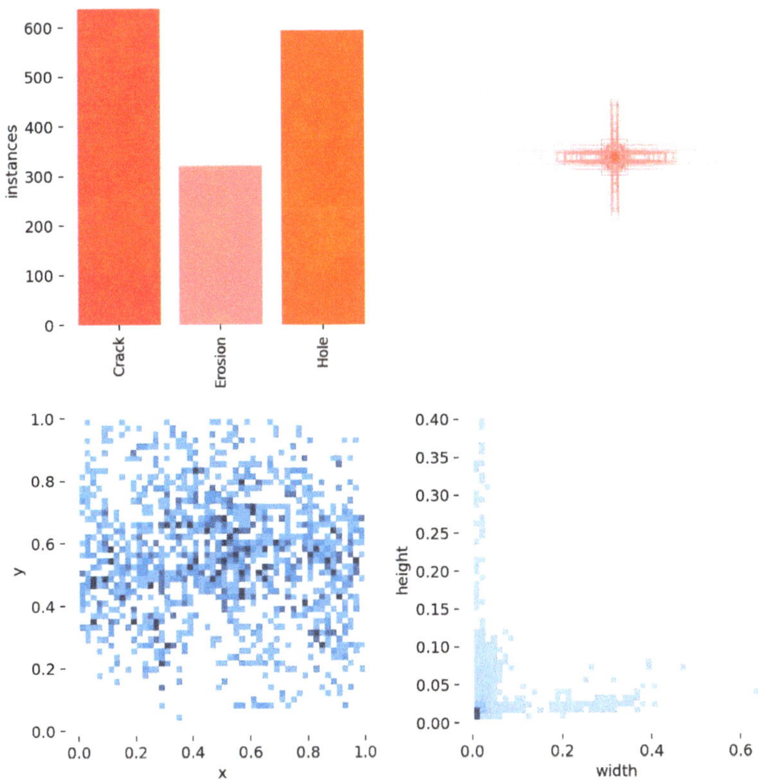

Figure 13. This figure showcases the label distribution and bounding box dimensions for YOLOv8. The (**top left**) bar chart shows the frequency of different defect types (cracks, erosion, holes) on the x-axis and their instances on the y-axis. The (**bottom left**) scatter plot shows the spatial distribution of defects with x- and y-coordinates representing the position on the WTB surface. The (**bottom right**) scatter plot shows the relationship between bounding box width and height with the width on the x-axis and height on the y-axis. The (**top right**) visualization depicts the anchor boxes, which are preset dimensions used during model training to identify objects of varying shapes and sizes. This visualization emphasizes the model's capacity to adjust to diverse defect dimensions, which is critical for accurate localization and classification in practical applications.

In conclusion, YOLOv8 was proven to be effective for defect detection in thermal images of WTBs, exhibiting promising results across various performance metrics. The analysis underscores the need for continued refinement in model training, especially to reduce confusion between classes and improve the detection of cracks and background instances. The results advocate the potential of YOLOv8 in the domain of wind turbine maintenance and the larger context of renewable energy sustainability.

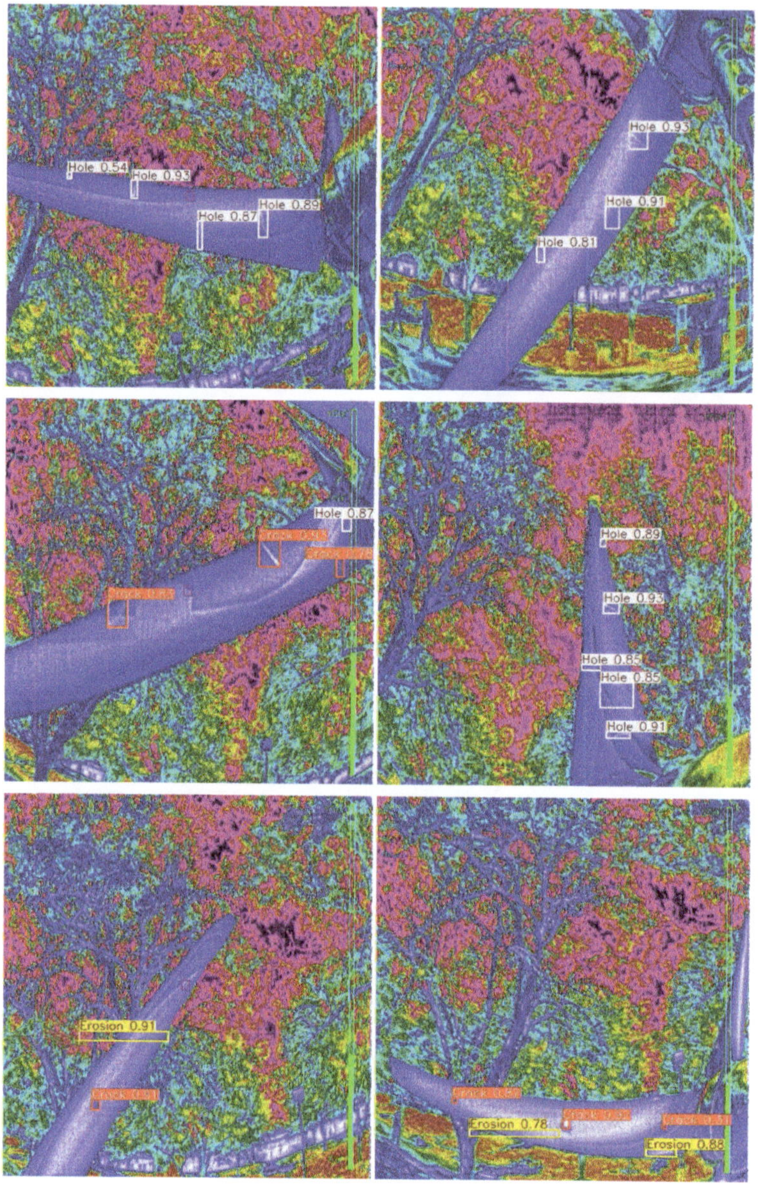

Figure 14. Examples of defect detection using YOLOv8 on thermal images of wind turbine blades (WTBs). The first row corresponds to Blade 1, the second row to Blade 2, and the third row to Blade 3. Detected defects include holes, erosion, and cracks, demonstrating accurate localization and classification.

3.3. YOLOv9 Results

Continuing the advancements in defect detection, YOLOv9 was evaluated under the same experimental conditions as YOLOv8. The model was tasked with identifying three classes of defects—holes, erosion, and cracks—in thermal images of WTBs.

The YOLOv9 model exhibited significant improvements in detecting cracks, with a detection accuracy of 75%, compared to YOLOv8's 58%. This enhancement suggests that YOLOv9 was more adept at distinguishing crack defects, likely due to advanced feature extraction techniques. Erosion detection remained consistently high, with an accuracy of 85%, similar to that of YOLOv8, indicating that both models were equally proficient in identifying this defect type. However, while the detection of holes improved to 65%, there remained considerable confusion with the background, as evidenced by a 48% misclassification rate. This ongoing challenge emphasizes the necessity for further model refinement to better distinguish between holes and background elements. Additionally, the background class continued to show confusion with both crack and hole detections, suggesting that additional data or targeted training enhancements could significantly benefit model performance.

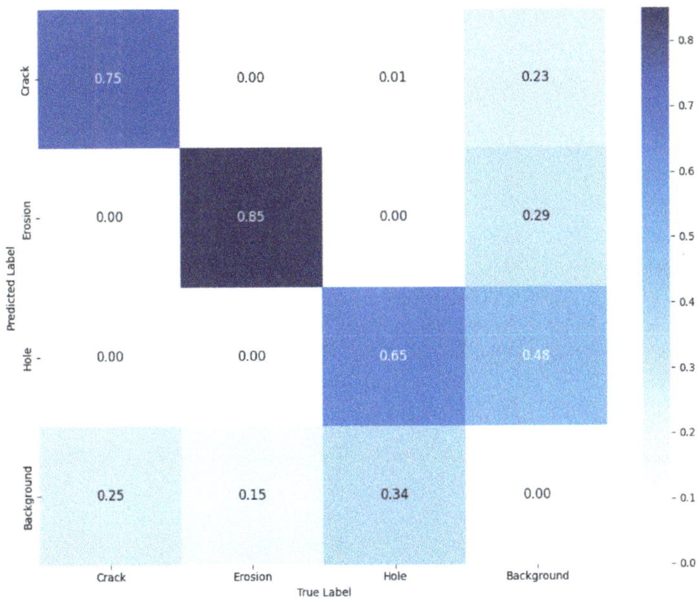

Figure 15. Normalized confusion matrix for YOLOv9.

The statistical significance of the YOLOv9 results is supported by narrower confidence intervals for crack and erosion detection, indicating that these results are more reliable and consistent compared to those obtained with YOLOv8. The marked improvement in crack detection accuracy is particularly notable, as it reflects the model's enhanced capability in identifying defects that previously presented challenges. However, the wider confidence interval observed for hole detection, despite the accuracy improvement, points to variability that may require further training data or model adjustments to achieve more stable performance. These findings highlight the ongoing need for refinement in detecting defects that share similar thermal characteristics with background noise.

Precision and recall are essential metrics for evaluating model performance. Figure 16 illustrates YOLOv9's precision and recall across different confidence thresholds. These metrics will be compared with YOLOv8's performance, highlighting areas of improvement or potential regression.

The F1–confidence curve for YOLOv9 is illustrated in Figure 17, demonstrating the harmonic mean of precision and recall at various confidence levels. This curve will be compared to YOLOv8's to evaluate if the newer model achieved a better balance between precision and recall.

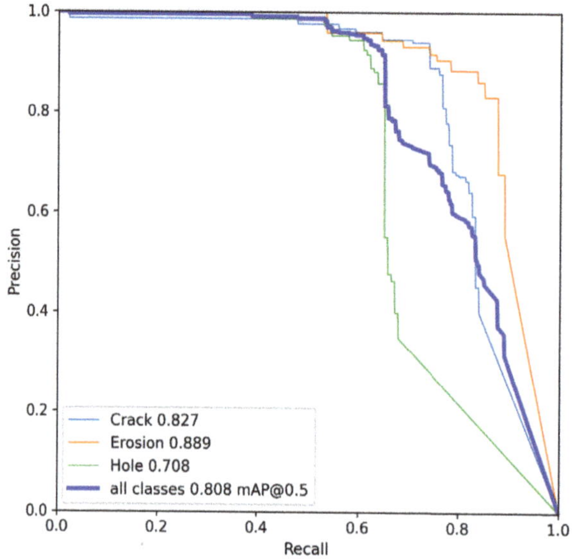

Figure 16. Precision–recall curve for YOLOv9. Visualization is enhanced with clear, non-overlapping labels to ensure optimal readability.

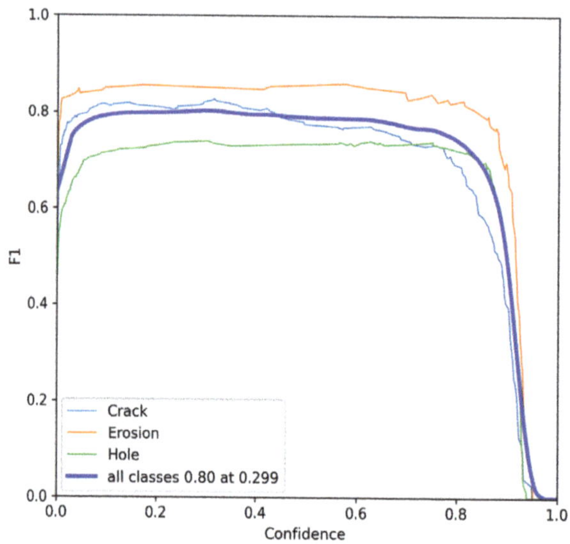

Figure 17. F1–confidence curve for YOLOv9.

YOLOv9's training and validation loss over time, along with precision and recall metrics, are depicted in Figure 18. These metrics indicate the model's learning efficiency and can be used to compare the convergence and reliability of defect localization against the previous model iteration.

Understanding the learning process of YOLOv9 involves analyzing the distribution of dataset labels and the quality of bounding box annotations, as depicted in Figures 19 and 20. These visualizations provide insights into any enhancements that YOLOv9 may offer in terms of defect detection and localization, particularly in comparison to its predecessor, YOLOv8.

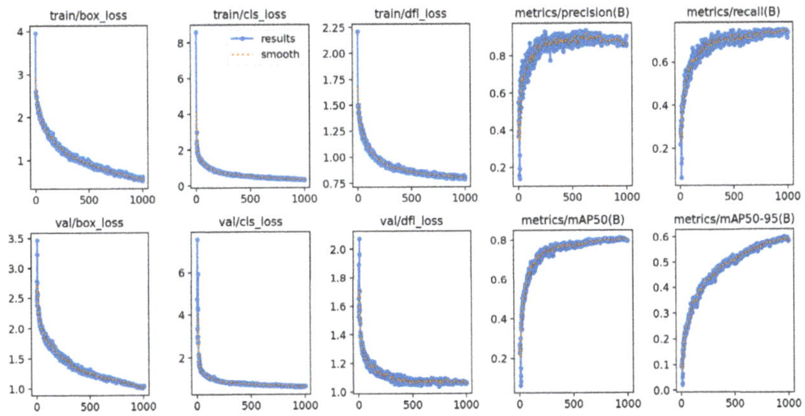

Figure 18. Loss and performance metrics for YOLOv9.

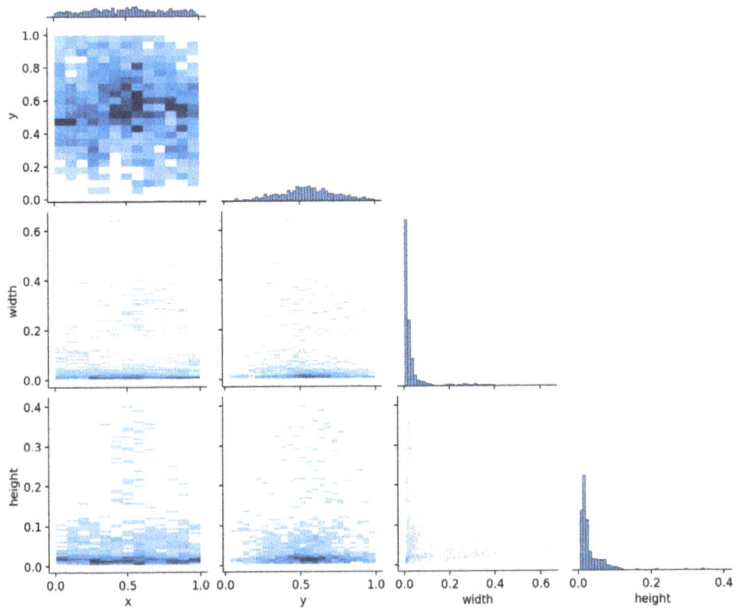

Figure 19. The label correlogram for YOLOv9, illustrating correlations and interactions between different defect categories. This aids in identifying complex patterns in defect occurrences, crucial for enhancing detection algorithms and training methodologies.

Figure 19 presents a correlogram that explores the inter-relationships between different types of defects identified in the dataset. This visualization helps to understand co-occurrence and potential dependencies among various defect categories, crucial for refining the model's training strategies and achieving robust defect detection. The x- and y-axes represent the bounding box attributes (x, y, width, height). The color intensity indicates the density of occurrences, with darker colors representing higher densities. This visualization aids in identifying complex patterns in defect occurrences, crucial for enhancing detection algorithms and training methodologies.

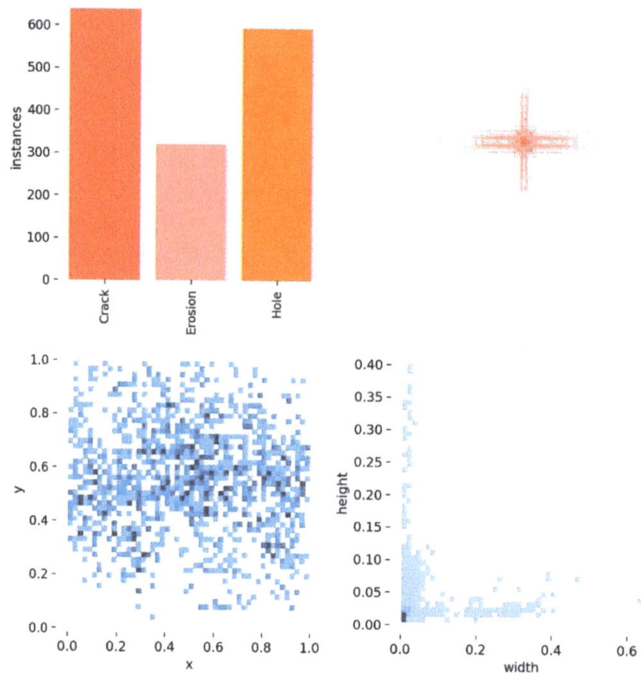

Figure 20. This figure highlights the label distribution and bounding box dimensions for YOLOv9, emphasizing the model's advancements in capturing defect sizes and accurately localizing them on turbine blades, thereby showcasing the improvements over YOLOv8 in handling defect dimensions and distribution.

Figure 20 displays the distribution of labels within the dataset and the dimensions of the bounding boxes used. This figure is key to assessing the precision with which the model identified and classified defects, showcasing improvements in how the model handled various defect sizes and their placements on the WTBs. The top left bar chart shows the frequency of different defect types (cracks, erosion, holes) on the x-axis, with the number of instances on the y-axis, indicating that cracks were the most common defect type, followed by holes and erosion. The bottom left scatter plot shows the spatial distribution of defects with x- and y-coordinates representing the position on the WTB surface. The bottom right scatter plot displays the relationship between the bounding box width and height, with width on the x-axis and height on the y-axis, illustrating the range of defect sizes. The top right part of the figure depicts the anchor boxes used by the YOLOv9 model, which are preset dimensions optimized to match the common sizes and aspect ratios of defects. This visualization emphasizes the model's advancements in capturing defect sizes and accurately localizing them on turbine blades, thereby showcasing the improvements over YOLOv8 in handling defect dimensions and distribution.

Furthermore, Figure 21 presents examples of defect detection using YOLOv9. This figure showcases the model's capability to accurately localize and classify defects in thermal images of WTBs, with a focus on identifying holes, erosion, and cracks.

These visual aids serve as a basis for a detailed discussion in the Results Section, where the two versions of the YOLO model will be critically analyzed, and conclusions will be drawn about their respective performances in defect detection tasks.

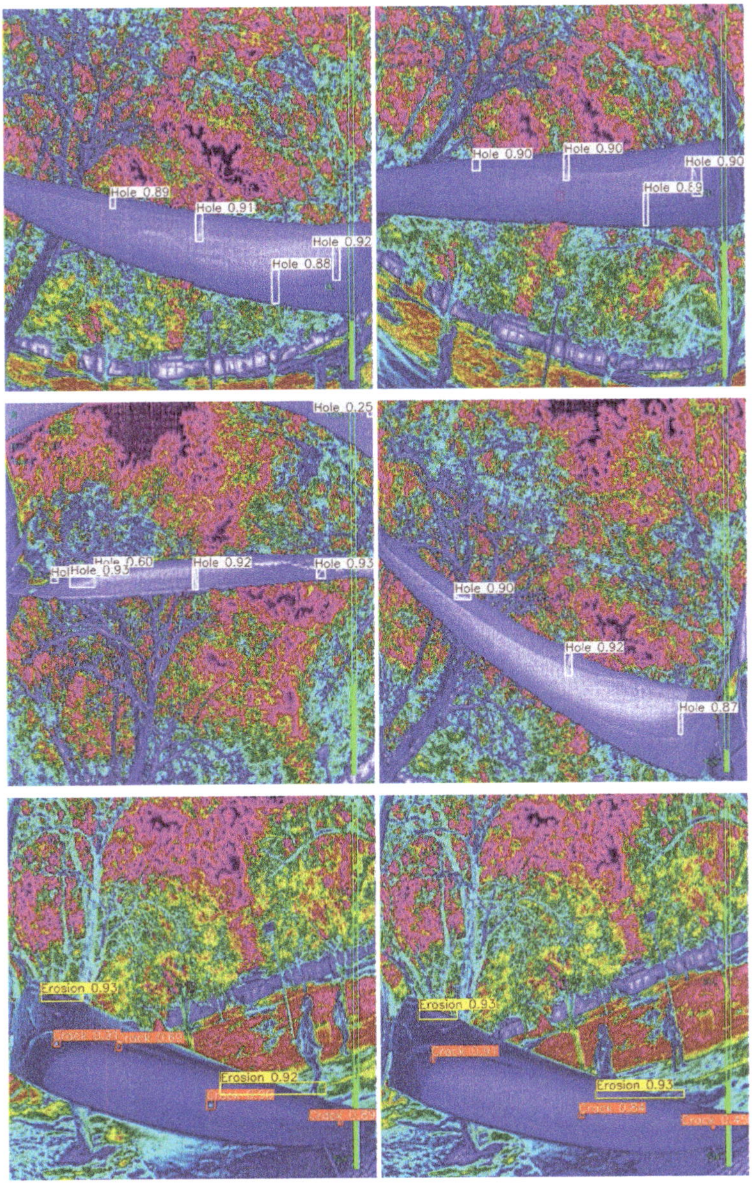

Figure 21. Examples of defect detection using YOLOv9 in thermal images of wind turbine blades (WTBs). The first row corresponds to Blade 1, the second row to Blade 2, and the third row to Blade 3. Detected defects include holes, erosion, and cracks, demonstrating accurate localization and classification.

3.4. Comparative Analysis of YOLOv8 and YOLOv9 Performance

A comprehensive analysis was conducted to compare the performances of YOLOv8 and YOLOv9 in the context of detecting defects in the created thermal images of WTBs. Both models were evaluated based on their ability to classify and localize three types of defects: cracks, erosion, and holes.

The normalized confusion matrices for both models provide a clear insight into their classification accuracy. As observed in Figures 8 and 15, YOLOv9 demonstrated a slight improvement in accurately classifying the crack and erosion defects, with a notable reduction in false positives for the background class, which is a significant challenge in object detection.

The precision–recall curves (Figures 9 and 16) show that YOLOv9 improved the precision across all classes, especially at higher recall levels. This indicates a robustness in YOLOv9's ability to maintain high precision without sacrificing recall, thereby providing a more balanced detection capability.

The F1 score, which combined both precision and recall into a single metric, peaked at a higher confidence threshold for YOLOv9 (Figure 17) compared to YOLOv8 (Figure 10). This suggests that YOLOv9 was more effective in balancing the trade-off between precision and recall, thus providing a better overall performance.

An analysis of the loss and performance metrics revealed that YOLOv9 converged more quickly and with a lower loss compared to YOLOv8 (Figures 11 and 18). This rapid convergence is indicative of YOLOv9's improved learning efficiency, which can be attributed to the architectural enhancements and optimized training protocols.

When examining the dataset distribution and bounding box annotations (Figures 12 and 19), YOLOv9 shows better generalization across varying sizes and shapes of defects. This indicates an improved capability in learning from the dataset and suggests potential for better performance in diverse real-world scenarios.

In conclusion, YOLOv9 outperformed YOLOv8 in every key performance metric, indicating that the improvements in architecture and training methods have translated into a more robust and accurate object detection model. YOLOv9's advancements underscore its suitability for real-time applications where precision and reliability are paramount.

4. Conclusions and Future Work

This research contributes to the evolving field of wind turbine maintenance, harnessing advanced image processing and machine learning techniques to enhance the accuracy and efficiency of inspections. By integrating cutting-edge thermal imaging technologies with the latest real-time object detection models YOLOv8 and YOLOv9, we established a robust framework for detecting and localizing defects in wind turbine blades. This approach not only improves the precision of defect assessments but also optimizes maintenance strategies, leading to increased turbine reliability and longevity. The deployment of the FLIR Vue TZ20 camera, chosen for its drone compatibility, underscores our forward-thinking approach. Although drone integration was not completed in the current phase of research, this setup paves the way for future applications involving aerial inspections. Utilizing drones promises to revolutionize wind turbine inspections by providing comprehensive aerial data acquisition, enabling access to difficult-to-reach areas and significantly reducing inspection times and costs. Furthermore, the creation of a specialized thermal image dataset tailored for wind turbine blades represents a pivotal step toward refining the training and effectiveness of machine learning models in real-world scenarios. This dataset enhances our understanding of defect characteristics under various operational conditions, facilitating more accurate predictions and insights.

Specific Recommendations and Broader Implications. Our future research will focus on three key objectives: (1) fully integrating drone technology into our inspection framework, including the development of advanced control systems for stable and precise thermal imaging during flight; (2) expanding and diversifying our thermal image dataset by incorporating data from larger, operational wind turbines under varied environmental conditions to enhance model robustness; and (3) continuously improving the YOLO models to increase their accuracy and reliability in real-world applications. These steps will address the challenges and limitations identified in the current phase, particularly in detecting defects that have similar thermal characteristics to background noise. The successful implementation of these recommendations will advance wind turbine maintenance practices and contribute significantly to the broader renewable energy sector by improving the reliability and efficiency of wind energy systems, thus supporting the global transition towards more sustainable energy sources.

Challenges and Limitations of Drone Integration. While integrating drones into our inspection protocol shows great promise, it also presents several technical and operational challenges that must be addressed in future work. These include ensuring the stability of the drone during flight, particularly in high-wind conditions typical of wind turbine sites, and optimizing the coordination between the drone's movement and the thermal camera's data capture to avoid motion blur and other imaging distortions. Additionally, developing autonomous navigation and obstacle avoidance systems will be critical to enable safe and efficient drone operations near wind turbines. Overcoming these challenges will be essential to fully realize the potential of drone-based inspections and achieve the goal of autonomous, real-time wind turbine maintenance.

Through this project, we not only built upon existing knowledge but also laid the groundwork for future innovations in renewable energy maintenance. Our continued efforts will focus on enhancing model accuracy, expanding dataset dimensions, and integrating drone technology to fully realize the autonomous operational management of wind turbines.

Author Contributions: Conceptualization, M.M. and M.S.; Software, M.M.; Validation, M.M. and M.S.; Formal analysis, M.M. and M.S.; Investigation, M.S., M.A.S.M. and A.C.S.; Resources, M.S. and M.A.S.M.; Data curation, M.M. and M.S.; Writing—original draft, M.M.; Writing—review & editing, M.S., M.A.S.M. and A.C.S.; Visualization, M.S.; Supervision, M.S.; Project administration, M.S.; Funding acquisition, M.S., M.A.S.M. and A.C.S. All authors have read and agreed to the published version of the manuscript.

Funding: This work was supported by the Utah System of Higher Edication (grant number: 20210016UT).

Data Availability Statement: The data presented in this study are available on request from the corresponding author.

Acknowledgments: The authors sincerely thank the Office of the Commissioner of Utah System of Higher Education (USHE) for supporting this research under the Deep Technology Talent Initiative Grant 20210016UT. We also extend our gratitude to Mohammad Shekaramiz for his oversight and management of the project prior to his untimely passing in December 2024. Additionally, we acknowledge the ECE undergraduate students Benjamin Collier, Kaden Clements, and Hannah Chappell for collaborating on and annotating the dataset.

Conflicts of Interest: The authors declare no conflicts of interest. The funders had no role in the design of the study; in the collection, analyses, or interpretation of data; in the writing of the manuscript; or in the decision to publish the results.

Abbreviations

Abbreviation	Definition
AI	Artificial Intelligence
CNN	Convolutional Neural Network
COCO	Common Objects in Context
DTAGAN	Dual-Threshold Attention-Guided GAN
FLIR	Forward-Looking InfraRed
GAN	Generative Adversarial Network
GELAN	Generalized Efficient Layer Aggregation Network

GRU	Gated Recurrent Unit
IR	Infrared
IRT	Infrared thermography
LCOE	Levelized Cost of Energy
MIR	Mid-infrared
NDT	Non-destructive testing
OCT	Optical Coherence Tomography
PCA	Principal Component Analysis
PGI	Programmable Gradient Information
PV	Photovoltaic
RoI	Region of interest
SVM	Support Vector Machine
UAV	Unmanned Aerial Vehicle
WTB	Wind turbine blade
YOLO	You Only Look Once

References

1. Memari, M.; Shakya, P.; Shekaramiz, M.; Seibi, A.C.; Masoum, M.A. Review on the Advancements in Wind Turbine Blade Inspection: Integrating Drone and Deep Learning Technologies for Enhanced Defect Detection. *IEEE Access* **2024**, *12*, 33236–33282. [CrossRef]
2. Li, W.; Pan, Z.; Hong, N.; Du, Y. Defect detection of large wind turbine blades based on image stitching and improved Unet network. *J. Renew. Sustain. Energy* **2022**, *15*, 013302. [CrossRef]
3. Clero, K.; Nadour, M.; Ed-Diny, S.; Achalhi, M.; Cherkaoui, M.; Ait Abdelali, H.; El Fkihi, S.; Benzakour, I.; Rziki, S.; Tagemouati, H.; et al. A Review of the Use of Thermal Imaging and Computer Vision for Pattern Recognition. *Comput. Sci. Inf. Technol.* **2023**, *13*, 121–136. [CrossRef]
4. Yang, R.; He, Y.; Mandelis, A.; Wang, N.; Wu, X.; Huang, S. Induction Infrared Thermography and Thermal-Wave-RADAR Analysis for Imaging Inspection and Diagnosis of Blade Composites. *IEEE Trans. Ind. Inform.* **2018**, *14*, 5637–5647. [CrossRef]
5. Smith, J.; Carter, D. Application of UAVs in detecting subsurface defects in wind turbine blades using thermal imaging. *Wind Energy* **2019**, *22*, 123–130.
6. Jones, A.; Taylor, M. Deep learning for real-time detection of defects in wind turbine blades via thermal imaging. *Renew. Energy* **2020**, *145*, 2291–2301.
7. Galleguillos, C.; Zorrilla, A.; Jimenez, A.; Diaz, L.; Montiano, Á.; Barroso, M.; Viguria, A.; Lasagni, F. Thermographic non-destructive inspection of wind turbine blades using unmanned aerial systems. *Plast. Rubber Compos.* **2015**, *44*, 98–103. [CrossRef]
8. Doroshtnasir, M.; Worzewski, T.; Krankenhagen, R.; Rollig, M. On-site inspection of potential defects in wind turbine rotor blades with thermography. *Wind Energy* **2016**, *19*, 1407–1422. [CrossRef]
9. Traphan, D.; Herráez, I.; Meinlschmidt, P.; Schlüter, F.; Peinke, J.; Gülker, G. Remote surface damage detection on rotor blades of operating wind turbines by means of infrared thermography. *Wind Energy Sci.* **2018**, *3*, 639–650. [CrossRef]
10. Chen, X.; Shihavuddin, A.; Madsen, S.H.; Thomsen, K.; Rasmussen, S.; Branner, K. AQUADA: Automated Quantification of Damages in Composite Wind Turbine Blades for LCOE Reduction. *Wind Energy* **2021**, *24*, 535–548. [CrossRef]
11. Xu, L.; Hu, J. A Method of Defect Depth Recognition in Active Infrared Thermography Based on GRU Networks. *Appl. Sci.* **2021**, *11*, 6387. [CrossRef]
12. Wang, K.; Zhang, J.; Ni, H.; Ren, F. Thermal Defect Detection for Substation Equipment Based on Infrared Image Using Convolutional Neural Network. *Electronics* **2021**, *10*, 1986. [CrossRef]
13. Jiang, C.; Ren, H.; Ye, X.; Zhu, J.; Zeng, H.; Nan, Y.; Sun, M.; Ren, X.; Huo, H. Object Detection from UAV Thermal Infrared Images and Videos Using YOLO Models. *Int. J. Appl. Earth Obs. Geoinf.* **2022**, *112*, 102912. [CrossRef]
14. Redmon, J.; Divvala, S.; Girshick, R.; Farhadi, A. You only look once: Unified, real-time object detection. In Proceedings of the IEEE Conference on Computer Vision and Pattern Recognition, Las Vegas, NV, USA, 27–30 June 2016; pp. 779–788.
15. Wang, C.; Gu, Y. Research on infrared nondestructive detection of small wind turbine blades. *Results Eng.* **2022**, *15*, 100570. [CrossRef]
16. Yu, J.; He, Y.; Zhang, F.; Sun, G.; Hou, Y.; Liu, H.; Li, J.; Yang, R.; Wang, H. An infrared image stitching method for wind turbine blade using UAV flight data and U-Net. *IEEE Sens. J.* **2023**, *23*, 8727–8736. [CrossRef]
17. Chen, X.; Sheiati, S.; Shihavuddin, A. AQUADA PLUS: Automated damage inspection of cyclic-loaded large-scale composite structures using thermal imagery and computer vision. *Compos. Struct.* **2023**, *318*, 117085. [CrossRef]

18. de Oliveira, A.K.V.; Bracht, M.K.; Aghaei, M.; Rüther, R. Automatic fault detection of utility-scale photovoltaic solar generators applying aerial infrared thermography and orthomosaicking. *Sol. Energy* **2023**, *252*, 272–283. [CrossRef]
19. Attallah, O.; Ibrahim, R.A.; Zakzouk, N.E. CAD system for inter-turn fault diagnosis of offshore wind turbines via multi-CNNs & feature selection. *Renew. Energy* **2023**, *203*, 870–880.
20. Shao, H.; Li, W.; Cai, B.; Wan, J.; Xiao, Y.; Yan, S. Dual-threshold attention-guided GAN and limited infrared thermal images for rotating machinery fault diagnosis under speed fluctuation. *IEEE Trans. Ind. Inform.* **2023**, *19*, 9933–9942. [CrossRef]
21. Goodfellow, I.; Pouget-Abadie, J.; Mirza, M.; Xu, B.; Warde-Farley, D.; Ozair, S.; Courville, A.; Bengio, Y. Generative adversarial networks. *Commun. ACM* **2020**, *63*, 139–144. [CrossRef]
22. Tanda, G.; Migliazzi, M. Infrared thermography monitoring of solar photovoltaic systems: A comparison between UAV and aircraft remote sensing platforms. *Therm. Sci. Eng. Prog.* **2024**, *48*, 102379. [CrossRef]
23. Petersen, C.R.; Fæster, S.; Bech, J.I.; Jespersen, K.M.; Israelsen, N.M.; Bang, O. Non-destructive and contactless defect detection inside leading edge coatings for wind turbine blades using mid-infrared optical coherence tomography. *Wind Energy* **2023**, *26*, 458–468. [CrossRef]
24. Zhao, X.; Zhao, Y.; Hu, S.; Wang, H.; Zhang, Y.; Ming, W. Progress in Active Infrared Imaging for Defect Detection in the Renewable and Electronic Industries. *Sensors* **2023**, *23*, 8780. [CrossRef]
25. Wang, Z.; Jiang, N.; Wen, R.; Sun, B. The Segmentation of Wind Turbine Defect Based on UAV Infrared Image. In Proceedings of the 4th Asian Quantitative InfraRed Thermography Conference, Abu Dhabi, United Arab Emirates, 30 October–3 November 2023.
26. Li, X.; He, Y.; Wang, H.; Sun, G.; Yu, J.; Du, X.; Hu, M.; Wang, Y. Thermal Inspection of Subsurface Defects in Wind Turbine Blade Segments Under the Natural Solar Condition. *IEEE Trans. Ind. Electron.* **2023**, *1*, 11488–11497. [CrossRef]
27. Zhou, W.; Wang, Z.; Zhang, M.; Wang, L. Wind Turbine Actual Defects Detection Based on Visible and Infrared Image Fusion. *IEEE Trans. Instrum. Meas.* **2023**, *72*, 3509208. [CrossRef]
28. Zheng, S.; Li, L.; Chu, N.; Zhou, C.; Mohammad-Djafari, A. Abnormal temperature detection of blower components based on infrared video images analysis. *IEEE Sens. J.* **2024**, *24*, 1919–1928. [CrossRef]
29. Jia, X.; Chen, X. AI-based optical-thermal video data fusion for near real-time blade segmentation in normal wind turbine operation. *Eng. Appl. Artif. Intell.* **2024**, *127*, 107325. [CrossRef]
30. Sheiati, S.; Chen, X. Deep learning-based fatigue damage segmentation of wind turbine blades under complex dynamic thermal backgrounds. *Struct. Health Monit.* **2024**, *23*, 539–554. [CrossRef]
31. Du, Y.; Zhou, S.; Jing, X.; Peng, Y.; Wu, H.; Kwok, N. Damage detection techniques for wind turbine blades: A review. *Mech. Syst. Signal Process.* **2020**, *141*, 106445. [CrossRef]
32. Aminzadeh, A.; Dimitrova, M.; Meiabadi, M.S.; Karganroudi, S.S.; Taheri, H.; Ibrahim, H.; Wen, Y. Non-Contact Inspection Methods for Wind Turbine Blade Maintenance: Techno—Economic Review of Techniques for Integration with Industry 4.0. *J. Nondestruct. Eval.* **2023**, *42*, 54. [CrossRef]
33. Redmon, J.; Farhadi, A. YOLO9000: Better, faster, stronger. In Proceedings of the IEEE Conference on Computer Vision and Pattern Recognition, Honolulu, HI, USA, 21–26 July 2017; pp. 7263–7271.
34. Redmon, J.; Farhadi, A. YOLOv3: An incremental improvement. In *Computer Vision and Pattern Recognition*; Springer: Berlin/Heidelberg, Germany, 2018.
35. Bochkovskiy, A.; Wang, C.Y.; Liao, H.Y.M. YOLOv4: Optimal speed and accuracy of object detection. *arXiv* **2020**, arXiv:2004.10934.
36. Shen, F.; Zeng, G. Weighted residuals for very deep networks. In Proceedings of the 2016 3rd International Conference on Systems and Informatics (ICSAI), Shanghai, China, 19–21 November 2016; pp. 936–941.
37. Wang, C.Y.; Liao, H.Y.M.; Wu, I.H.; Chen, P.Y.; Hsieh, J.W.; Yeh, I.H. CSPNet: A New Backbone that can Enhance Learning Capability of CNN. In Proceedings of the IEEE/CVF Conference on Computer Vision and Pattern Recognition Workshops, Seattle, WA, USA, 14–19 June 2020; pp. 390–391.
38. Yao, Z.; Cao, Y.; Zheng, S.; Huang, G.; Lin, S. Cross-Iteration Batch Normalization. In Proceedings of the 2021 IEEE/CVF Conference on Computer Vision and Pattern Recognition (CVPR), Nashville, TN, USA, 20–25 June 2021; pp. 12326–12335.
39. Ultralytics. YOLOv5: Object Detection at 640 × 640. Online Resource. 2020. Available online: https://github.com/ultralytics/yolov5 (accessed on 20 January 2025).
40. Tan, M.; Le, Q.V. EfficientNet: Rethinking Model Scaling for Convolutional Neural Networks. In Proceedings of the 36th International Conference on Machine Learning, Long Beach, CA, USA, 10–15 June 2019; pp. 6105–6114.
41. Li, C.; Wang, X.; Zhang, Y.; Chen, H.; Liu, J.; Zhao, T.; Xu, F.; Yang, L.; Sun, Q.; Wu, M. YOLOv6: An Incremental Improvement of YOLOv5 for Object Detection. Online Resource. 2023. Available online: https://github.com/meituan/YOLOv6 (accessed on 1 May 2024).
42. Reis, D.; Kupec, J.; Hong, J.; Daoudi, A. Real-Time Flying Object Detection with YOLOv8. *arXiv* **2023**, arXiv:2305.09972.
43. Wang, C.-Y.; Bochkovskiy, A.; Liao, H.-Y. M. YOLOv7: Trainable Bag-of-Freebies Sets New State-of-the-Art for Real-Time Object Detectors. In Proceedings of the IEEE/CVF Conference on Computer Vision and Pattern Recognition, Vancouver, BC, Canada, 17–24 June 2023; pp. 7464–7475.

44. Wang, C.Y.; Yeh, I.H.; Liao, H.Y.M. YOLOv9: Learning What You Want to Learn Using Programmable Gradient Information. *arXiv* **2024**, arXiv:2402.13616.
45. Ultralytics. Ultralytics Version 8.2.0. 2024. Available online: https://github.com/ultralytics (accessed on 1 May 2024).
46. Lin, T.Y.; Maire, M.; Belongie, S.; Bourdev, L.; Girshick, R.; Hays, J.; Perona, P.; Ramanan, D.; Zitnick, C.L.; Dollár, P. Microsoft COCO: Common Objects in Context. *arXiv* **2014**, arXiv:1405.0312.

machines

MDPI

Article

Artificial-Intelligence-Based Condition Monitoring of Industrial Collaborative Robots: Detecting Anomalies and Adapting to Trajectory Changes

Samuel Ayankoso [1,†,*], Fengshou Gu [1,†], Hassna Louadah [2,†], Hamidreza Fahham [2,†] and Andrew Ball [1,†]

1 Centre for Efficiency and Performance Engineering, University of Huddersfield, Huddersfield HD1 3DH, UK; a.ball@hud.ac.uk (A.B.)
2 Institute of Railway Research, University of Huddersfield, Huddersfield HD1 3DH, UK; h.louadah@hud.ac.uk (H.L.); h.faham@hud.ac.uk (H.F.)
* Correspondence: samuel.ayankoso@hud.ac.uk
† These authors contributed equally to this work.

Abstract: The increasing use of collaborative robots in smart manufacturing, owing to their flexibility and safety benefits, underscores a critical need for robust predictive maintenance strategies to prevent unexpected faults/failures of the machine. This paper focuses on fault detection and employs multivariate operational data from a universal robot to detect anomalies or early-stage faults using test data from designed anomalous conditions and artificial-intelligence-based anomaly detection techniques called autoencoders. The performance of three autoencoders, namely, a multilayer-perceptron-based autoencoder, convolutional-neural-network-based autoencoder, and sparse autoencoder, was compared in detecting anomalies. The results indicate that the autoencoders effectively detected anomalies in the examined complex and noisy datasets with more than 93% overall accuracy and an F1 score exceeding 96% for the considered anomalous cases. Moreover, the integration of trajectory change detection and anomaly detection algorithms (i.e., the dynamic time warping algorithm and sparse autoencoder, respectively) was proposed for the local implementation of online condition monitoring. This integrated approach to anomaly detection and trajectory change provides a practical, adaptive, and economical solution for enhancing the reliability and safety of collaborative robots in smart manufacturing environments.

Keywords: industrial collaborative robot; anomaly detection; trajectory change; autoencoder; dynamic time warping; online monitoring

1. Introduction

There is a tremendous rise and increasing interest in collaborative robots or cobots in the manufacturing industry. Cobots are created for tasks involving either human–robot interaction or proximity between the two (without needing safety cages) [1]. As we transition towards an era of smart manufacturing, cobots are being adopted in big and Small and Midsize Enterprise (SME) industries due to the benefits they have over traditional robots [2]. These benefits include their compact size in terms of weight and payload, ease of movement and installation, versatility for reprogramming to perform various tasks, and safe operation alongside humans [3]. Due to repetitive use and degradation of parts, these robots, like many other machines, are not free from faults, and their breakdown can have negative consequences on production. Therefore, implementing an effective maintenance strategy before the occurrence of faults is crucial for maintaining the reliability and safety of such industrial robots.

There have been so many changes in the way industrial/commercial machines are maintained over the years. Several factors have influenced such transitions including the need to reduce cost and machine downtime, increase safety, and adopt modern technologies [1]. As a result, machine maintenance has evolved from reactive to preventive maintenance, and from preventive to predictive maintenance, as discussed in [4,5] and illustrated in Figure 1.

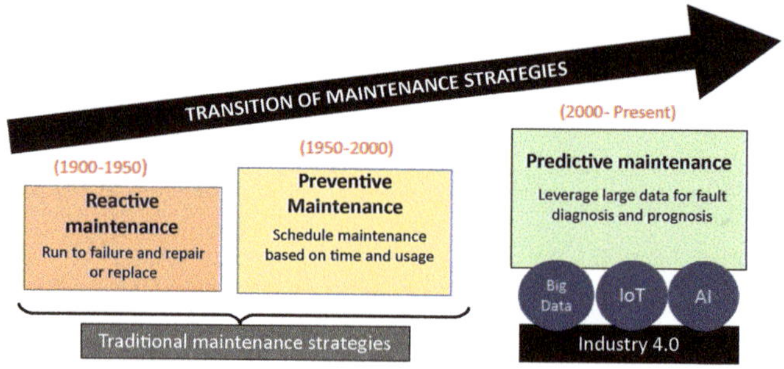

Figure 1. The transition of maintenance strategies.

Predictive maintenance (PM) is a maintenance philosophy targeted at improving the reliability and safety of industrial machines while driving down costs and reducing downtime. Due to the current phase of the industrial revolution (i.e., Industry 4.0), PM is crucial to smart machines, and its core drivers are Big Data, the Internet of Things (IoT), and Artificial Intelligence (AI) [1]. PM is enabled by Prognostics and Health Management (PHM). PHM is an engineering field that addresses the efficient detection, diagnosis, and prediction of faults in industrial machines [6].

PHM strategies are developed through models based on system physics, machine learning (ML), or deep learning (DL) [7]. The choice of any of the three models depends on the system's complexity, data availability, and the need for interpretability. Physics-based approaches offer high accuracy and reliability when dealing with a system whose underlying dynamic behavior and failure mechanisms are well understood. In contrast, ML and DL are data driven; ML requires manual feature engineering with the help of signal processing and expert knowledge to extract relevant patterns, while DL is capable of learning features automatically from raw or unprocessed data [6]. DL benefits from transfer learning, allowing models trained on large datasets to be adapted to new, smaller tasks, as demonstrated in [8]. Although DL often lacks transparency, explainable AI aims to address this issue by making DL model decisions more interpretable, improving trust in critical applications [9].

Different sensors are used in the PHM of industrial robots [10]. These sensors can be placed in the joints, end effector, or robot environment. The common inbuilt sensors are encoders and current transducers for measuring joint positions and motor input currents. The external sensors, on the other hand, include lasers, acoustic sensors, and cameras, but they are difficult to employ for PHM/condition monitoring in an industrial environment [11]. In addition, robot joints often do not come equipped with built-in vibration sensors or acoustic sensors, and they are usually installed in an intrusive way and at an extra cost. For example, a vibration sensor was externally installed in [12], while an acoustic one was installed on the joint surface in [13]. Built-in sensors are the most reliable option for diagnosing mechanical faults in motors and gears, as demonstrated through the use of joint internal current sensors for the six axes of a Hyundai robot in [14]. The data from the same sensors were leveraged in their subsequent works [15,16]. Cobots are designed with a variety of internal sensors that enhance their condition monitoring, as presented in [17] for a universal robot and [18] for a voraus robot.

The two major problems with the PHM of industrial collaborative robots are the lack of or insufficient fault data for training DL models and the inability of the models to generalize and adapt to different tasks and changes in operating conditions. First, it is difficult to acquire fault data, as the average mean time before the failure of industrial robots is tens of thousands of hours, as reported in [19]. Generally, it is very expensive to induce faults in industrial robots, and, even if carried out, without in-depth knowledge about the system's fault types and degrees, the observed faults may differ from those that occur in real cobots. Furthermore, though the utilization of ML and DL for fault detection and diagnosis has garnered significant attention in the last decade, few related works can be found on the implementation of these data-driven methods. Additionally, most fault detection or diagnosis models are typically trained based on a specific trajectory, which is often not applicable to different operational scenarios or tasks where trajectory change occurs [20,21].

In this study, the detection of artificial anomalies in cobots is investigated. Anomaly detection is crucial for industrial robots to identify abnormal or unusual behaviors early, which can signal potential issues before they escalate into faults or failures. Better put, anomalies represent deviations from normal operating behavior but do not always indicate faults, as they could stem from small variations or changes in operating conditions rather than actual system malfunctions. The multivariate operational data obtainable from a universal robot (UR5e) were employed as a means of detecting anomalies from two groups of designed anomalous conditions. The first group of anomalies were achieved by placing different weights on the robot's shoulder arm while a vibrator with different speeds was used to create the second class of anomalies around the elbow joint. The contributions of this work include:

1. Comparison of three state-of-the-art anomaly detection techniques, which are the multi-layer-perceptron-based autoencoder (MLP-AE), convolutional-neural-network-based autoencoder (CNN-AE), and sparse autoencoder (SAE), in detecting anomalies associated with the overall operation of a cobot;
2. Integration of trajectory change and anomaly detection algorithms for local implementation of online condition monitoring.

Our results show that AEs are advanced unsupervised deep learning techniques, and they are very useful for anomaly detection using complex and noisy datasets (with no manual extraction of health indicators). Additionally, our work provides insights into the potential of combining both anomaly and trajectory change detection models for the practical, adaptive, and cost-efficient online condition monitoring of cobots in smart manufacturing settings. The rest of this paper is structured as follows: The related work is summarized in Section 2, while the anomaly and trajectory change detection models are introduced in Sections 3.1–3.3. In Sections 3.4–3.6, the test setup and theoretical analysis of the experimented anomalies, as well as data visualization and preprocessing, are discussed. The model training and evaluation are covered in Section 4, and Section 5 is about online condition monitoring. The Section 6 is the conclusion and future recommendations of this work.

2. Related Work

Fault detection, diagnosis, and fault prognosis are three important parts of a complete PHM system. In fault detection, the machine's condition is monitored to detect any deviation from its normal operating condition. Fault diagnosis deals with fault type or class and severity detection in machines or components. In contrast, fault prognosis focuses on estimating the remaining useful life (RUL) of such components. Table 1 shows a comparison of the three stages of the PHM process, and more details about each are available in [6,22]. In addition, a complete PHM process is described in Figure 2 which usually starts with data acquisition, and each cycle ends with decision making.

The major challenge in cobot condition monitoring research is the lack/unavailability of fault data because it is costly to induce faults, and different operating conditions of the robot need to be experimentally considered. For example, in [16], a robust fault diagnosis model was developed for one of the servo motors of a six-axis industrial robot under various operating conditions. The operating conditions of the robot included different speeds (from 10 to 100% with 10% increment), loads (0 g, 500 g, 1000 g, 1500 g, 2000 g, 2500 g, 3000 g, and 3500 g), and motions (simple and complex). During the simple motion, all the robot joints were fixed except the third joint, which was allowed to move −50 +50 +50 −50. The motion of a welding task was adopted for the complex motion, in which all the joints of the robot moved independently. Moreover, while designing the experiment, two states were considered; the first was when the robot was operating normally, and the second state was when a bearing fault (inner race fault artificially created by grinding) was added to the third joint servo motor. The fault diagnosis of the harmonic reducer of an industrial robot was considered in [23] using time-domain vibration signals. A one-dimensional convolutional neural network with a matrix kernels framework was proposed and trained offline. The test stand consisted of three (3) fault-free industrial robots, six faulty harmonic reducers, and PV-571 orthogonal direction (XYZ) vibration sensors. The fault dataset was acquired by replacing the two joints of each of the fault-free industrial robots. The drawback of the two works ([16,23]) is the fact that it is expensive to obtain data relating to the joint's motor or gear fault of the robot.

Table 1. Comparison of the differences between fault detection, fault diagnosis, and fault prognosis in the context of condition-based and predictive maintenance for industrial collaborative robots.

Aspect	Fault Detection	Fault Diagnosis	Fault Prognosis
Objective	Identify the occurrence of fault or anomaly	Determine faulty conditions and their degree	Predict when failure or major fault would occur based on current condition
Data required	Healthy state sensor data	Healthy and fault states sensor data	Historical sensor data
Method	Residual models and threshold-based algorithms	Classification models	Time-series predictive models
Output	Alarm abnormal conditions	Show the class, type, or degree of fault that has occurred	The remaining useful life of machines/parts before failure
Examples	Abnormal trajectories, overheating, excessive vibrations	Identify failed motor or gear, sensor failure, and communication issue	Predict when to replace a cobot arm based on wear and usage patterns

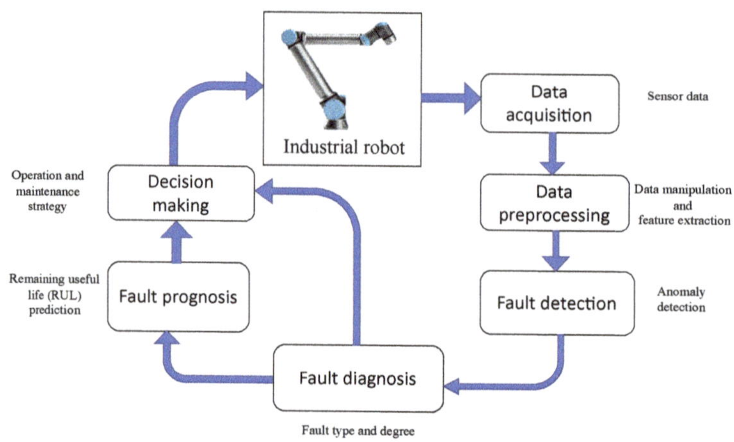

Figure 2. Cobot PHM process.

Given the challenge mentioned earlier, a potential solution is to create high-fidelity models capable of simulating faults, such as the framework outlined in [24] and the multibody model developed in [25]. Another solution is to focus on fault detection, as it has proven to be a more practical and cost-effective approach. The development of a fault detection model does not necessarily require fault data, and there is reduced expertise needed in its development compared to in fault diagnosis and prognosis.

There are a few existing studies on fault detection for collaborative robots to explore. First, a comparison of different anomaly detection methods for an industrial collaborative robot was performed in [17]. Fifteen anomaly detection methods were examined in the pick-and-place task using a cylindrical object and experimenting with different anomalies such as pushing joint 3 or 5 during the task, changing the moved object weight (from 0.2 kg to 0.5 kg, 1 kg, and 2.5 kg), object drop, applying a twisting force to the robot's Tool Center Point (TCP), and others. Some of the methods compared were the K-nearest neighbor technique, Principal Component Analysis (PCA), deep AE, variational AE, LSTM-based AE, and others. Their analysis showed that many methods gave high detection accuracy; however, the model accuracy was reduced when subjected to new program branches, non-deterministic trajectories, and deliberate contact with the environment. This means the developed detection models will underperform when there is a trajectory change, payload, or contact with the environment.

A framework that can inform an operator of anomalous trajectories was developed for a seven-axis collaborative robot. A two-level framework was proposed in [20]. The first level is where fault detection is performed through an algorithm, while the second level is a diagnostic check by a human operator to isolate the fault. The proposed algorithm for fault detection has three steps: (1) trajectory clustering; (2) health state identification; and (3) functional failure clustering. Three anomalous conditions were artificially induced to ascertain the effectiveness of the proposed framework, which were end effector overload, excessive friction in a joint, and random variations in the test data. The gap in the authors' work is that four trajectories and three failures were investigated, and they do not cover all the possible cases that the robot can be subjected to during operation. This suggests the fault detection system will be unreliable when an unknown trajectory is used.

In [26], the authors proposed a model-based method for the condition monitoring of cobots. Rubber bands of varying elasticity were used to simulate different degrees of force on one of the robot joints (specifically, the third joint). The physical model of the UR5 collaborative robot was used to obtain signals, such as position and current, reflecting the robot's operational behavior under diverse test conditions. These signals were compared with actual signals from the robot during operation. Abnormal joint conditions were effectively observed when calculating the residual values between the target and actual signals. The findings indicate that the residual of the current signal exhibits an increasing trend as the magnitude of abnormal forces increases across different trajectories, distinguishing normal and abnormal states.

In our previous work [27], anomalous conditions were artificially introduced to a cobot by attaching different weights on its third joint. Two unsupervised learning models, namely, PCA and an SAE, were developed to identify such anomalies using multivariate data obtained from a universal robot (UR5e). These models were employed to reconstruct the normal or baseline data. The evaluation of the PCA model involved examining the Hotelling T^2 and Q-statistic metrics based on the reconstruction error, while the Mahalanobis distance was utilized for the SAE model to identify abnormal conditions. The outcomes demonstrate that both methods are effective at sensitively detecting the investigated anomaly. However, PCA exhibits greater computational efficiency. Due to the potential of AEs, further research is needed to explore SAE and other variations of AEs, while also considering another group of anomalies.

An adaptive anomaly detection scheme was proposed in [21] by the authors to provide a robust detection system that works when there is a change in the data pattern. The scheme has two stages: the first is the concept drift (CD) detection stage, and the second stage is anomaly detection (AD). The CD model was trained with multivariable data such as current, position, speed, and torque data from the robot, while the AD model was trained using torque data. During the online deployment of the scheme, the CD and AD models were retrained when a concept drift was detected; otherwise, AD was performed. A probabilistic encoder was used for the CD detector, while a CNN encoder structure was developed for the anomaly detector. The scheme was validated using normal and fault data from an Omron TM5-900 cobot. A major issue identified about the introduced concept was the computation requirement of the two models. The authors plan to combine the two ideas (CD and AD) in a single model in their future work. In addition, the fault in the robot motor or gear system was simulated by manipulating the torque data (10% increase).

Research Challenges and Proposed Solutions

A summary of existing challenges related to cobot anomaly detection includes (1) model performance degradation under a change in operating condition (i.e., speed or trajectory change) and environmental changes (e.g., external push), (2) limited coverage of fault scenarios during experiments targeted at fault diagnosis, (3) the computational complexity of anomaly detection models and very few works on the online deployment of detection schemes.

To solve the challenges highlighted above, two classes of algorithms are proposed in this study. The first class is for anomaly detection, while the second is for trajectory change detection. The sensitivity of three AE-based anomaly detection algorithms to two different anomalies is compared. Moreover, we introduce, analyze, and implement a flowchart to integrate the anomaly and trajectory change detection models into the real-time monitoring of cobots.

This work shows that slight variations (anomalies) in robot operating data can be detected by combining the Mahalanobis distance with few-layer autoencoder models. Another novel aspect of this study is the implementation of the trajectory adaptive condition monitoring framework, which improves the reliability of the anomaly detection models.

3. Methodology

The Section 3 explains the basics of AI-based anomaly detection and introduces three different algorithms for detecting anomalies. It also presents an algorithm for detecting changes in the robot's trajectory. Additionally, this section covers the test setup, the effects of the designed anomalies, and how the data were visualized and preprocessed.

3.1. AI-Based Anomaly Detection

The framework for anomaly detection is shown in Figure 3, which expressly depicts the training and testing phases of an AI-based detection model called an autoencoder (AE). This unsupervised learning AI method comprises three components, an encoder, latent space, and a decoder, in that order, as illustrated in Figure 4.

An AE is a type of neural network that learns to compress input data into a smaller representation (encoding) and then reconstructs the original input from this representation (decoding) through training. During training, the normal data alone are used, and they serve as both input and a target. Subsequently, the trained model undergoes testing using normal and anomalous data, with a higher reconstruction error in this scenario indicating abnormalities within the data [28,29].

Despite being unsupervised, the weights are updated iteratively using optimization algorithms like Adam optimization or stochastic gradient descent, minimizing the difference between the input and the reconstructed output through backpropagation [30]. After presenting the general architecture of an AE, the three specific types of AEs utilized in this study will be examined.

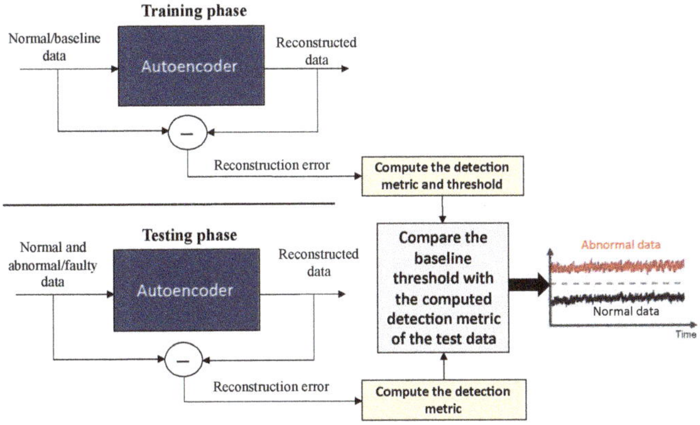

Figure 3. Anomaly detection framework.

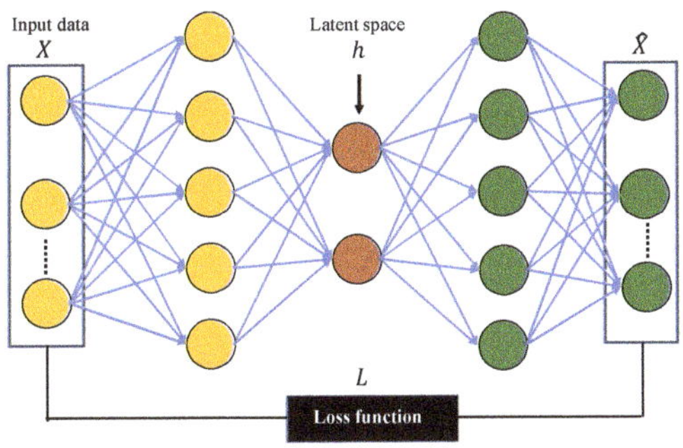

Figure 4. Autoencoder architecture.

3.1.1. MLP-Based Autoencoder

The MLP-AE is a type of AE architecture that utilizes multi-layer perceptrons (MLPs) or an FNN as its building blocks—that is, the encoder and decoder parts. An MLP consists of multiple layers of neurons where each neuron in one layer is connected to every neuron in the subsequent layer, forming a dense, fully connected network [31,32]. Mathematically, the operation within a single neuron is expressed as follows:

$$f_i = f_{act}\left(\sum_{i=1}^{n} w_i x_i + b\right), \tag{1}$$

where f_i is the neuron output upon activation, $f_{act}(.)$ represents the activation function (common examples include sigmoid, tanh, ReLU, and softmax functions [33]), w_i is the weight associated with the i-th input, b is the bias, and n is for the number of inputs (x). MLP-AEs are flexible and can be adapted to various tasks and datasets by adjusting the number of layers, the number of neurons in each layer, and the activation functions used.

3.1.2. CNN-Based Autoencoder

This is a type of AE architecture that incorporates convolutional and max pooling or upsampling layers in its encoder and decoder structure. The convolution operation involves applying a filter, also known as a kernel, to an input image or time-series data to highlight patterns and features present in the input [29,31,34]. The convolution operation is expressed mathematically as follows:

$$f_{ij}^{l} = f_{act}\left(\sum_{i}^{i+m-1} k_j^l * x_i^{l-1} + b_j^l\right), \tag{2}$$

where f_{ij}^{l} and x_i^l are the output and input of the convolution, $*$ represents the convolution operator, i is the sliding index, which is determined by the input (x_i^l) size, k_j^l is the kernel or filter of the j-th neuron in layer l, m is the filter size, and b_j^l is the scaler bias of layer l.

A max pooling operation is a downsampling operation used in the encoder part and after the convolutional layers to decrease the spatial dimensions of the input data, while an upsampling operation is used in the decoder part to increase the spatial dimensions of feature maps. The CNN-AE architecture is particularly well suited for tasks involving spatial or sequential data such as images, time series, or sequences.

3.1.3. Sparse Autoencoder

This is a type of AE with the ability to enforce sparsity in the learned representations through an improvement of the standard loss function such as the mean squared error (MSE) [35]. Technically, overfitting is prevented by adding a regularization term to the loss function, and it also helps in a more efficient use of computational resources. The modified loss function has $L2$ sparsity regularization terms, hence the reason for the name SAE [36].

$$L = \underbrace{\frac{1}{N}\sum_{n=1}^{N}\sum_{k=1}^{K}(x_{kn} - \widehat{x}_{kn})^2}_{mse} + \underbrace{\gamma * \Omega_{weights}}_{L2\ regularization} + \underbrace{\mu * \Omega_{sparsity}}_{sparsity\ regularization}, \tag{3}$$

where L is the improved loss function, γ is the coefficient of the $L2$ regularization, and μ is the coefficient of the sparsity regularization.

3.2. *Anomaly Scores and Detection Metrics*

Anomaly scores are numerical values used to quantify the degree of abnormality. There are various ways of computing anomaly scores, one of which is the maximum mean error of the AE model. Given the consideration of multivariate data, the Mahalanobis distance (MD) serves as a key metric for computing the distance between multivariable time-series data [37]. It is preferred over the Euclidean distance due to its ability to incorporate data correlations.

$$MD = \sqrt{(x-\mu)C^{-1}(x-\mu)^T}, \tag{4}$$

where MD is the Mahalanobis distance of data x to reference data y. μ and C represent the mean and covariance of the reference or baseline data.

The threshold of the model can be calculated through the 3-limits of the baseline MD result.

$$MD_\alpha = \mu_{MD} + 3\sigma_{MD}, \tag{5}$$

where μ_{MD} is the mean of MD, and σ_{MD} is the standard deviation of MD.

The confidence limit of the model is $MD \leq MD_\alpha$.

After computing the anomaly scores and threshold, there are other metrics to evaluate the performance of anomaly detection algorithms. These include accuracy, precision, recall, F1 score, false-positive rate, false-negative rate, receiver operating characteristic area under the curve, etc.

3.3. *Trajectory Change Detection Models*

Robot trajectory can change based on the task; however, this change leads to variations in sensor readings, motor responses, and overall system behavior. Hence, fault/anomaly detection and diagnosis models trained on previous trajectory patterns may struggle to adapt and accurately detect faults or anomalies when faced with new trajectories. The speed at which the robot operates and the payload carried by the robot also affect the overall dynamics; however, the scope of this work is limited to drift caused by a change in trajectory. There are a few algorithms that can be used to detect a change in robot trajectories. These include dynamic time warping and kernel density estimation. In this work, dynamic time warping (DTW) is used because it is robust to time variation and non-linear temporal behavioral differences between two trajectories. DTW is a powerful detection algorithm which calculates the similarity between robot trajectories and identifies any significant deviations or changes [38]. The steps of the DTW algorithm are as follows:

1. Define two trajectories' data (Q_1 and Q_2), representing the robot's movements in the X, Y, Z coordinates over time:

$$Q_1 = \begin{bmatrix} q_{x,1} & q_{y,1} & q_{z,1} \\ q_{x,2} & q_{y,2} & q_{z,2} \\ \vdots & \ddots & \vdots \\ q_{x,n} & q_{y,n} & q_{z,n} \end{bmatrix}, \tag{6}$$

$$Q_2 = \begin{bmatrix} q_{x,1} & q_{y,1} & q_{z,1} \\ q_{x,2} & q_{y,2} & q_{z,2} \\ \vdots & \ddots & \vdots \\ q_{x,m} & q_{y,m} & q_{z,m} \end{bmatrix}, \tag{7}$$

   where n and m are the number of data points in each trajectory.
2. Create a matrix CD of size $n \times m$, in which the elements in the ith row and the jth row correspond to the distance between the ith elements in Q_1 and the jth elements in Q_2. This involves computing and storing the Euclidean distance between points X, Y, and Z in Q_1 and Q_2 within a distance matrix.
3. Create the cumulative distance matrix. First, the top-left corner of the matrix is initialized as $CD[1,1]$ because the DTW algorithm needs to align the start of the two sequences:

$$CD[1,1] = D[1,1]. \tag{8}$$

   Then, the other elements in the matrix are calculated by as follows:

$$CD[i,1] = CD(i-1,j) + D[i,1]\,, \tag{9}$$

$$CD[1,j] = CD(1,j-1) + D[1,j]\,, \tag{10}$$

$$CD[i,j] = D(i,j) + min(CD[i-1,j], CD[i,j-1], CD[i-1,j-1]). \tag{11}$$

   The final DTW distance is $CD[n,m]$, which is the lower-right element of the matrix CD. The normalized final distance, CD_{final}, is

$$CD_{final} = \frac{CD[n,m]}{(n+m)}, \tag{12}$$

   The smaller the value of CD_{final}, the higher the similarity between Q_1 and Q_2. A trajectory change happens if the threshold is $CD_{th} < CD_{final}$.

3.4. Test Setup

The industrial robot utilized for this study was a UR5e collaborative robot manufactured by Universal Robots [39]. To enable seamless communication and data exchange between the universal robot and external apps, a Real-Time Data Exchange (RTDE) framework was employed [40]. RTDE utilizes appropriate networking protocols and APIs such as an Ethernet connection to provide a reliable and high-bandwidth communication channel between the external app program (e.g., a Python script) and the UR5e robot controller, thus facilitating efficient data exchange, reducing latency, and enabling fast and accurate control of the robot's movements. An overview of the setup for data exchange in the UR5e robot is depicted in Figure 5.

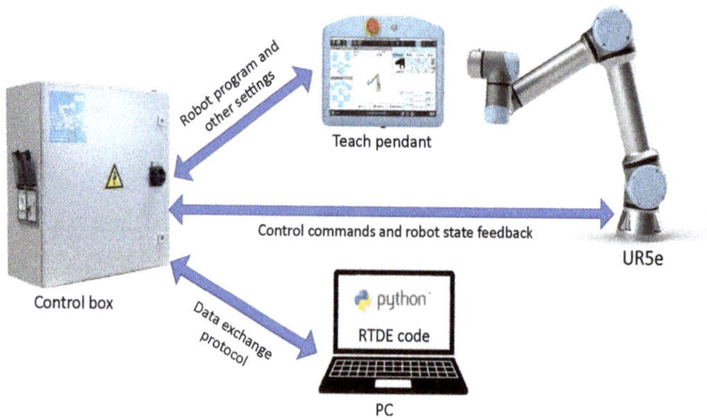

Figure 5. Overview of the data exchange setup for UR5e robot.

By leveraging the capabilities of RTDE, the data-gathering process was customized and adapted according to the requirements of our application. This allowed us to collect a wide range of variables, including joint temperatures, current, voltages, velocities, positions, moments, and TCP information, in real time. During the experiments conducted, the collection frequency was set to every 0.04 s within the RTDE program to ensure the data were acquired at a high resolution. Ideally, this frequency would correspond to a sampling rate of 25 Hz. However, due to the concurrent execution of other processes within the program, the actual data collection frequency was approximately 22 Hz. Despite this slight deviation from the desired frequency, the collected data provide valuable insights into the robot's performance and behavior. Two sets of experiments were carried out, and the same trajectory was followed during each experiment. The two groups of experiments were conducted as follows.

3.4.1. Anomaly 1 Experiments

Five tests were performed for the group 1 experiments, as summarized in Table 2. The first test of experimental set 1 was the baseline—in this case, no external weight was added to the shoulder arm of the robot—whereas, in the remaining four tests in this group, a 3D-printed part with holes (for attaching different numbers and sizes of bolts) was fitted to the shoulder arm of the robot as shown in Figure 6a. The weight of the 3D-printed part was 150, and this was increased by changing the number and size of bolts on the 3D-printed part. The weights used for the four cases were 150 g, 330 g, 450 g, and 750 g, respectively.

Table 2. Group 1 experiments for the first class of anomalies.

Experiment	Weight	Label
1	0 g	normal
2	150 g	abnormal
3	330 g	
4	450 g	
5	750 g	

3.4.2. Anomaly 2 Experiments

In the second group of experiments, summarized in Table 3, normal and abnormal operational data were collected when a vibrator was placed on the elbow joint of the robot. The vibrator was powered by a lithium battery and controlled using an ESP32 microcontroller, as illustrated in Figure 6b. The first test in experimental set 2 was the baseline, in which the vibrator was not turned on. Two more tests were performed in experimental set 2 where the vibrator was set to half and full speed, respectively (based on the PWM cycle in the control code).

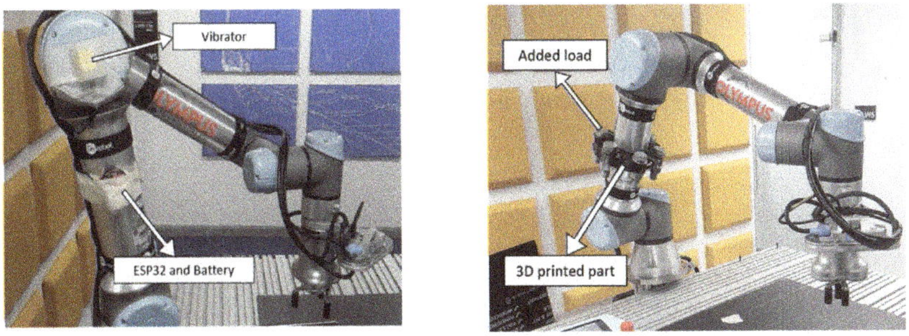

(**a**) Anomaly 1 setup. (**b**) Anomaly 2 setup.

Figure 6. Universal robot (UR5e) with two different anomalies.

Table 3. Group 2 experiments for the second class of anomalies.

Experiment	Speed	Label
1	0 speed	normal
2	50% speed	abnormal
3	00% speed	

In the two groups/sets of experiments conducted, the robot moved through a path consisting of 6 points, as observable in the 3D trajectory plot in Figure 7a. As shown in Figure 7b, another trajectory named TCP translational motion 2 was designed using the teach pendant programming interface and used for the acquisition of the test dataset employed in validating the trajectory change detection model. The two trajectories were designed to be arbitrary rather than tailored to a specific task like the robot picking up and placing an object. However, a key consideration in designing the trajectory paths using the teach pendant was to ensure that all six joints were involved in the movement. In both trajectories (TCP translational motions 1 and 2), none of the joints remained static, allowing for comprehensive observation of the impact of the investigated anomalies across all the joints. The maximum speed and acceleration of the motion were 100 mm/s and 400 mm/s^2, respectively. In total, 93 variables were acquired from the robot. Furthermore, the total number of data points in each experiment was 4000, and each experiment lasted for about 185–186 s/$\approx$ 3 min. Out of the 93 variables, 30 were used in training the models studied in this work. The signals/variables eliminated were either the target signals or they were variables with very small residuals when the plots of each signal based on different anomalies were produced. Additionally, the joint temperatures were not considered in the anomaly detection model development because we noticed that these variables increased based on the time and the order of the experiment.

Among the selected variables, four of them, namely, motor input current, joint angular position, TCP position, and TCP force/torque, are plotted in Figures 8 and 9 to show the effect of the normal and abnormal conditions on those robot variables. By visual inspection of both figures, no significant difference can be seen between the data variables obtained under normal and abnormal conditions. The reason for this can be linked to the closed-loop feedback control system of the robot, which quickly corrected deviations. Notwithstanding, the signals will be statistically analyzed in Section 3.6 to show the variations caused to each signal by the two sets of anomalies.

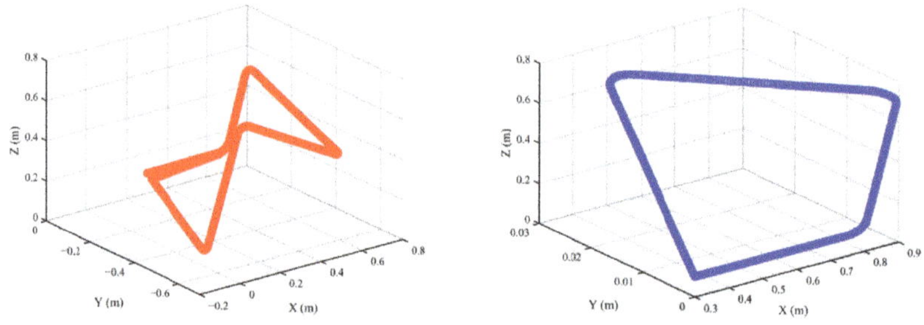

(**a**) TCP translational motion 1. (**b**) TCP translational motion 2.

Figure 7. Combined view of TCP translational motions.

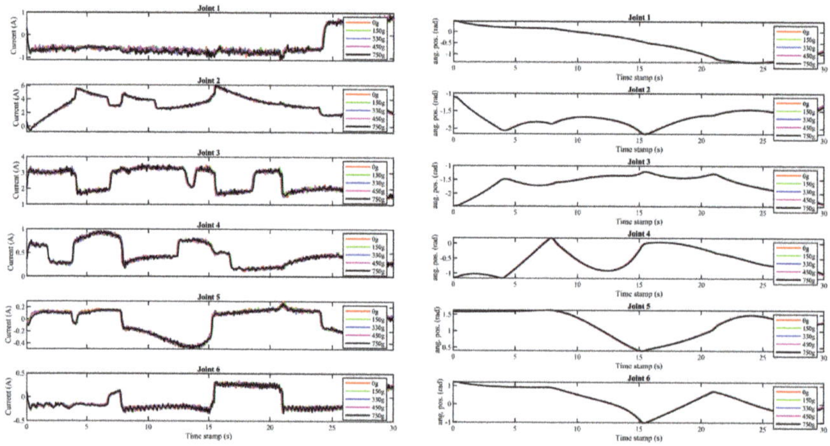

(**a**) Motor input current. (**b**) Angular position.

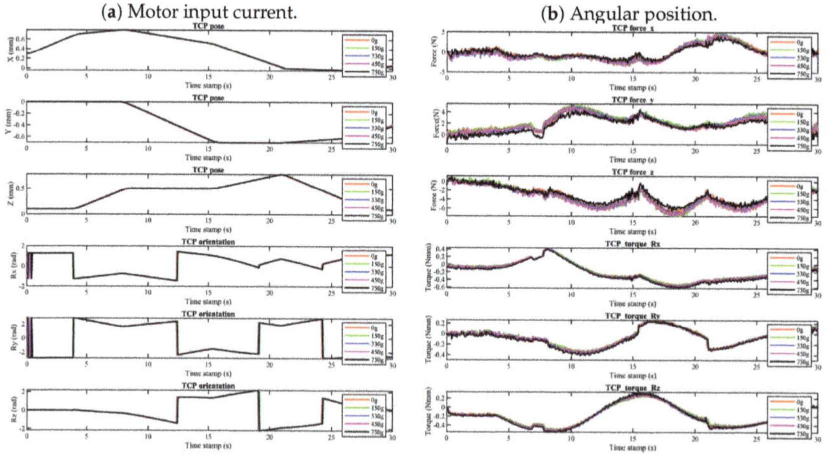

(**c**) TCP position and orientation. (**d**) TCP force/torque.

Figure 8. Plots of some cobot variables for the normal condition and first anomalies.

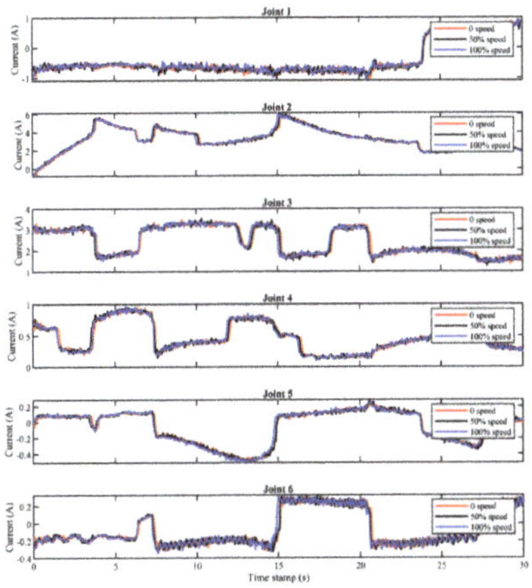

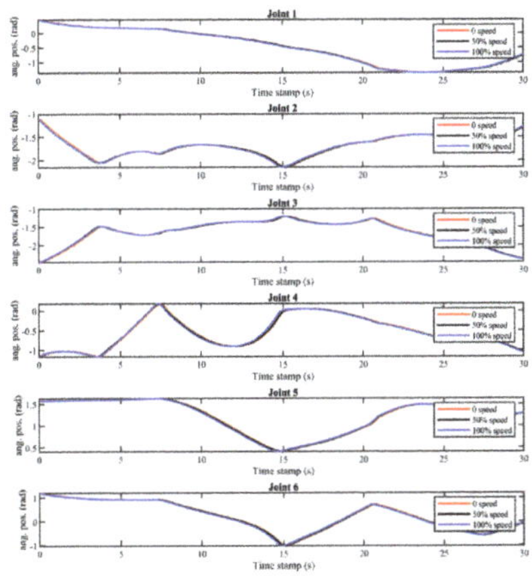

(**a**) Motor input current. (**b**) Angular position.

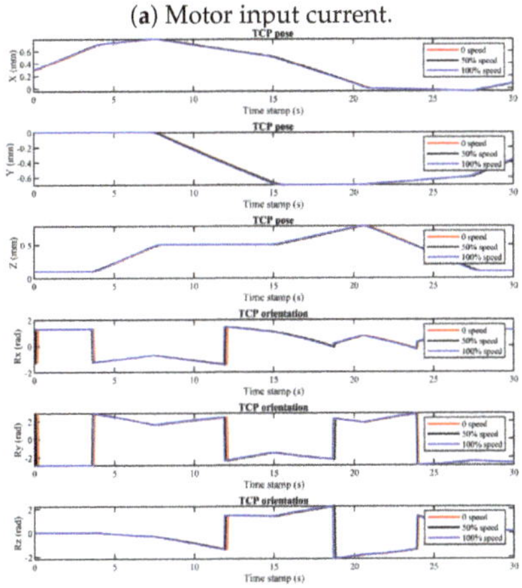

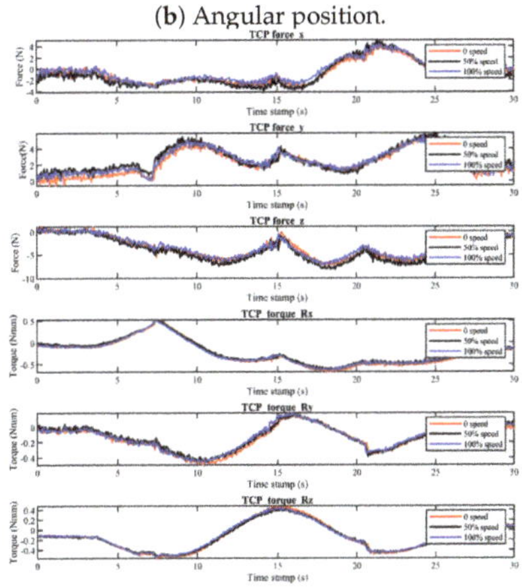

(**c**) TCP position and orientation. (**d**) TCP force/torque.

Figure 9. Plots of some cobot variables for the normal condition and second anomalies.

3.5. Theoretical Analysis of the Impact of Designed Anomalies

The theoretical explanation of the effects induced by the two anomalies is presented in this section. Generally, the motion of any industrial robot/manipulator solely depends on the kinematic model, dynamic model, and control algorithm, and these three parts are interconnected. As a result, we can calculate some elements in the dynamic equation from the kinematic knowledge, and the robot can be controlled to the desired trajectory using the dynamic information. The dynamic model of a robot is usually expressed as

$$M(q)\ddot{q} + C(q,\dot{q})\dot{q} + G(q) = \tau, \tag{13}$$

where $M(q) \in \mathbb{R}^{n\times n}$ is the inertia matrix; $C(q,\dot{q}) \in \mathbb{R}^{n\times n}$ is the matrix of Coriolis and Centrifugal terms; $G(q) \in \mathbb{R}^{n\times 1}$ denotes the gravitational force vector; $q, \dot{q}$, and $\ddot{q} \in \mathbb{R}^{n\times n}$ are the joint's position, velocity, and acceleration, respectively; and τ is the vector of the total joint torques (which is dependent on the motor torque, friction, and other external torques).

When friction(τ_f) is accounted for, the equation becomes

$$M(q)\ddot{q} + C(q,\dot{q})\dot{q} + G(q) = \tau - \tau_f. \tag{14}$$

Anomaly 1 involved applying varying weights to the link near joint 3. This increased the torque (τ) and input current required to move the joint. Joint faults, often caused by component degradation and increased friction, are a significant concern. The added weight increases the inertia (M) of the link; thus, more force is required to accelerate or decelerate it. Moreover, as the torque increases, the frictional torque is increased (τ_f), additional strain is put on the motor and gearbox, and the robot's accuracy and precision in reaching desired positions are affected based on the control algorithm used.

The second group of anomalies involved introducing a vibrator with two distinct vibrating speeds to joint 3. In practice, the simulated anomalies created with the vibrator could represent mechanical faults in the joint's motor or gears, which are common fault sources in robotics. The vibrator induces vibrations in the joint, resulting in displacement from its intended position, which affects the accuracy of the joint's positioning. Consequently, the joint's torque and current are changed (increased or decreased) to compensate for the anomalous condition. If the vibration intensity is significant, it may induce oscillations or resonance effects in the mechanical parts of the joint. Additionally, the mechanical stress induced by the vibration could accelerate wear on components such as bearings and gears.

3.6. Data Visualization and Preprocessing

For easy visualization of the variation in our dataset, the difference between the mean of each abnormal signal and that of the corresponding normal signal was computed with and without normalization for each group of experiments. The mathematical equation for the normalization of each of the robot variables ($y_{\mathrm{k,\,norm}}$) is

$$y_{\mathrm{k,norm}} = \frac{y_k - \mathrm{mean}(y_k)}{\mathrm{std}(y_k)}, \tag{15}$$

where k is an indicator of the data variable. Figure 10a shows the mean error plot of each signal for the different weights (i.e., weights 1–4 in the test cases) with respect to the baseline data, while Figure 10b shows the mean error plots after the signals were normalized. In the first plot, the TCP forces show high magnitude and variation compared to the other variables. A similar figure is shown in Figure 11a,b for the second anomaly experimental dataset. By examining the figure, it is clear that the mean errors of the TCP force variables are more dominant, and more variables appeared after normalization in Figure 11b. Furthermore, the data were also smoothed to remove noise from the measured signals. The smoothening function takes the median of data variables over a w-length sliding window through the following equation:

$$y_s[i] = \frac{1}{s} \sum_{i-\frac{s}{2}}^{i+\frac{m}{2}} y[i+j], \tag{16}$$

where $y_s[i]$ is the smooth value at point i, $y[i+j]$ represents the original data points within the window centered at position i, and s is the span of the moving average window. The sliding window length used in this work was 50 based on the result of comparing different window lengths in [27].

In the two figures (Figures 10 and 11), the variables on the x-axis are *jointi pos*, *jointi vel*, *jointi cur*, *TCP pos*, and *TCP force*. *Jointi* pos, *jointi vel*, and *jointi cur* represent the i-th joint position, velocity, and current, respectively, while *TCP pos* and *TCP force* represent the Tool Center Point position and force/torque in x, y, and z positions and Rx, Ry, and Rz orientation.

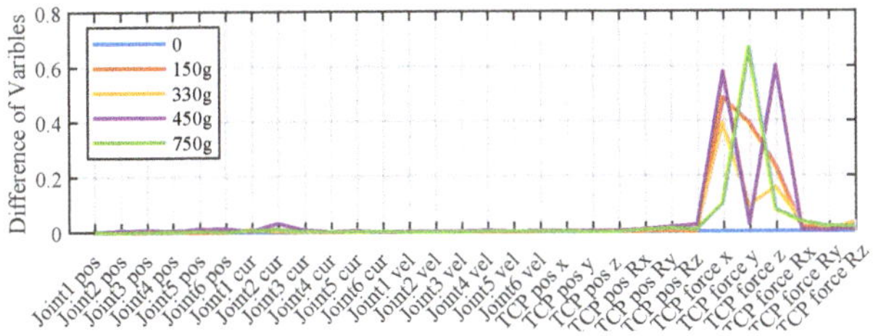

(**a**) Raw data.

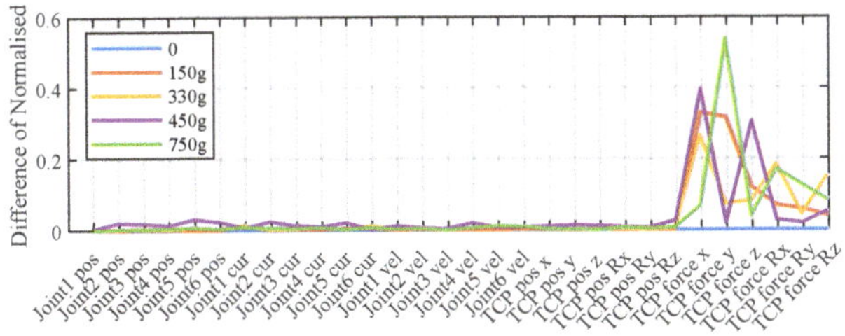

(**b**) Normalized raw data.

Figure 10. Anomaly 1 experimental dataset visualization.

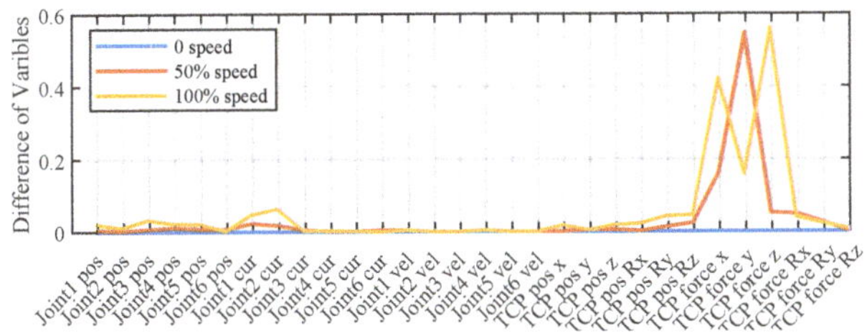

(**a**) Raw data.

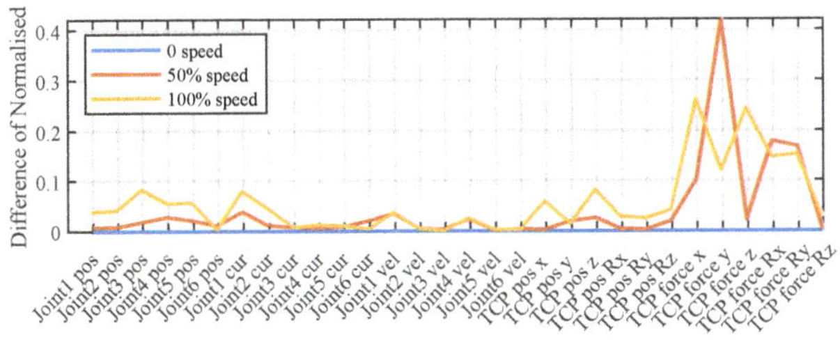

(**b**) Normalized raw data.

Figure 11. Anomaly 2 experimental dataset visualization.

4. Model Training and Results

In this section, the training process and results of the anomaly and trajectory change detection models are discussed.

4.1. Anomaly Detection Models

The architecture of the three AEs discussed in Section 3.1 was created, and the models were trained using the normal data from the two groups of experiments that were previously normalized and smoothed. This means six models were trained simultaneously:

three different AE models for group 1 normal data and another three unique AE models for group 2. The MLP-AE architecture consisted of four and three layers in the encoder and decoder parts, respectively, as shown in Figure 12a, while the CNN-AE architecture was made up of four layers in both the encoder and decoder parts, as illustrated Figure 12b. The SAE architecture was only composed of one layer in both the encoder and the decoder parts, as shown in Figure 12c. The training epochs set for the MLP-AE, CNN-AE, and SAE were 50, 10, and 50, respectively, when the Adam optimizer was used. The time it took to train each model is summarized in Table 4. It was observed that the MLP-AE took the least amount of time, followed by the SAE, while the CNN-AE took a relatively longer time.

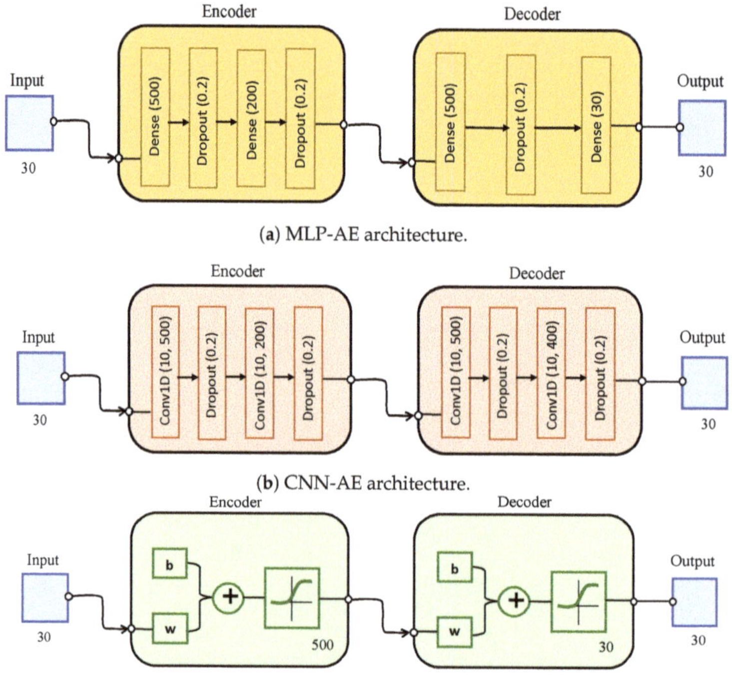

(**a**) MLP-AE architecture.

(**b**) CNN-AE architecture.

(**c**) SAE architecture.

Figure 12. Autoencoders' architecture.

Table 4. Training time of the anomaly detection models.

	Training Time (s)	
Model	**Group 1 Experiment**	**Group 2 Experiment**
MLP-AE	6.10	6.23
CNN-AE	60.78	60.59
SAE	11.64	12.85

In Figures 13 and 14, the *MD* between the baseline residual of the SAE model and its mean is shown. The detection threshold was calculated using the three-sigma value of *MD*, as depicted by the broken red lines in both plots. In the first subplot of Figure 13, the *MD* values (the solid black line) are below the threshold except for the starting and end points, where non-zero *MD* values can be observed. In subplots 2 to 5 of Figure 13, the distance between the residual of the different weights (150 g, 330 g, 450 g, and 750 g) and the mean of the baseline residual is plotted, and it can be seen that the *MD* values are mostly above the threshold except for a few points. This means that the SAE model could detect the anomaly caused by the weights (weights 1–4). Similarly, the plot obtained for anomaly 2 (Figure 14) has the same interpretation as that of anomaly 1 (Figure 13). It can be observed that the *MD* values (solid blue lines) are mostly below the threshold for the baseline case, while the *MD* values are mostly above the threshold for the anomalous conditions where the vibrator speed was set to 50 and 100%, respectively.

Furthermore, the accuracy and F1 score of each experimental set with respect to the three models (MLP-AE, CNN-AE, and SAE) were evaluated for the anomalous cases, and the results are shown in Tables 5 and 6. Accuracy measures how well the model correctly predicts both normal and anomalous instances out of the total instances, while the F1 score reflects the balance between precision and recall. Across all experiments, the three models achieved an accuracy greater than 93% for the anomalous cases, with the lowest F1 score exceeding 96%. Moreover, in both tables, it can be observed that the CNN-AE had the overall best accuracy. However, SAE was used for the local condition monitoring implementation in Section 5 because it had the simplest architecture. Since anomaly detection is similar to binary classification, with only two data classes, it tends to result in higher accuracy. To prevent overfitting, we used smaller-layer models, applied early stopping, and included a regularization term in the SAE.

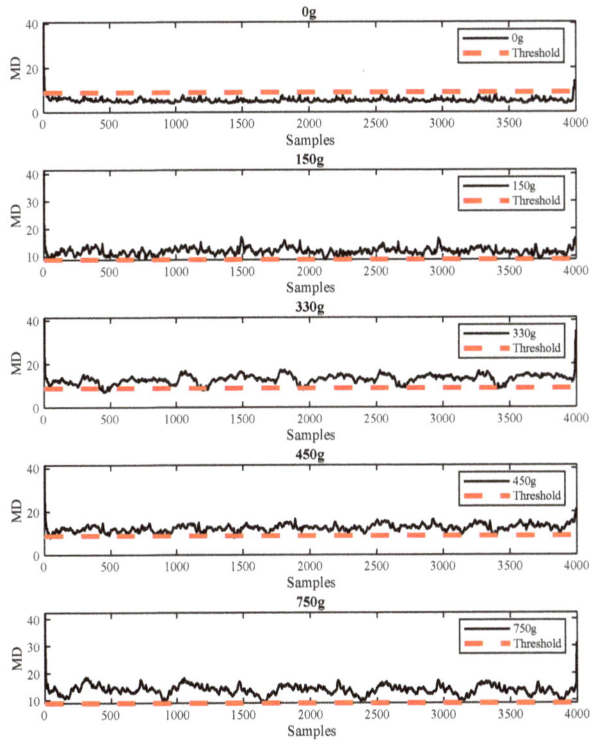

Figure 13. *MD* plot of the baseline and different weights of the arm load (w = 50).

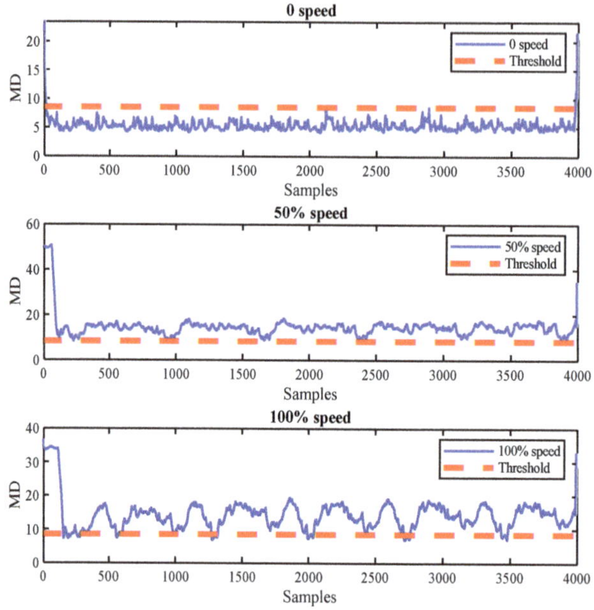

Figure 14. *MD* plot of the baseline and different speeds of the vibrator (w = 50).

Table 5. Performance metrics of the anomaly detection models with group 1 anomalies.

	MLP-AE		CNN-AE		SAE	
S/No.	**Accuracy**	**F1 Score**	**Accuracy**	**F1 Score**	**Accuracy**	**F1 Score**
150 g	100%	100%	100%	100%	96.78%	98.36%
330 g	96.13%	98.02%	100%	100%	97.48%	98.72%
450 g	99.45%	99.72%	100%	100%	99.60%	99.80%
750 g	100%	100%	100%	100%	99.80%	99.90%

Table 6. Performance metrics of the anomaly detection models with group 2 anomalies.

	MLP-AE		CNN-AE		SAE	
S/No.	**Accuracy**	**F1 Score**	**Accuracy**	**F1 Score**	**Accuracy**	**F1 Score**
50% speed	99.90%	99.95%	100%	100%	99.90%	99.95%
100% speed	95.82%	97.87%	98.50%	99.24%	93.50%	96.64%

4.2. Trajectory Change Detection Model

The cobot TCP position in the X, Y, and Z directions was used in the trajectory change detection model development. The DTW algorithm was employed in estimating the cumulative distance between trajectory 1 data for the baseline case and the weight 4 anomalous condition. The distance between the two trajectory data was found to be 8.5609×10^{-4}. Based on the distance, a threshold was then set as 0.005 (based on the observed range of variation) to detect a change in trajectory. Afterward, we used the same algorithm to determine the cumulative distance between TCP translational motion 1 (i.e., the trajectory displayed in Figure 7a) and the new motion shown in Figure 7b. We then compared the cumulative distance to the threshold set, and the result obtained shows that the distance between the test trajectory and the previous one was higher than the set threshold, thus indicating that a trajectory change was detected.

5. Online Monitoring

This section presents a framework for the online condition monitoring of cobots, and the framework was also implemented.

5.1. Monitoring Framework

One of the key benefits of using cobots, as opposed to traditional industrial robots, lies in their flexibility. This flexibility allows cobots to handle multiple tasks simultaneously, both in small- and large-scale industries. Hence, we have designed a robust and trajectory change adaptive framework to accommodate this versatility, as illustrated in Figure 15.

The framework assumptions are the following:

1. The data read by the cobot are for one cycle of a particular collaborative task.
2. The robot has primary (trajectory 1) and secondary collaborative tasks (i.e., other trajectories, 2, 3, 4, etc.).
3. It is assumed that the detection model used for each trajectory is pre-trained, and a new model is selected from the saved secondary tasks' models when a trajectory change occurs.
4. The online monitoring framework will alert the maintenance operator either when an anomaly is detected or when there is a case of a completely new trajectory without a corresponding pre-trained model.

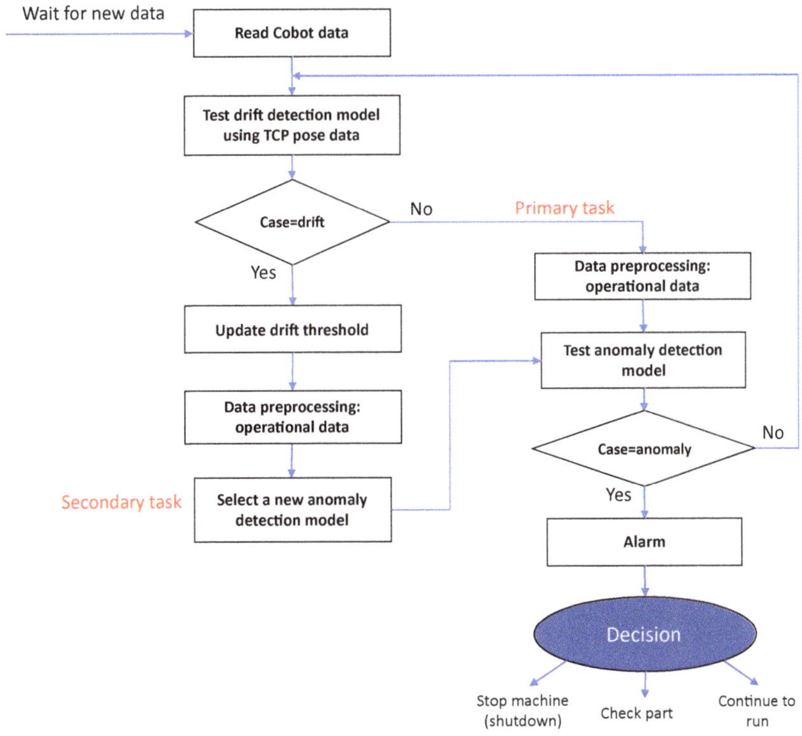

Figure 15. A flowchart for robust and drift-resilient cobot online condition monitoring framework.

5.2. Online Monitoring Implementation

In this section, we will discuss the implementation of the online monitoring framework. The implementation was carried out locally with a personal computer, and the MATLAB app designer was used to design a graphical user interface (GUI) for the online condition monitoring of the UR5e cobot. The online monitoring process includes real-time data acquisition, preprocessing, and the detection of anomalies as the robot performed different operations. The code begins by specifying the folder where the robot data are stored. Then, the data variables of interest are read, and some program-related variables are also initialized, including a variable where the list of processed data files can be tracked and another variable for setting flags for the loop control. The user interface shown in Figure 16 is updated with messages indicating that the robot monitoring has begun and the type of task or trajectory by the robot is being checked (see this update in Figure 17a). Afterward, a while loop runs continuously to monitor the robot. This loop is designed to keep running while the program is active. Within the loop, the code performs all or some of the following steps for each data file:

- Checks if the file has been processed before.
- If it is a new data file, it reads the data from the Excel file and stores them.
- Smooths and normalizes the data.
- Calculates the DTW distance between the new trajectory data and a baseline trajectory.
- Determines if a trajectory change is detected and updates the UI.
- Employs the SAE for anomaly detection and updates the UI.

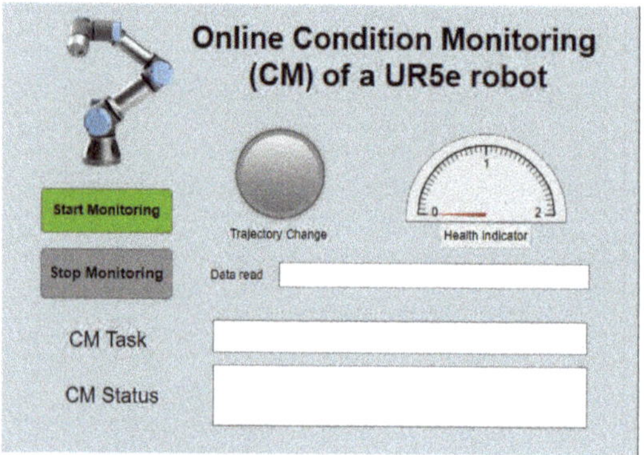

Figure 16. Online monitoring GUI.

Depending on whether a trajectory change is detected, the code updates the user interface and performs anomaly detection:

- If no change is detected, the accuracy of the detection model is checked, and the UI is updated accordingly.
- If a change is detected, the anomaly detection model is used to check the accuracy of a different trajectory, and the UI is also updated.

In Figure 17b, an example case of an anomaly and no trajectory change is displayed in the GUI. The code keeps track of processed files by adding them to the list of processed files. Additionally, a timer is used in the code to determine if new data have been detected within a specified timeout period. If no new data are detected within the timeout, the loop stops, and a message is displayed in the user interface (see Figure 17c). As shown in Figure 17d, the program can also stop when the 'Stop Monitoring' push button is pressed. In summary, the code is designed to continuously monitor data from an industrial robot, compare its trajectory with a baseline, and detect anomalies in the robot's behavior.

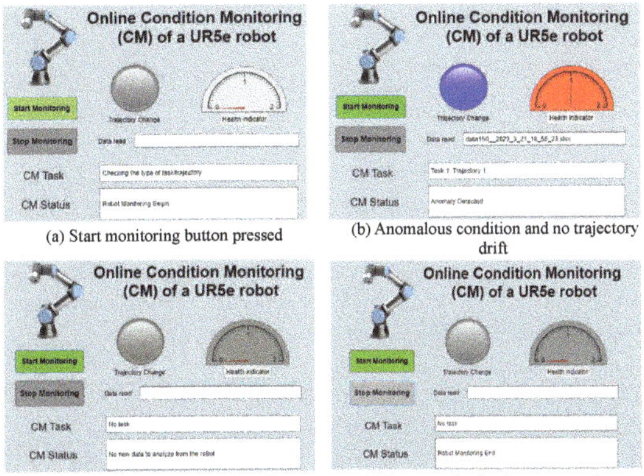

(a) Start monitoring button pressed (b) Anomalous condition and no trajectory drift (c) No batch data from the robot after some time (d) Stop monitoring button pressed

Figure 17. Online monitoring GUI for different test scenarios.

6. Conclusions

This research presents the condition monitoring of cobots using their multivariable operational data. It was found that acquiring fault data for developing fault diagnosis and prognosis models which can capture various faults and estimate the remaining useful life (RUL) of robots and their components is costly. As a result, fault/anomaly detection is a more practical and cost-effective approach. The literature review also revealed that the performance of anomaly detection models is affected by changes in trajectory. Therefore, both anomaly detection and trajectory change detection models were explored in this study.

The performance of three AEs (MLP-AE, CNN-AE, and SAE) was compared in terms of their anomaly detection capabilities. Additionally, DTW was employed for trajectory change detection. The results showed that the AEs were highly effective, achieving an overall accuracy of over 93% and an F1 score exceeding 96% for different anomalous conditions. The DTW algorithm also demonstrated satisfactory performance in distinguishing between two different trajectories.

Furthermore, a framework was presented for the online deployment of two digital models. These models included a trajectory change model and an anomaly detection model, which were trained offline using data collected during the cobot's primary task (trajectory 1). The assumptions crucial for the framework's operation were examined. Finally, the online monitoring framework was implemented locally and displayed the operation through a user interface. The algorithms and framework have the potential to be used for cobots and other industrial robots performing predefined manufacturing tasks in an industrial setting. In our future work, we aim to enhance the online monitoring algorithm by incorporating a model that is trained online once a trajectory change is detected, without the need to use pre-trained anomaly detection models. This will foster the online training and update of the trajectory change and anomaly detection algorithms to create a more robust condition monitoring framework. Additionally, the anomaly detection models developed in this study are for a particular operating condition (i.e., defined trajectory, tool speed, and acceleration). The model cannot be generalized for other trajectories. Hence, a residual-based model leveraging the robot mathematical model will be explored as part of our future research.

Author Contributions: S.A.: methodology, experiments, data analysis, coding, visualization, writing—original draft, writing—review and editing. F.G.: methodology, experiments, data analysis, coding, validation, visualization, writing—review and editing, resources, supervision. H.L.: methodology, experiments, visualization, writing—review and editing. H.F.: methodology, experiments, writing—review and editing, resources. A.B.: supervision, resources. All authors have read and agreed to the published version of the manuscript.

Funding: This research received no external funding.

Data Availability Statement: Data will be made available on request.

Conflicts of Interest: The authors declare no conflicts of interest.

References

1. Aliev, K.; Antonelli, D. Proposal of a Monitoring System for Collaborative Robots to Predict Outages and to Assess Reliability Factors Exploiting Machine Learning. *Appl. Sci.* **2021**, *11*, 1621. [CrossRef]
2. Knudsen, M.; Kaivo-oja, J. Collaborative Robots: Frontiers of Current Literature. *J. Intell. Syst. Theory Appl.* **2020**, *3*, 13–20. [CrossRef]
3. Lee, J.; Ni, J.; Singh, J.; Jiang, B.; Azamfar, M.; Feng, J. Intelligent Maintenance Systems and Predictive Manufacturing. *J. Manuf. Sci. Eng.* **2020**, *142*, 110805. [CrossRef]
4. Achouch, M.; Dimitrova, M.; Ziane, K.; Sattarpanah Karganroudi, S.; Dhouib, R.; Ibrahim, H.; Adda, M. On Predictive Maintenance in Industry 4.0: Overview, Models, and Challenges. *Appl. Sci.* **2022**, *12*, 8081. [CrossRef]
5. Montero Jiménez, J.J.; Vingerhoeds, R.; Grabot, B.; Schwartz, S. An ontology model for maintenance strategy selection and assessment. *J. Intell. Manuf.* **2023**, *34*, 1369–1387. [CrossRef]
6. Biggio, L.; Kastanis, I. Prognostics and Health Management of Industrial Assets: Current Progress and Road Ahead. *Front. Artif. Intell.* **2020**, *3*, 578613. [CrossRef]
7. Kumar, P.; Khalid, S.; Kim, H. Prognostics and Health Management of Rotating Machinery of Industrial Robot with Deep Learning Applications—A Review. *Mathematics* **2023**, *11*, 3008. [CrossRef]
8. Raouf, I.; Kumar, P.; Lee, H.; Kim, H.S. Transfer Learning-Based Intelligent Fault Detection Approach for the Industrial Robotic System. *Mathematics* **2023**, *11*, 945. [CrossRef]
9. Ucar, A.; Karakose, M.; Kırımça, N. Artificial Intelligence for Predictive Maintenance Applications: Key Components, Trustworthiness, and Future Trends. *Appl. Sci.* **2024**, *14*, 898. [CrossRef]
10. Sabry, A.H.; Ungku Amirulddin, U.A.B. A review on fault detection and diagnosis of industrial robots and multi-axis machines. *Results Eng.* **2024**, *23*, 102397. [CrossRef]
11. Galan-Uribe, E.; Amezquita-Sanchez, J.P.; Morales-Velazquez, L. Supervised Machine-Learning Methodology for Industrial Robot Positional Health Using Artificial Neural Networks, Discrete Wavelet Transform, and Nonlinear Indicators. *Sensors* **2023**, *23*, 3213. [CrossRef] [PubMed]
12. Han, H.; Lin, Y.; Gu, L.; Xu, Y.; Gu, F. Vibration Analysis Based Condition Monitoring for Industrial Robots. In *Proceedings of the IncoME-V & CEPE Net-2020*; Zhen, D., Wang, D., Wang, T., Wang, H., Huang, B., Sinha, J.K., Ball, A.D., Eds.; Springer International Publishing: Cham, Switzerland, 2021; pp. 186–195. [CrossRef]
13. Yun, H.; Kim, H.; Jeong, Y.H.; Jun, M.B.G. Autoencoder-based anomaly detection of industrial robot arm using stethoscope based internal sound sensor. *J. Intell. Manuf.* **2023**, *34*, 1427–1444. [CrossRef]
14. Rohan, A.; Raouf, I.; Kim, H.S. Rotate Vector (RV) Reducer Fault Detection and Diagnosis System: Towards Component Level Prognostics and Health Management (PHM). *Sensors* **2020**, *20*, 6845. [CrossRef]
15. Raouf, I.; Lee, H.; Kim, H.S. Mechanical fault detection based on machine learning for robotic RV reducer using electrical current signature analysis: A data-driven approach. *J. Comput. Des. Eng.* **2022**, *9*, 417–433. [CrossRef]
16. Lee, H.; Raouf, I.; Song, J.; Kim, H.S.; Lee, S. Prognostics and Health Management of the Robotic Servo-Motor under Variable Operating Conditions. *Mathematics* **2023**, *11*, 398. [CrossRef]
17. Graabæk, S.G.; Ancker, E.V.; Christensen, A.L.; Fugl, A.R. An Experimental Comparison of Anomaly Detection Methods for Collaborative Robot Manipulators. *IEEE Access* **2023**, *11*, 65834–65848. [CrossRef]
18. Brockmann, J.T.; Rudolph, M.; Rosenhahn, B.; Wandt, B. The voraus-AD Dataset for Anomaly Detection in Robot Applications. *IEEE Trans. Robot.* **2023**, *40*, 438–451. [CrossRef]
19. Majid, M.A.A.; Fudzin, F. Study on Robots Failures in Automotive Painting Line. *ARPN J. Eng. Appl. Sci.* **2017**, *12*, 62–67.
20. Polenghi, A.; Cattaneo, L.; Macchi, M. A framework for fault detection and diagnostics of articulated collaborative robots based on hybrid series modelling of Artificial Intelligence algorithms. *J. Intell. Manuf.* **2023**, *35*, 1929–1947. [CrossRef]
21. Kermenov, R.; Nabissi, G.; Longhi, S.; Bonci, A. Anomaly Detection and Concept Drift Adaptation for Dynamic Systems: A General Method with Practical Implementation Using an Industrial Collaborative Robot. *Sensors* **2023**, *23*, 3260. [CrossRef]
22. Qiu, S.; Cui, X.; Ping, Z.; Shan, N.; Li, Z.; Bao, X.; Xu, X. Deep Learning Techniques in Intelligent Fault Diagnosis and Prognosis for Industrial Systems: A Review. *Sensors* **2023**, *23*, 1305. [CrossRef]
23. Zhou, X.; Zhou, H.; He, Y.; Huang, S.; Zhu, Z.; Chen, J. Harmonic reducer in-situ fault diagnosis for industrial robots based on deep learning. *Sci. China Technol. Sci.* **2022**, *65*, 2116–2126. [CrossRef]
24. Samuel, A.; Eric, K.; Hassna, L.; Hamidreza, F.; Fengshou, G.; Andrew, B. A Hybrid Digital Twin Scheme for the Condition Monitoring of Industrial Collaborative Robots. *Procedia Comput. Sci.* **2024**, *232*, 1099–1108. [CrossRef]
25. Raviola, A.; Guida, R.; Bertolino, A.C.; De Martin, A.; Mauro, S.; Sorli, M. A Comprehensive Multibody Model of a Collaborative Robot to Support Model-Based Health Management. *Robotics* **2023**, *12*, 71. [CrossRef]

26. Han, H.; Huang, C.; Song, Y.; Li, D.; Hamidreza, F.; Fengshou, G.; Andrew, B. Residual-based fault detection of abnormal joint running state of industrial collaborative robot. In *Proceedings of the UNIfied Conference of DAMAS, IncoME and TEPEN Conferences (UNIfied 2023)*; Springer: Cham, Switzerland, 2024; Volume 1. [CrossRef]
27. Samuel, A.; Xiaoxia, L.; Hassna, L.; Hamidreza, F.; Fengshou, G.; Andrew, B. Sensitivity of PCA and Autoencoder-Based Anomaly Detection for Industrial Collaborative Robots. In *Proceedings of the UNIfied Conference of DAMAS, IncoME and TEPEN Conferences (UNIfied 2023)*; Springer: Cham, Switzerland, 2024; Volume 1. [CrossRef]
28. Tziolas, T.; Papageorgiou, K.; Theodosiou, T.; Papageorgiou, E.; Mastos, T.; Papadopoulos, A. Autoencoders for Anomaly Detection in an Industrial Multivariate Time Series Dataset. *Eng. Proc.* **2022**, *18*, 23. [CrossRef]
29. Chen, Z.; Yeo, C.K.; Lee, B.S.; Lau, C.T. Autoencoder-based network anomaly detection. In Proceedings of the 2018 Wireless Telecommunications Symposium (WTS), Phoenix, AZ, USA, 17–20 April 2018; pp. 1–5. [CrossRef]
30. Soydaner, D. A Comparison of Optimization Algorithms for Deep Learning. *Int. J. Pattern Recognit. Artif. Intell.* **2020**, *34*, 2052013. [CrossRef]
31. Ayankoso, S.; Olejnik, P. Time-Series Machine Learning Techniques for Modeling and Identification of Mechatronic Systems with Friction: A Review and Real Application. *Electronics* **2023**, *12*, 3669. [CrossRef]
32. Popescu, M.C.; Balas, V.E.; Perescu-Popescu, L.; Mastorakis, N. Multilayer Perceptron and Neural Networks. *World Sci. Eng. Acad. Soc. (WSEAS) Trans. Circuits Syst.* **2009**, *8*, 579–588.
33. Sharma, S.; Sharma, S.; Scholar, U.G.; Athaiya, A. Activation Functions in Neural Networks. *Int. J. Eng. Appl. Sci. Technol.* **2020**, *4*, 7. [CrossRef]
34. Indolia, S.; Goswami, A.; Mishra, S.; Asopa, P. Conceptual Understanding of Convolutional Neural Network- A Deep Learning Approach. *Procedia Comput. Sci.* **2018**, *132*, 679–688. [CrossRef]
35. Adcock, B.; Dexter, N. The gap between theory and practice in function approximation with deep neural networks. *SIAM J. Math. Data Sci.* **2021**, *3*, 624–655 [CrossRef]
36. Train an Autoencoder—MATLAB. Available online: https://uk.mathworks.com/help/deeplearning/ref/trainautoencoder.html (accessed on 20 April 2023).
37. Leys, C.; Klein, O.; Dominicy, Y.; Ley, C. Detecting multivariate outliers: Use a robust variant of the Mahalanobis distance. *J. Exp. Soc. Psychol.* **2018**, *74*, 150–156. [CrossRef]
38. Kloska, M.; Grmanova, G.; Rozinajova, V. Expert enhanced dynamic time warping based anomaly detection. *Expert Syst. Appl.* **2023**, *225*, 120030. [CrossRef]
39. Universal Robot UR5 Technical Specification. 2016. Available online: https://www.rarukautomation.com/wp-content/uploads/2022/11/universal-robot-ur5-technical-spec.pdf (accessed on 4 May 2023).
40. Real-Time Data Exchange (RTDE) Guide. 2024. Available online: https://www.universal-robots.com/articles/ur/interface-communication/real-time-data-exchange-rtde-guide/ (accessed on 24 April 2023).

machines

MDPI

Article

Multi-Sensor Information Fusion with Multi-Scale Adaptive Graph Convolutional Networks for Abnormal Vibration Diagnosis of Rolling Mill

Rongrong Peng [1,*], Changfen Gong [2] and Shuai Zhao [3]

[1] School of Mechanical and Vehicle Engineering, Nanchang Institute of Science and Technology, Nanchang 330108, China
[2] School of Education, Nanchang Institute of Science and Technology, Nanchang 330108, China; 15179122587@126.com
[3] School of Information and Artificial Intelligence, Nanchang Institute of Science and Technology, Nanchang 330108, China; zs1025198686@163.com
* Correspondence: 15294476178@163.com

Abstract: Graph data and multi-sensor information fusion have been integrated into the abnormal vibration type classification and the identification of the rolling mill for extracting spatial–temporal and robust features. However, most of the existing deep learning (DL) based methods exploit only single sensor information and Euclidean space data, which results in incomplete information contained in the features extracted by in-depth networks. To solve this issue, a multi-sensor information fusion with multi-scale adaptive graph convolutional networks (M^2AGCNs) framework is proposed to model graph data and multi-sensor information fusion in a unified in-depth network and then to achieve abnormal vibration diagnosis. First, convolutional neural networks (CNNs) were adopted for the deeper features of multi-sensor signals. And then, the extracted features were fed into the proposed feature-driven adaptive graph generation network to build graphs to extract spatial–temporal correlation between multi-sensor data. After that, the multi-scale graph convolutional networks (MSGCNs) were employed to aggregate and enrich several different receptive information to further improve valuable features. Finally, the extracted multi-sensor features were integrated into a unified network to achieve the abnormal vibration type classification and identification of the rolling mill. Meanwhile, we performed horizontal, vertical, and coupled abnormal vibration experiments, and then three different types of studies were conducted to illustrate the superiority and usefulness of this method in the paper and the feasibility of rolling mill abnormal vibration diagnosis. It can be seen from the results that the proposed M^2AGCNs can be able to achieve valuable feature extraction effectively from multi-sensor information and to obtain more excellent behavior of the abnormal vibration diagnosis of the rolling mill in comparison with the mainstream methods.

Keywords: rolling mill; abnormal vibration; monitoring and diagnosis; multi-sensor information fusion; multi-scale graph convolutional networks

1. Introduction

The rolling mills often work in complex and changing conditions, so there will inevitably occur abnormal vibration during long-term operations [1,2]. The phenomenon of abnormal vibration can cause certain damage to the product and rolling mill compo-

https://doi.org/10.3390/machines13010030

nents, which will cause damage to the rollers and lead to the rolling mill downtime for maintenance [3,4].

Early abnormal vibration diagnosis methods of the rolling mill mostly belonged to the model-based category, where a state observer or linear and nonlinear physical models are constructed, and then abnormal vibration diagnosis can be achieved by monitoring variations in the residuals. Zhang et al. proposed a multi-channel monitoring coefficients model to diagnose the abnormal vibration [5]. Du et al. designed a dynamic comprehensive model based on the strain gradient elasticity theory to predict vibration [6]. However, these methods are time-consuming and labor-intensive in the face of industrial big data. Recently, with the advancement in sensor technology and the improvement in data storage volume, data-driven machine learning methods have become a research spot due to their powerful nonlinear data extraction and fitting capabilities. Liu et al. introduced an ensemble learning model based on multi-source information to predict abnormal vibration [7]. Lu et al. discussed a data-driven method, such as a neural network-based method and support vector regression, to predict and analyze the vibration of the cold rolling mill [8]. Pan et al. combined modified nonlocal means and modified empirical wavelet transform for data-driven vibration identification [9]. Li et al. proposed an improved hidden semi-Markov model to monitor the health states of the rolling mill [10]. Currently, intelligent diagnosis techniques for rolling mills have made great progress in feature extraction thanks to the boom in deep learning (DL) theory and computational resources [11–14]. The emerging deep neural networks, such as convolutional neural networks (CNNs), auto-encoders (AEs), deep belief networks (DBNs), recurrent neural networks (RNNs), and long short-term memory (LSTM), can automatically extract features from the original signals. Zhang et al. designed an improved residual network to extract the raw cutting force and vibration signals to achieve the condition monitoring of the rolling mill [15]. Shah et al. suggested the stacked LSTM method to diagnose the health states of mechanical equipment [16]. Xu et al. designed the novel transfer learning framework online to better analyze the health states of the rolling mill [17]. Zhao et al. adopted the advanced signals processing methods and then combined them with multivariate composite multi-scale weighted permutation entropy function (MCMWPE) to locate and identify the health states of the rolling mill [18]. DL methods are designed for Euclidean data, which deal with a regular spatial structure and so cannot easily capture the spatial–temporal relationships between different data. The existence of non-Euclidean graph data cannot be exploited by traditional DL methods because of restrictions in computational criteria and network architecture, thereby greatly limiting the application of DL methods.

To solve the above limitation, graph neural networks (GNNs) have rapidly gained attention due to their wide applications and powerful performance. GNNs can perform DL representation learning by grasping information from input node and structure features as shown in Figure 1. Zhang et al. extracted singular values from original signals which are input to construct graphs to diagnose the health states of the roller bearings [19]. Yang et al. used the spatial–temporal correlation of the data to form graphs for achieving the fault diagnosis of the rotating machinery [20]. Liu et al. introduced and trained a SuperGraph method to extract valuable features of vibration signals to realize fault diagnosis [21].

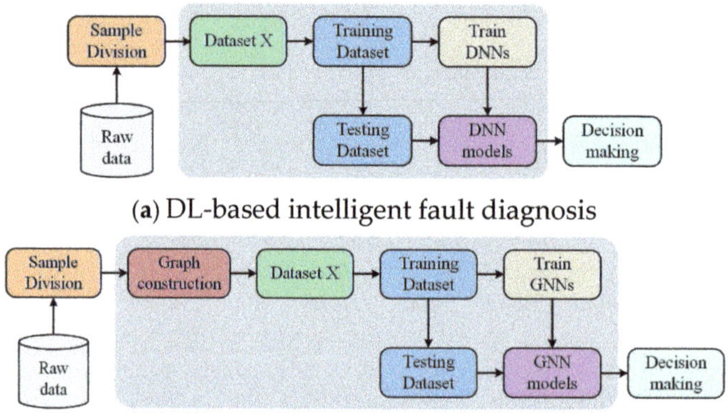

(**a**) DL-based intelligent fault diagnosis

(**b**) GNN-based intelligent fault diagnosis

Figure 1. The framework of DL-based methods and GNN-based methods.

Multi-sensor information compared with single-sensor information can not only contain redundant and complementary information, but also can produce more robust and reliable state representations of a rolling mill. Multi-sensor information fusion methods can be divided into three types: data-level fusion, feature-level fusion, and decision-level fusion as shown in Figure 2. Herein, data-level fusion integrates raw data from multiple sources, feature-level fusion combines the extracted features to enhance representation, and decision-level fusion aggregates individual decisions to improve the accuracy and robustness of fault diagnosis. Among them, the feature-level fusion method can reduce the amount of data to be processed and then improve the real-time performance after extracting the original multi-sensor signals. Hence, this feature-level fusion method is considered more effective than the other two fusion methods. Liu et al. [22] designed a multi-sensor information fusion diagnosis framework for rolling bearings. The framework fuses the information of different frequency components in multi-sensors utilizing a weighted fusion rule based on fuzzy entropy and performs fault classification with multi-core attentional CNNs based on them. Huo et al. [23] fused thermal imaging with vibration signals for the fault diagnosis of rotating machinery under a strong noise background. Huang et al. [24] presented a residual gated dynamic sparse network to fuse multi-sensor information of gearboxes for identifying their health status.

Although the DL-based abnormal vibration diagnosis methods combined with multi-sensor information fusion have made great progress in the rolling mill diagnosis field, there are still some limitations and research gaps [25–27]. Firstly, traditional operations are easily applied to Euclidean data, but they make it difficult to model graph data in non-Euclidean space. Meanwhile, traditional DL methods make it difficult to extract the spatial–temporal relationships before data. Secondly, the graph generation method for non-Euclidean data is too mechanical to fully capture features from input data. Thirdly, abnormal vibration diagnosis methods on the basis of a multi-sensor data fusion technique present complex and difficult challenges due to the non-stationary properties of fault vibration signals.

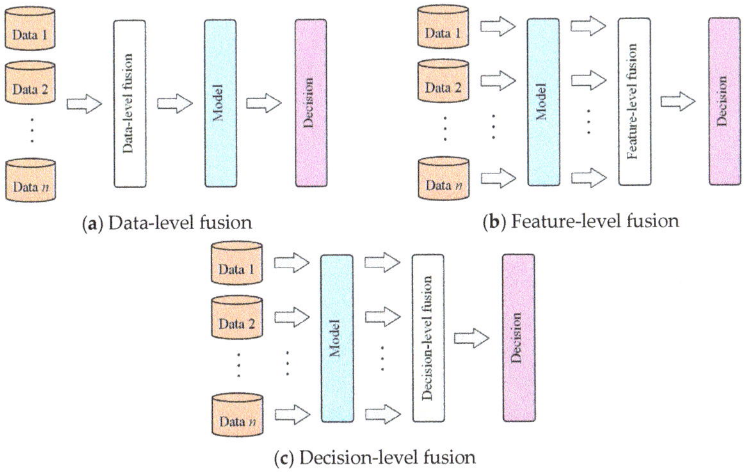

Figure 2. The framework of multi-sensor information fusion methods.

To address the above limitations, multi-scale adaptive graph convolutional networks on the basis of a multi-sensor data fusion technique have been proposed for the abnormal vibration diagnosis of the rolling mill. In M^2AGCNs, the abnormal vibration labels of the rolling mill and multi-sensor original signals are associated and then modeled through the multi-scale graph convolutional networks (MSGCNs) and adaptive functional graph construction (AFGC) in a unified deep network framework. First, the multi-sensor features are extracted from the input multi-sensor signals using the feature extraction CNNs. After that, an adaptive graph construction layer is introduced to learn the data structure through the features extracted by a CNN and then the graphs are constructed by mining data features and correlations. Then, the MSGCNs can model structural information at different scales of the instance graphs of different sensors and then fuse these multi-scale features to the output. Finally, the feature-level fusion framework is applied to fuse redundant, valuable, and spatial–temporal features by modeling multi-source sensor signals to gain abnormal vibration type classification and the identification of the rolling mill.

The primary contributions of this academic paper can be identified as follows:

(1) The AFGC is designed to build the graphs based on the features extracted by CNNs between multi-sensors, where feature vectors can be converted to an adjacency matrix adaptively.
(2) The MSGCNs contain receptive fields of different scales to enrich the representation features of the original vibration signals under multiple sensors.
(3) The multi-sensor information fusion framework is established for realizing the abnormal vibration diagnosis of the rolling mill, including the corresponding feature fusion functions and general framework output formula.

The rest of this paper can be structured as follows: Section 2 introduces the related works of GNNs. In Section 3, the proposed AFGC, MSGCNs, and the M^2AGCNs are illustrated. Section 4 details the architecture and sensor locations of the laboratory equipment of the rolling mill and then shows the information-collecting device. Section 5 introduces the collection process and division method of the experimental datasets and then verifies the superiority and dominance of the proposed method by three case studies of the abnormal vibration of the rolling mill. Then, the discussions of the proposed method is presented in Section 6. Finally, Section 7 depicts the conclusions and forthcoming work under these experiments of this paper.

2. Graph Convolutional Networks (GCNs)

Specifically, the GCNs, as one of the most classical branches of graph neural networks (GNNs), has become gained significant success in the field of DL, which essentially extracts the spatial–temporal features based on graphs by the defined convolution operation, as seen in Figure 3. Hence, the undirected graph is a fundamental component of GCNs. It can be defined as $G = \{V, E, A, F\}$, where V denotes the set of n-vertices, E is the set of edges, $A \in R^{n \times n}$ expresses the adjacency matrix, and $F \in R^{n \times m}$ represents the feature matrix of the nodes.

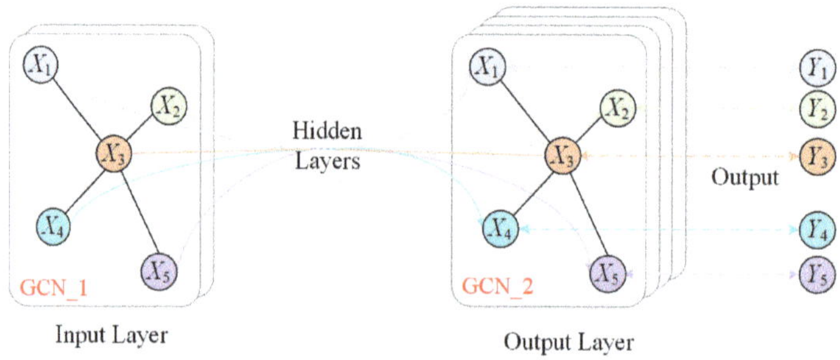

Figure 3. The classical structure of the GCNs.

The normalized graph Laplacian matrix is introduced by the adjacency matrix, which serves mainly to study the mapping properties of the graph, which can be computed by

$$L = I_N - D^{-\frac{1}{2}} A D^{\frac{1}{2}} = U \Lambda U^T \quad (1)$$

where U and Λ refers to the matrix of eigenvectors and a diagonal matrix with eigenvalues corresponding to the normalized graph Laplacian L, respectively. D denotes a degree diagonal matrix.

Remarkably, Khan et al. [28] defined the graph convolution operation $\langle * \rangle$ applied to the Fourier domain of the graph Laplacian to further calculate eigen-decomposition values, which can be expressed as

$$g_\theta * F = U g_\theta(\Lambda) U^T F \quad (2)$$

where θ is the learnable modal parameter.

Next, based on the previous theory, Defferrard et al. [29] proposed the ChebNet by using the K-localized convolution to reduce the computational volume of Equation (2), which can be written as

$$g_\theta \star F \approx \sum_{k=0}^{K} \theta_k T_k(\hat{L}) F \quad (3)$$

where $T_k(\hat{L})$ denotes the Chebyshev polynomials written by $T_k(x) = 2x T_{k-1}(x) - T_{k-2}(x)$ in which $T_0(x)$ and $T_1(x)$ is the 1 and x, respectively. $\hat{L} = \frac{2}{\lambda_{\max}} L - I_N$ is the refined diagonal matrix, with the largest eigenvalue $\lambda_{\max}$. θ_k represents the Chebyshev coefficient vector.

Finally, Kipf and Welling [30] proposed that K could be designed as one to alleviate the problem of the layer-wise convolution operation and then $\lambda_{\max}$ could be approximated as 2. Hence, Equation (3) can be simplified as

$$g_{\theta'} \star F \approx \theta'_0 F + \theta'_1 (L - I_N) F = \theta'_0 F - \theta'_1 D^{-\frac{1}{2}} A D^{\frac{1}{2}} F \quad (4)$$

where θ'_0 and θ'_1 represent two free parameters.

Then, to further reduce the number of free parameters, the above equation can be expressed by applying $\theta = \theta'_0 = -\theta'_1$ as

$$g_\theta \star F \approx \theta\left(I_N + D^{-\frac{1}{2}} A D^{\frac{1}{2}}\right) F \quad (5)$$

Finally, the renormalization trick of $I_N + D^{-\frac{1}{2}} A D^{\frac{1}{2}} \to \widetilde{D}^{-\frac{1}{2}} \widetilde{A} \widetilde{D}^{\frac{1}{2}}$ is applied to solve the problems of exploding or vanishing gradients and numerical instabilities, and the graph convolutional of the feature maps can be defined as

$$Y = \widetilde{D}^{-\frac{1}{2}} \widetilde{A} \widetilde{D}^{\frac{1}{2}} X W \quad (6)$$

where $Y \in N^{n \times n}$ denotes the final output and $W \in N^{m \times s}$ represents the trainable model parameters. $\widetilde{A}$ and $\hat{D}$ denotes the $A + I_N$ and $\sum_j \widetilde{A}_{ij}$.

3. The Proposed Method

3.1. Adaptive Functional Graph Construction (AFGC)

Functional connectivity is innovatively constructed based on the dependencies or correlations of the features extracted by the CNNs. Time-frequency relationships [20] and feature distances [31] are the two dominant methods to determine the constructions of the graph. However, these methods can only generate graphs for specific-sensor data and lack self-learning and generalization capabilities. Second, their node relationships are represented inadequately due to the manual intervention. Hence, we propose a new feature-driven graph construction method for adaptive functional connectivity. The feature-driven method can adaptively construct the graphs for data with different sensors based on different fault features between the nodes as shown in Figure 4. The specific expression of the AFGC is explained as follows.

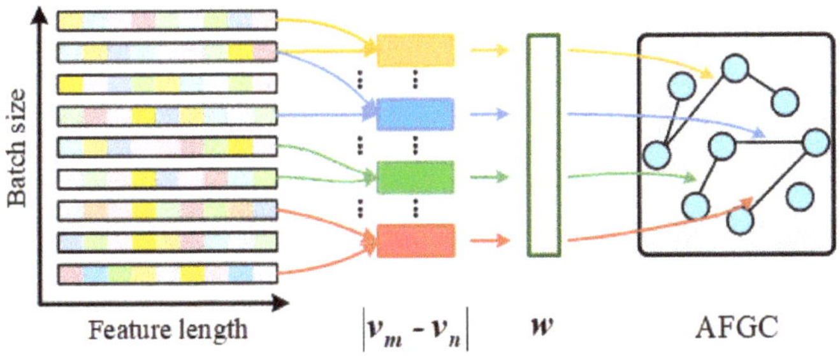

Figure 4. The adaptive graph construction method.

Given the features $V_i = (v_1^i, v_2^i, \ldots, v_N^i)^{\mathrm{T}} \in \mathbb{R}^{N \times D}$ extracted by the embedding layer, we then define an adaptive connection function $f(v_m, v_n), (m, n \in \{1, 2, \ldots, N\})$ to represent the adjacency matrix based on the nodes $v_m \in \mathbb{R}^{1 \times D}$ and $v_n \in \mathbb{R}^{1 \times D}$. We implement the $f(v_m, v_n)$ over the weight layer containing the learnable parameter $w = (w_1, w_2, \ldots, w_D)^{\mathrm{T}} \in \mathbb{R}^{D \times 1}$. We form the new adaptive adjacency matrix A_{mn}^{AFGC} as follows:

$$A_{mn}^{AFGC} = f(v_m, v_n) = \frac{\exp(\sigma(w^{\mathrm{T}} |v_m - v_n|))}{\sum_{n=1}^{N} \exp(\sigma(w^{\mathrm{T}} |v_m - v_n|))} \quad (7)$$

where $\sigma(\cdot)$ denotes a sigmoid activation function to guarantee that A_{mn}^{AFGC} is nonlinear and non-negative. The $\exp(\cdot)$ operation is adopted to normalize the A_{mn}^{AFGC} to improve the parameter training speed.

We designed the new loss function $\mathcal{L}_{AFGC}$ to update the learnable weight vector w by minimizing it, which can be written as

$$\mathcal{L}_{AFGC} = \sum_{m,n=1}^{N} \lambda \|v_m - v_n\|_2^2 \times A_{mn}^{AFGC} + \beta \left\| A_{mn}^{AFGC} \right\|_F^2 \quad (8)$$

where λ and β are the weight and sparsity parameters. The former term denotes the weight connections between the nodes of graph A_{mn}^{AFGC} and the latter term means the regularization.

3.2. Multi-Scale Graph Convolutional Networks (MSGCNs)

The main idea of the MSGCNs is to contain the multi-scale receptive fields to aggregate and enrich several different information of the original features. The aggerate process of the MSGCNs is displayed in Figure 5, assuming that the node feature $X \in \mathbb{R}^{m \times n}$, where m is the number of nodes and n denotes the features.

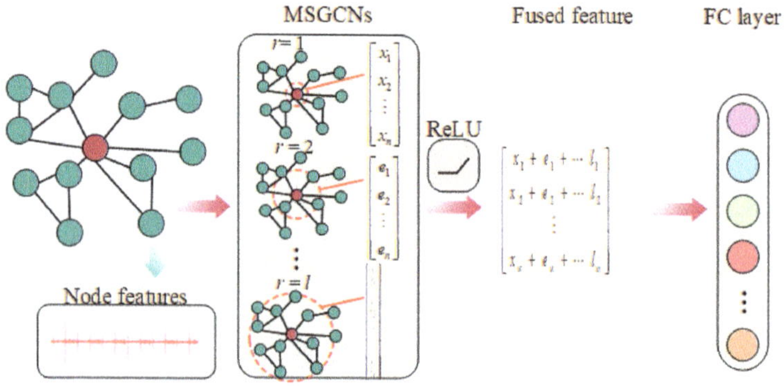

Figure 5. The framework of the MSGCNs.

The MSGCNs first execute multiple spectral convolution operations through different receptive sizes (e.g., r = 1, 2, . . ., l is applied in this paper), and then fuse and stack the features extracted by different receptive sizes to obtain more complementary and powerful feature representations. Therefore, the operation process of the MSGCNs can be defined as

$$\boldsymbol{H} \approx \sigma\left(\left[\sum_{k_1=0}^{K_1} \theta_{k_1} T_{k_1}(\hat{\boldsymbol{L}})\boldsymbol{X}; \sum_{k_2=0}^{K_2} \theta_{k_2} T_{k_2}(\hat{\boldsymbol{L}})\boldsymbol{X}; \cdots ; \sum_{k_l=0}^{K_l} \theta_{k_l} T_{k_l}(\hat{\boldsymbol{L}})\boldsymbol{X}\right]\right) \in \mathbb{R}^{m \times 3l} \quad (9)$$

where $\boldsymbol{H}$ denotes the fused multi-scale features, $\sigma(\cdot)$ means the activation function named rectified linear unit (ReLU), $[\cdot]$ is the concatenation algorithm, and l represents the number of the multiple receptive sizes. $\theta_{k_1}, \theta_{k_2}, \ldots, \theta_{k_l}$ are the learnable parameters of the MSGCNs. $K_1, K_2, \ldots, K_l$ denote the different spectral convolutional kernel sizes. $k_1, k_2, \ldots, k_l$ represent the kth-order Chebyshev polynomials in the Laplacian spectral domain.

3.3. Multi-Sensor Information Fusion Framework

Compared with single-sensor data, the advantage of multi-sensor data is that they often comprise more valuable, robust, and complementary data about the abnormal vibration of the rolling mill. Meanwhile, multi-sensor data can also improve the accuracy of the diagnosis models to improve the reliability and confidence of the model decision and enhance the generalization and adaptability. Hence, performing discriminative feature extraction is a very essential step in the multimodal data field. We adopted adaptive graph construction for multi-source sensors, and then used different MSGCNs for different graphs to extract multi-scale features, respectively. Finally, the proposed M^{2}AGCNs fused the features extracted from different MSGCNs at the feature level to implement multi-sensor feature fusion by the original signals. Most importantly, compared to the other fault diagnosis methods, the proposed MSAGCNs can not only extract multi-scale and multi-level features of data in non-Euclidean space, but also realize the multi-spatial information fusion from multi-source sensors. The multi-sensor information processing frame can be shown in Figure 6.

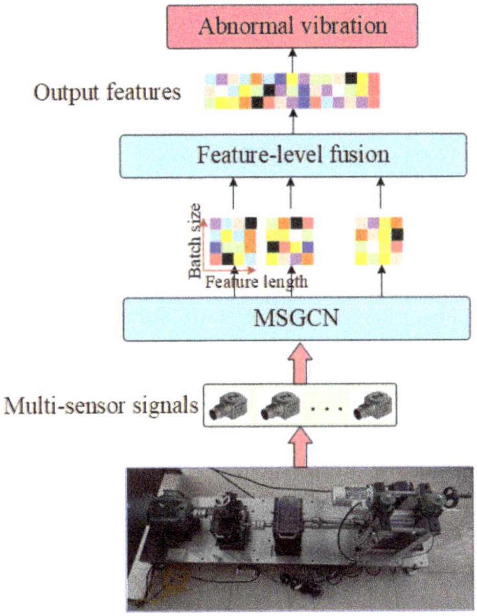

Figure 6. The multi-sensor information fusion framework.

The proposed MSGCNs are used to process the multi-sensor data and then extract reliable, overall, and powerful features to fusion and then obtain the desired classification and identification of the abnormal vibration results of the rolling mill based on the resulting features. First, the forward learning process of the proposed M^{2}AGCNs is composed of the individual features extracted by the different MSGCNs in the early stage, which can be defined as

$$f(x,y) = F(x,\theta) \tag{10}$$

where x denotes the original single signals, and θ represents the learnable model parameters.

Then, the feature-level fusion (FF) layer is employed to process the original signals for further fusing the different features extracted from the different MSGCNs, which can be expressed as follows:

$$h(x,y) = \left[\sum_{i=1}^{n} f(x,y),\theta\right] \tag{11}$$

where θ means the FF layer parameter.

Then, to extract and learn deeper and abstract features further, the output of the FF layer is fed into the fully connected (FC) layer as the input, which can be expressed as

$$g(x,y) = f(h(x,y),\theta) \tag{12}$$

where θ means the FC layer learning parameter.

Next, the final feature vectors are obtained and input into the classification layer, and the outcome of the proposed M^{2}AGCNs through the Soft-max function is described as

$$O^L = \frac{\exp[g(x,y,\theta^t_{FC,1},k)]}{\sum_{j=1}^{K}\exp[g(x,y,\theta^t_{FC,1},j)]} \tag{13}$$

where O^L means the output of the whole framework, and K denotes the number of categories of abnormal vibration types.

The essence of the forward propagation is to fuse the different features obtained from multi-sensor signals and then to output the abnormal vibration diagnosis results of the rolling mill. Finally, the parameters of the proposed M^{2}AGCNs can be updated iteratively by the following defined loss function, thus it can be written as

$$L(h(x,y,\theta),z) = -\frac{1}{m}\sum_{i=1}^{m}\sum_{j=1}^{K} z_j \ln g(x,y,\theta,k) \tag{14}$$

where z means the true abnormal vibration type and m is the sample number.

The M^{2}AGCN method for detecting the abnormal vibration types of the rolling mill is shown in this section. Figure 7 demonstrates the entire process from data acquisition to final vibration recognition and then complies with the following steps provided in this paper.

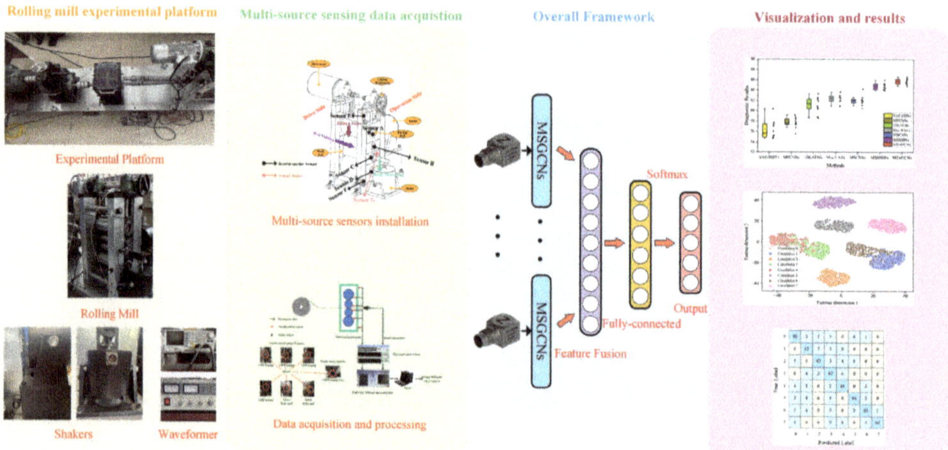

Figure 7. The overall framework of this paper. Herein, the yellow block shows the rolling mill experimental platform. The green block represents the locations of multi-sensor and data acquisition and processing systems. The blue block indicates the overall framework of the proposed method. The pink block describes the visualization results.

Step 1: First, we locate multiple sensors I~VI on the rolling mill stand and three case studies are then conceived in order to produce different abnormal vibration types.

Step 2: Second, we use the data acquisition system to collect multi-sensor signals of the abnormal vibration of different locations based on the rolling mill.

Step 3: After that, the whole parameters of the proposed M^{2}AGCN method are trained and updated in order to learn robust and valuable multiple features through the multiple signals.

Step 4: Next, through three abnormal vibration case studies, we evaluate the suggested M^{2}AGCNs compared with the state-of-the-art (SOTA) methods based on different types of abnormal vibration of the rolling mill and then show the effectiveness and advancement of this method in this paper.

4. Rolling Mill Laboratory Equipment and Information Collecting Device

The abnormal vibration signals were collected by multiple accelerometers from a rolling mill laboratory equipment, which includes two backup rollers of 120 mm and two work rollers of 60 mm, respectively, as shown in Figure 8. The rolling mill experimental platform contains three key components: (i) an experimental bench, which comprises four main parts—a rolling mill, a planetary gearbox, a reduction gearbox, and a driving motor; (ii) a multi-sensor information-collecting device as exhibited in Figure 8b; and (iii) a computer console as seen in Figure 8c.

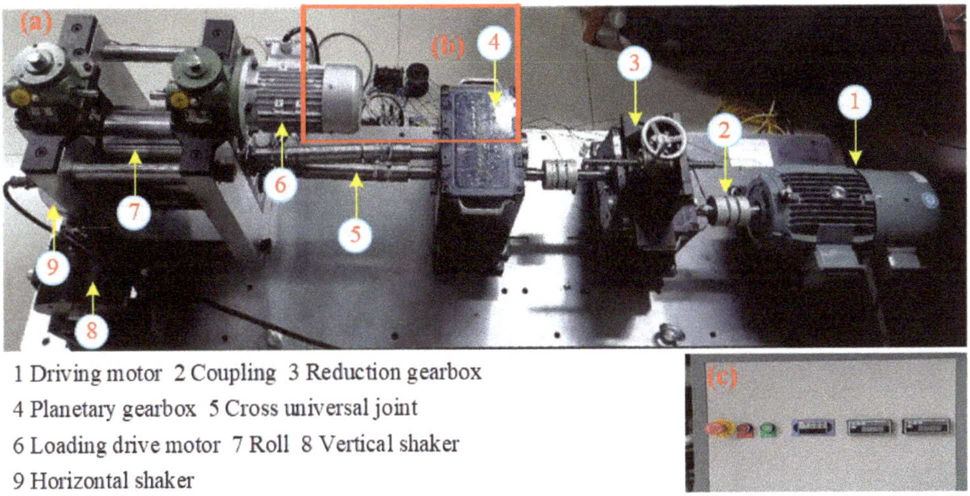

Figure 8. Rolling mill vibration experimental platform. (**a**) rolling mill experimental platform (**b**) multi sensor information data acquisition system (**c**) control platform.

To fully acquire multiple complementary and redundant vibration signals from six sensors, we designed the arrangements of the sensors in detail. The elaborate sketch of the rolling mill and the specific placements of multi-source sensors are displayed in Figure 9.

Various sensors are mounted in the Y-direction at different locations on the rolling mill stand to acquire various vibration signals. It can be seen that Sensor I is located at the lateral shaft end seat of the upper backup roller; Sensor II is situated at the side-end of the upper work roll to further collect original signals; Sensor III is located outside the bearing of the lower work roller. Sensor IV is located on the outer side of the rolling mill stand of the lower backup roll. Sensors V and VI are located on the outer side of the upper and lower part of the rolling mill frame, respectively.

Figure 10 displays a diagram of the actual installation of several sensors, the collection process of the multi-sensor data, and the treatment procedure. In addition, the signal waveforms of the sensor data are simultaneously visualized on the on-site computer, while the information is recorded in digital storage.

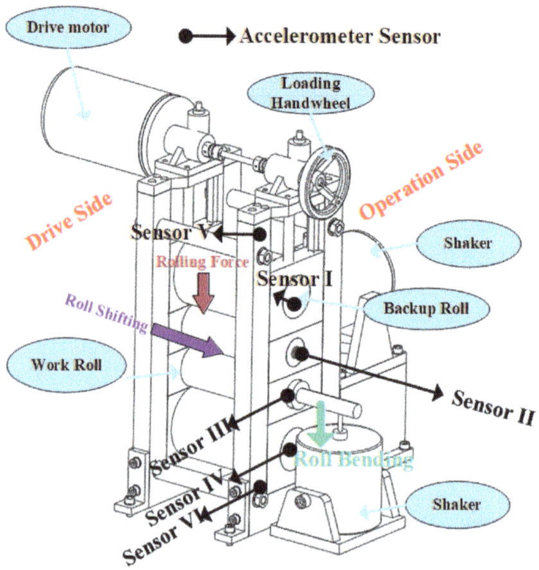

Figure 9. The sketch of the rolling mill and the detailed locations of sensors.

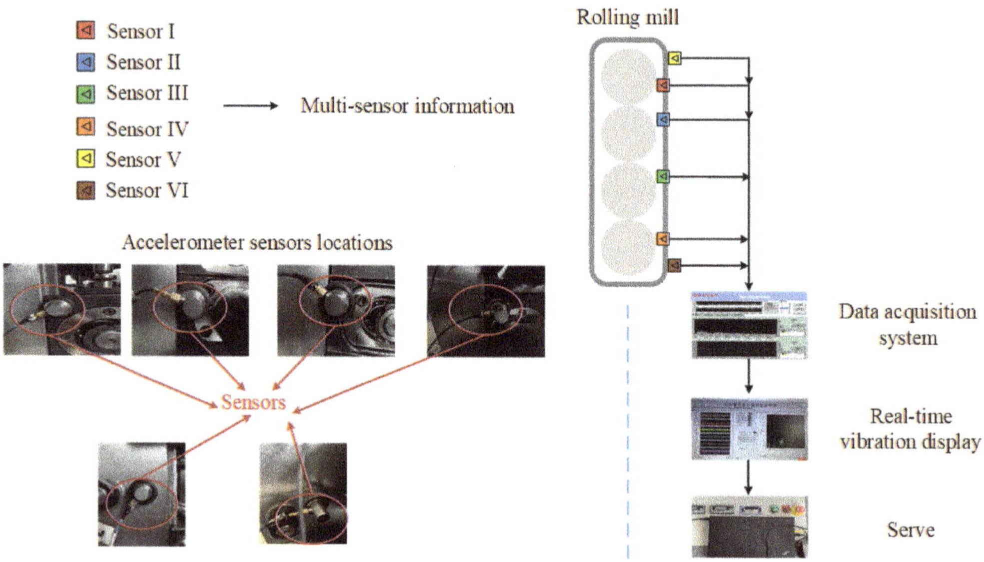

Figure 10. The information collection process.

5. Case Study

5.1. Experimental Designs and Data Processing

The various abnormal vibration experiments of the rolling mill were performed in December 2023 and then the collection process lasted for two months. The sampling frequency of all the sensors was set as 10,240 Hz. Then, the multi-sensor signals collection process could be divided into three parts: (1) horizontal abnormal vibration, (2) vertical abnormal vibration, and (3) coupling abnormal vibration. We added two shakers on the base in the radial and axial ends of the stand of the rolling mill laboratory equipment to excite the stand vibration. In other words, they are employed to model the horizontal abnormal vibration, vertical abnormal vibration, and coupling abnormal vibration produced by the rolling process in a real environmental condition, as shown in Figure 11. Meanwhile, the process of collecting multi-sensor signals for the three different case studies is the same. The speed of the rollers used to collect the multi-sensor signals was set at 600 rpm, as this speed was chosen to simulate the operating conditions typical of the rolling mill. The multi-sensor signals collected for these case studies lasted ten seconds, a duration chosen to capture sufficient data for fault analysis while accounting for the operational constraints of the rolling mill.

(**a**) The axial shaker (**b**) The radial shaker

Figure 11. The two shakers.

The multi-sensor signals collected from three case studies are normalized using the Z-score normalization technique. Next, the original vibration signals are divided and segmented using a sliding window length of 1024 to form subsamples without overlap. The proposed method and SOTA methods are written in Python 3.8, and then three case studies are carried out on a Windows 11 operating system, an Intel I7-11800H CPU (Intel, Santa Clara, CA, USA), a GeForce RTX 1060 GPU (NVIDIA, Santa Clara, CA, USA), and a Pytorch machine learning framework.

5.2. Case Study 1: Horizontal Abnormal Vibration

In this case study, four abnormal vibration types (i.e., 1, 10, 30, and 50 Hz) are stimulated by the axial shaker. At the same time, two load pressures (i.e., 0 and 1) are designed and then each workload can be regarded as a working state. Therefore, for vibration types combined with load pressures, there are eight types of rolling mill abnormal vibration conditions in total. Next, there are 500 samples for each rolling mill's abnormal vibration type and then 2000 samples in total for each working condition. In addition, 80% of the samples were randomly selected in each rolling mill abnormal vibration condition as the training dataset and the remaining 20% were used as the testing set. Consequently, for each rolling mill's abnormal vibration condition, a total of 400 samples were used for training the methods and 100 samples for testing to evaluate the methods. Specifically, the detailed description of the number of training and testing samples of the single sensor can be described in Table 1.

Table 1. Rolling mill health state datasets under the horizontal abnormal vibration.

Sensors	Load	Horizontal Frequency (Hz)	Label	Training Datasets	Testing Datasets
Sensor I~VI	0	1	0	400	100
		10	1	400	100
		30	2	400	100
		50	3	400	100
	1	1	4	400	100
		10	5	400	100
		30	6	400	100
		50	7	400	100

The time domain diagrams of the multi-sensor signals under the horizontal abnormal vibration frequency of 30 Hz and 0 load can be partly shown in Figure 12.

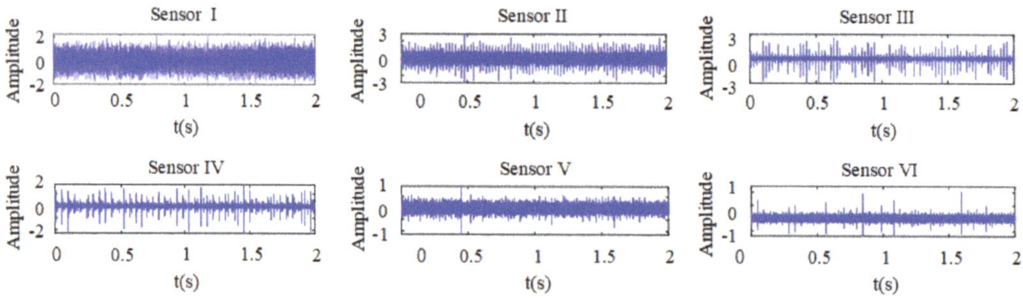

Figure 12. The multi-sensor signals under the horizontal abnormal vibration.

To further sufficiently illustrate the superiority of the proposed method M^2AGCNs for abnormal vibration diagnosis tasks, six SOTA methods are introduced and implemented. It is noteworthy that the network learning rate of our proposed method is designed as 0.001, and the training epochs are determined at 300. Meanwhile, the MSGCNs consist of two multi-scale graph convolutional layers, with the filters set to (256, 1200) and (1200, 300), respectively. The batch size is 64 and the momentum is 0.999. Each method was run 10 times to ensure the reliability and stability of the results. The brief descriptions of these mainstream methods can be summarized below:

(1) SAE-DBNs [32]: Two-layer SAEs are employed to extract shallow features and then the fused features are input to the DBNs to produce results.
(2) MPCNNs [33]: The multi-channel parallel convolutional neural network combined with the attention mechanism based on the number of sensors can be constructed to perform the rolling bearing fault diagnosis. The number of neurons in the two convolutional layers is set to 64 and 32, respectively. The batch size and maximum epochs are configured as 64 and 50, respectively.
(3) 1DLSTMs [34]: The one-dimension convolutional LSTMs are proposed to capture the spatial–temporal features of multi-sensor signals to diagnose the bearings. This 1DLSTM is composed of three one-dimensional convolutional layers, one fully connected layer, one output layer, and three LSTM layers. Herein, the kernel stride of the three convolutional layers is set to (16, 1, 1), and the number of nodes in the three LSTM layers is configured as (256, 128, 64). Specifically, the batch size is set to 128, and the dropout rate is 0.5.

(4) Mix-VAEs [35]: The mixture of Gaussians and variational autoencoders (VAEs) are combined to capture multi-source features to achieve fault diagnosis. The Mix-VAEs architecture consists of four one-dimensional convolutional layers, four pooling layers, and one fully connected layer. The rectified linear unit (ReLU) is used as the activation function. The learning rate and momentum are set to 0.1 and 0.8, respectively. Additionally, the trade-off parameters for reconstruction loss and weight loss are set to 1 and 0.01, respectively.

(5) MSICNNs [36]: The improved one-dimension CNNs and two-dimension CNNs are adopted to extract multi-source heterogeneous information (acoustic signals and vibration signals) to achieve the health state diagnosis of the rolling mill. The learning rate is set to 0.001, and the training process is conducted for 300 epochs. MSICNNs consist of six convolutional layers, four pooling layers, one dropout layer, and one output layer.

(6) MSIDBNs [37]: The improved DBNs are proposed to sufficiently exploit the useful information from individual sensors, and then the MSIDBNs are employed to extract and fuse multi-sensor signals for health states classification of rolling mill. The network learning rate of MSIDBNs is set to 0.001, with a total of 300 training epochs. MSIDBNs consist of three improved RBMs, with network parameters set to 512, 256, and 128, respectively.

The diagnosis result and their corresponding bar chart of all the methods are demonstrated in Table 2 and Figure 13. From the horizontal abnormal vibration diagnosis results, it can be observed that the SAE-DBN method reveals the poorest diagnosis performance, with only 85.374% diagnosis accuracy. And then, the diagnosis performance of the MPCNN method only improves on the former method by 2.518% to 87.892%. This indirectly confirms the phenomenon that traditional DL methods do not achieve satisfactory diagnosis results by simply stacking. Compared with the former methods, the 1DLSTM and Mix-VAE methods achieved diagnosis performance of 89.433% and 90.628%, respectively. Though these two methods combined with other mathematics techniques can boost the recognition performance of the methods to some extent, the performance can not guarantee the safety requirements. Next, the MSCNNs and MSIDBNs can improve the underlying structure of the methods and thus improve the valuable feature extraction of multi-sensor signals and their diagnosis results are 91.439% and 93.542%, respectively. However, they cannot extract the spatial–temporal features and correlations of multi-sensor signals, and thus fail to further improve the model performance. Finally, the proposed M^2AGCNs achieved optimal performance; the proposed M^2AGCNs can exhibit a comparatively constant property with an average diagnosis accuracy of 94.254%. This substantiates the superiority of the proposed method in the horizontal abnormal vibration diagnosis tasks.

Table 2. The detailed diagnosis results under the horizontal abnormal vibration.

Dataset Name	Models	Min-Acc (%)	Max-Acc (%)	Avg-Acc (%)
Case study 1	SAE-DBNs	83.592	86.219	85.374
	MPCNNs	85.483	88.734	87.892
	1DLSTMs	88.395	90.281	89.433
	Mix-VAEs	89.718	91.724	90.628
	MSCNNs	90.336	92.568	91.439
	MSIDBNs	93.257	94.117	93.542
	M^2AGCNs	93.129	94.731	94.254

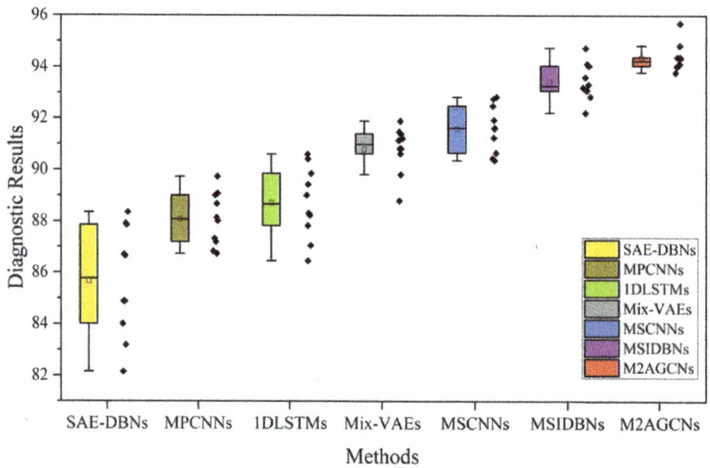

Figure 13. The avg-acc (%) box plot under the horizontal abnormal vibration.

To further research the classification accuracy of horizontal abnormal vibration diagnosis, the confusion matrix of the diagnosis results for different methods can be shown in Figure 14. In contrast, the proposed M^2AGCNs maintain a high classification level, achieving a recognition accuracy of over 90% on each abnormal vibration.

To present the identification of each health state more visually, the features extracted by each method at the 8 available labeled samples per class were visualized by the t-distributed stochastic neighbor embedding (t-SNE) technique, as depicted in Figure 15. The different colors and symbols represent different health states. As can be observed, among the features extracted by the proposed M^2AGCNs, those originating from different health states tend to diverge more from each other and have less confusion among them. Hence, the proposed M^2AGCNs can, thus, learn more discriminative features and accumulate diagnosis knowledge so as to achieve good generalization.

5.3. Case Study 2: Vertical Abnormal Vibration

In this case study, four different types of vertical abnormal vibration can be tested in the rolling mill. Meanwhile, two different roll load pressures (i.e., 0 and 1) are designed, so each roll load is viewed as the working condition. Hence, each working condition includes four vertical abnormal vibration types (i.e., 1, 10, 30, and 50 Hz). The Z-score normalization technique standardizes the gathered vibration signals of each data file. Subsequently, the same divide technique in Case Study 1 is adopted to intercept the subsamples. Thus, 500 samples are yielded for each health state and then 2000 samples are available for each working condition in total. In addition, 80% of the samples in each working condition are randomly selected as the training set, and the remaining samples are used as the testing set. More specifically, a detailed description of the number of training and testing samples for a single sensor can be described in Table 3.

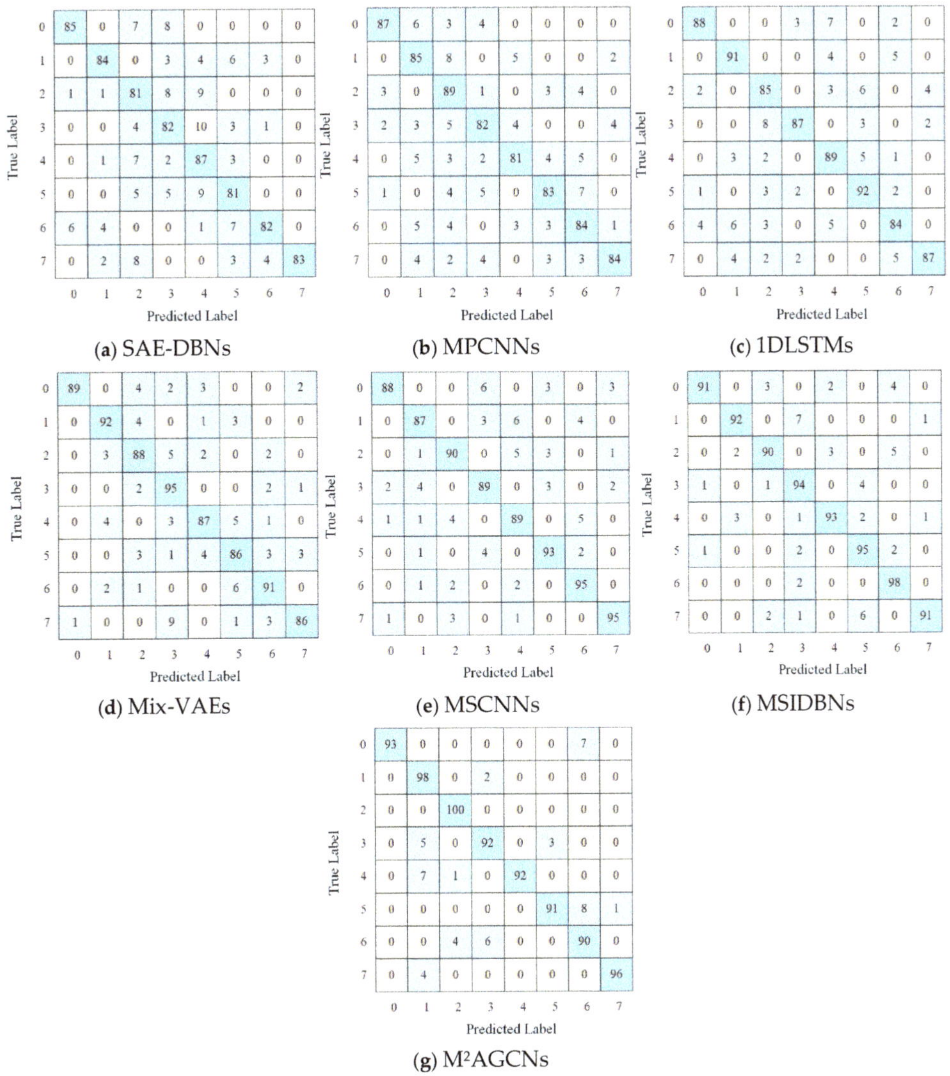

Figure 14. The confusion matrix of the diagnosis results for the horizontal abnormal vibration of the rolling mill.

Table 3. Rolling mill health state datasets under the vertical abnormal vibration.

Sensors	Load	Vertical Frequency (Hz)	Label	Training Datasets	Testing Datasets
Sensor I~VI	0	1	0	400	100
		10	1	400	100
		30	2	400	100
		50	3	400	100
	1	1	4	400	100
		10	5	400	100
		30	6	400	100
		50	7	400	100

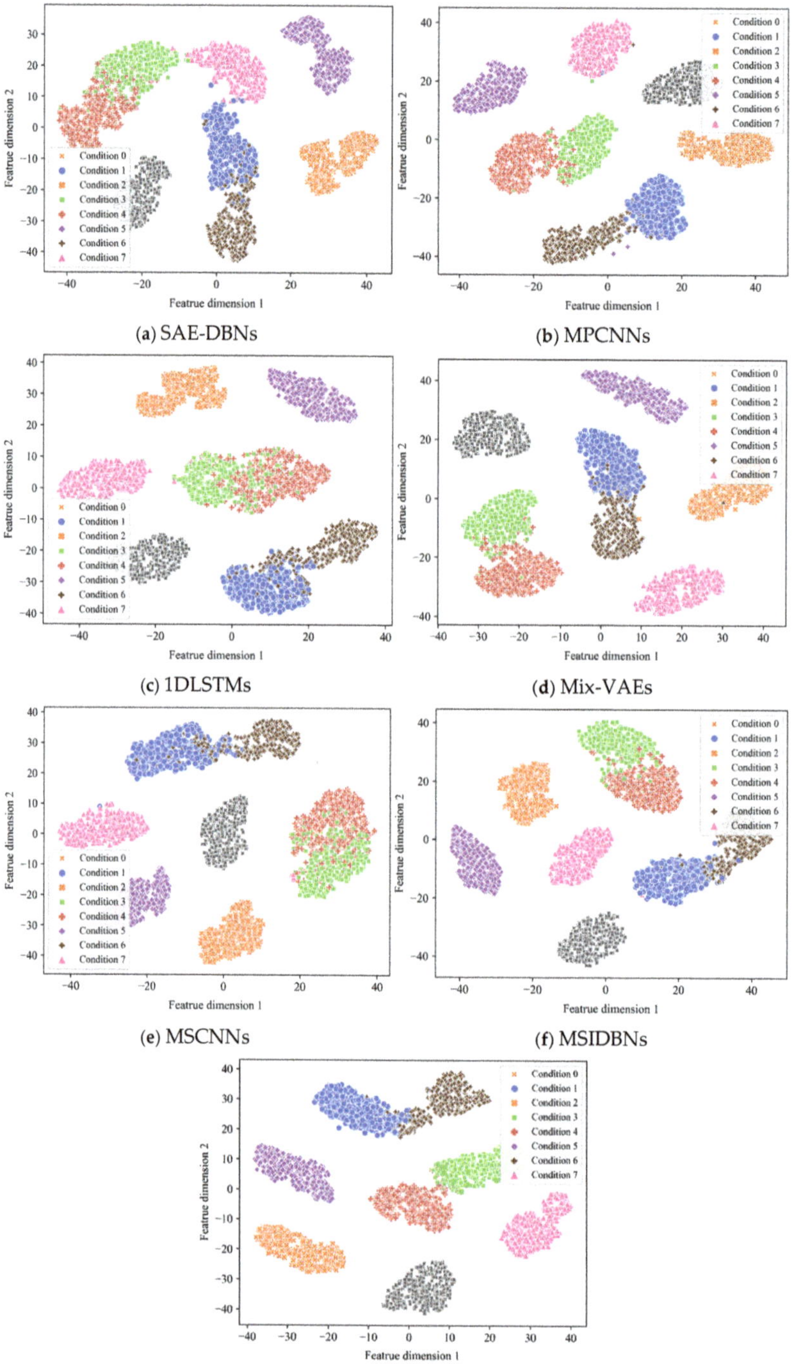

(**a**) SAE-DBNs (**b**) MPCNNs (**c**) 1DLSTMs (**d**) Mix-VAEs (**e**) MSCNNs (**f**) MSIDBNs (**g**) M²AGCNs

Figure 15. The visualization results of different methods under the horizontal abnormal vibration.

The time domain diagrams of the multi-sensor signals under the vertical abnormal vibration frequency of 30 Hz and 0 load can be partly shown in Figure 16.

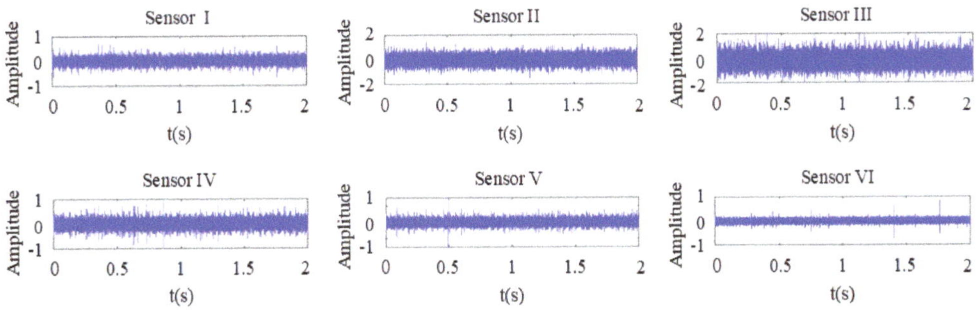

Figure 16. The multi-sensor signals under the vertical abnormal vibration.

The diagnosis results of the proposed M^2AGCNs and other comparative methods on the vertical abnormal vibration diagnosis task are demonstrated in Figure 17 and Table 4. Since multiple sensors were placed on the axial end stand of the rolling mill, it is more difficult for vertical abnormal vibration diagnosis for the multiple sensors to collect valuable information. As observed, the proposed M^2AGCNs still provide more efficient and stable diagnosis performance than the other SOTA methods for the more complex vertical abnormal vibration tasks. The average diagnosis accuracy of the proposed M^2AGCNs on the vertical abnormal vibration task was improved by 9.002% and 6.833% over the SAE-DBN method and the MPCNN method, respectively. In contrast to the 1DLSTM and Mix-VAE methods combined with math techniques, the proposed M^2AGCNs also achieved a more robust diagnosis performance of abnormal vibration, which rose by 5.394% and 4.292%, respectively. Finally, for the advanced MSCNN and MSIDBN methods, the proposed method can also achieve 91.734% diagnosis accuracy. Most importantly, the proposed method can achieve a high diagnosis performance, which mainly extracts the correlations between multi-sensor data and the multi-scale features.

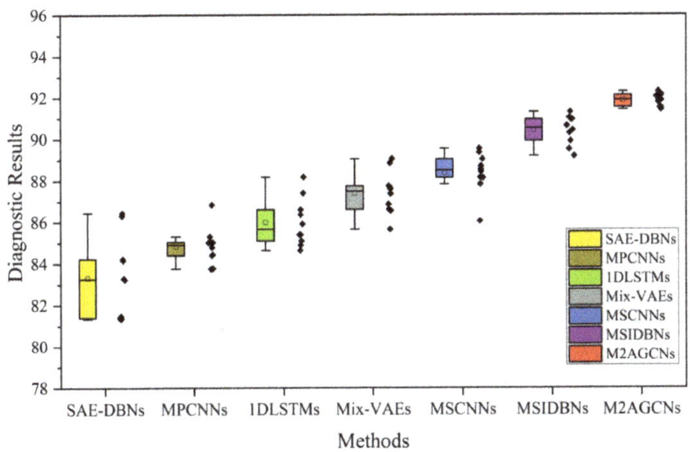

Figure 17. The avg-acc (%) box plot under the vertical abnormal vibration.

Table 4. The detailed diagnosis results under the vertical abnormal vibration.

Dataset Name	Models	Min-Acc (%)	Max-Acc (%)	Avg-Acc (%)
Case study 2	SAE-DBNs	80.375	83.747	82.732
	MPCNNs	82.954	85.431	84.901
	1DLSTMs	85.312	87.415	86.337
	Mix-VAEs	86.014	88.179	87.442
	MSCNNs	87.741	89.612	88.734
	MSIDBNs	87.730	91.194	90.456
	M^2AGCNs	90.339	92.586	91.734

Afterward, the t-SNE technique was employed to visualize the features extracted when vertical abnormal vibration occurs. The results of the feature visualization are depicted in Figure 18. To reflect the effectiveness of the M^2AGCN method and to further visualize the feature extraction capability on the basis of the multi-sensor original signals. It can be observed that the features extracted by these methods except the proposed M^2AGCNs cannot distinguish and visualize by t-SNE between the categories of the vertical abnormal vibration diagnosis task. Then, following the extraction of features by the proposed M^2AGCNs can be achieved to diagnose the abnormal vibration types.

To further reflect the vertical abnormal vibration type classification and identification results, we used the confusion matrix under multi-sensor information of the M^2AGCN method, which can be demonstrated in Figure 19. It can be concluded that type 6 of the vertical abnormal vibration has the highest diagnosis performance and type 7 of the abnormal vibration has the lowest diagnosis performance. Meanwhile, all the types of abnormal vibration have satisfactory diagnosis results.

To further investigate the sensitivity and robustness of the proposed method, receiver operating characteristic (ROC) curves are presented. It is worth noting that the ROC curves are generated from the predicted probability of the model for the sample category and the true label of the sample. The ROC curves of these methods can be shown in Figure 20. It is very clear from the ROC curves that the proposed method achieves a higher area under curve (AUC) with an AUC of about 0.9. This suggests that the proposed M^2AGCNs can effectively recognize various health states. Moreover, the micro-average index of 94.79% and the macro-average index of 94.61% further illustrate the stability and sensitivity of the proposed M^2AGCNs in industrial diagnosis.

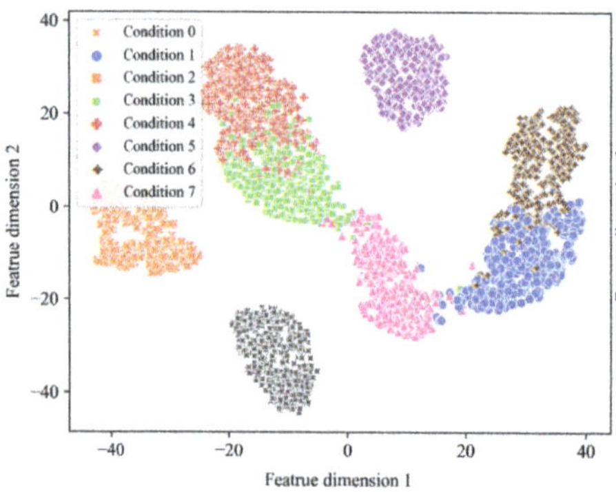

(a) SAE-DBNs

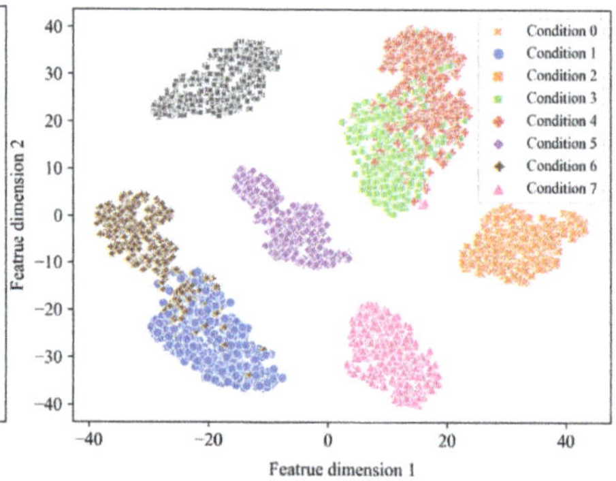

(b) MPCNNs

Figure 18. *Cont.*

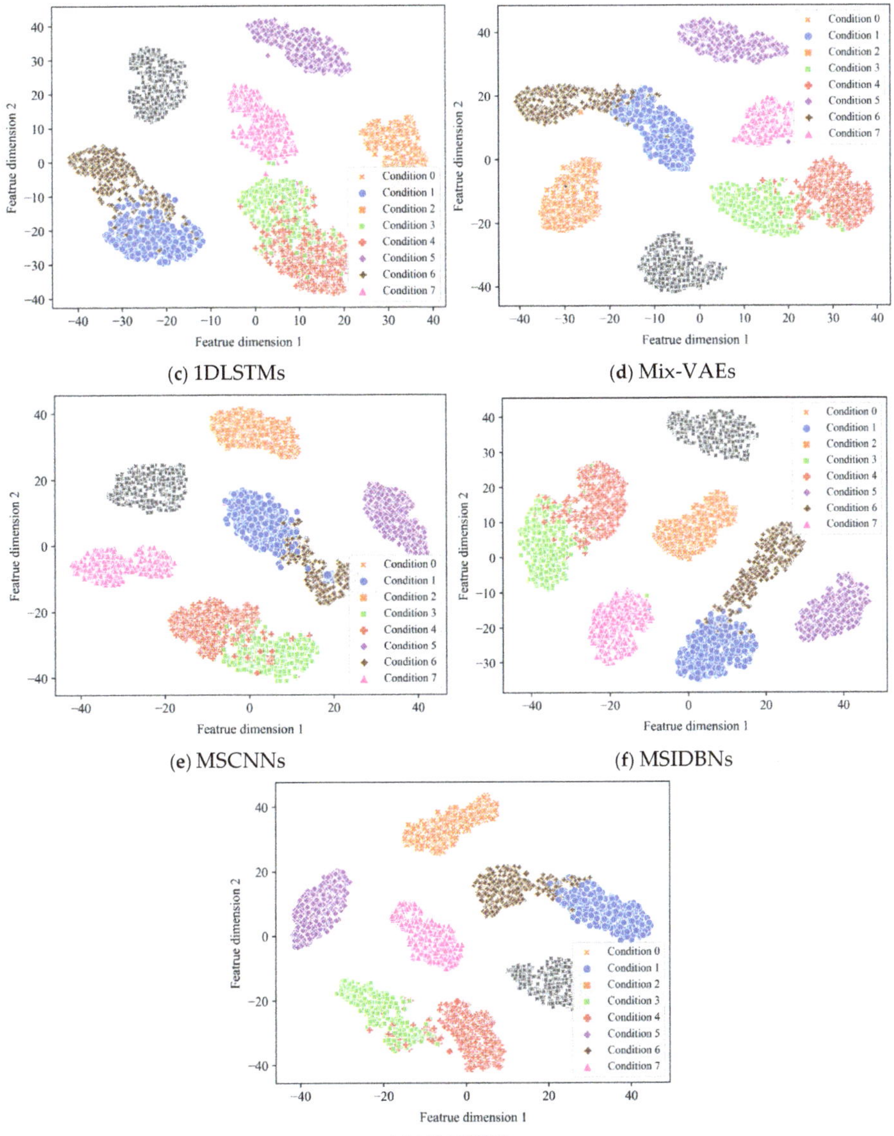

(c) 1DLSTMs (d) Mix-VAEs
(e) MSCNNs (f) MSIDBNs
(g) M²AGCNs

Figure 18. The visualized results of the different methods under the vertical abnormal vibration of the rolling mill.

5.4. Case Study 3: Coupling Abnormal Vibration

In this case study, four different types of coupling abnormal vibration can be stimulated in the rolling mill. The coupling abnormal vibration is mainly simulated by the excitation of two shakers on the base in the radial and axial ends of the stand of the rolling mill experimental platform at the same time. Meanwhile, two different roll load pressures (i.e., 0 and 1) are designed, so each roll load is viewed as the working condition. Hence, each working condition includes four coupling abnormal vibration types (i.e., 1, 10, 30, and 50 Hz). The Z-score normalization technique standardizes the gathered vibration signals of each data file. Subsequently, the raw vibration signal is segmented with the same divide technique to intercept the subsamples. Thus, 500 samples are yielded for each health state and then 2000 samples are available for each working condition in total. More specifically, a detailed description of the number of training and testing samples for a single sensor can be described in Table 5. The task of diagnosing the coupling abnormal vibration of the rolling mill is the most difficult because of the complexity and variability of coupling abnormal vibration situations and the fact that they are easily confused with other forms of vibration.

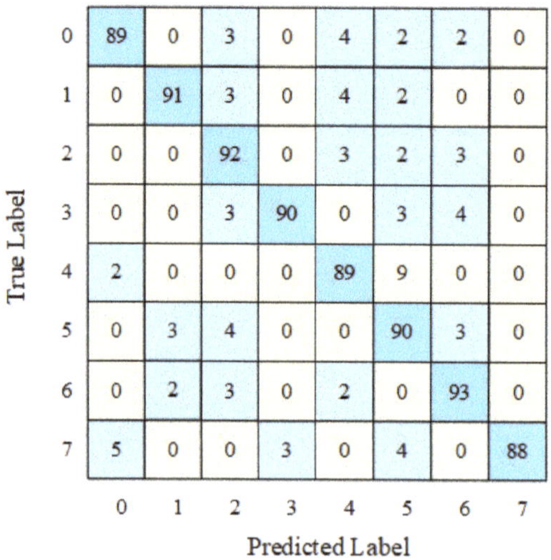

True \ Predicted	0	1	2	3	4	5	6	7
0	89	0	3	0	4	2	2	0
1	0	91	3	0	4	2	0	0
2	0	0	92	0	3	2	3	0
3	0	0	3	90	0	3	4	0
4	2	0	0	0	89	9	0	0
5	0	3	4	0	0	90	3	0
6	0	2	3	0	2	0	93	0
7	5	0	0	3	0	4	0	88

Figure 19. The confusion matrix of the proposed method under the vertical abnormal vibration.

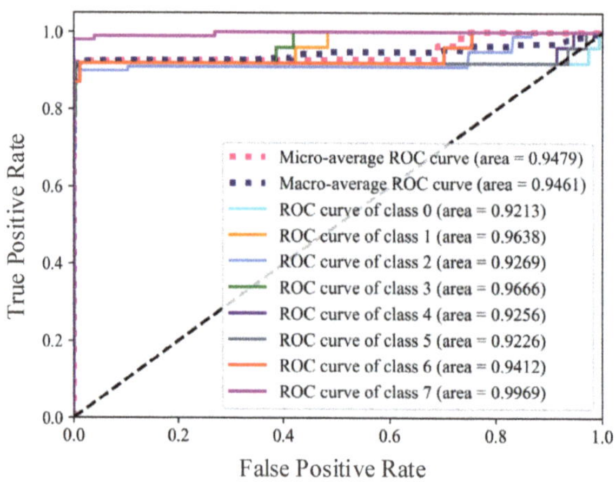

Figure 20. The ROC of the proposed method under the vertical abnormal vibration of the rolling mill.

Table 5. Rolling mill health state datasets under the coupling abnormal vibration.

Sensors	Load	Horizontal Frequency (Hz)	Vertical Frequency (Hz)	Label	Training Datasets	Testing Datasets
Sensor I~VI	0	1	1	0	400	100
		10	10	1	400	100
		30	30	2	400	100
		50	50	3	400	100
	1	1	1	4	400	100
		10	10	5	400	100
		30	30	6	400	100
		50	50	7	400	100

The time domain diagrams of the multi-sensor signals under the coupling abnormal vibration frequency of 30 Hz and 0 load can be partly shown in Figure 21.

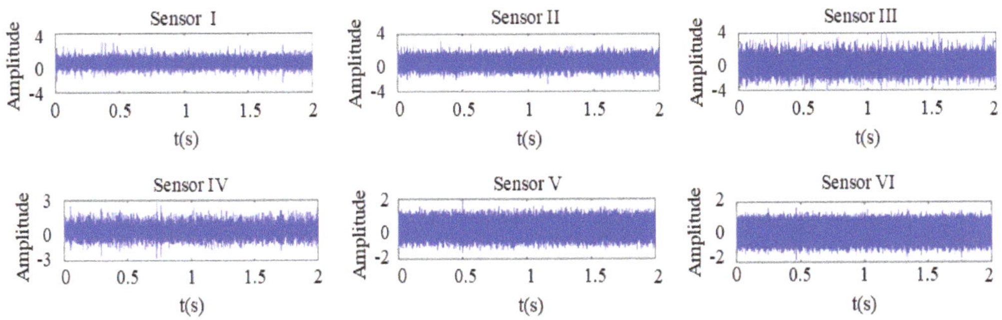

Figure 21. The multi-sensor signals under the coupling abnormal vibration of the rolling mill.

The diagnosis results and graph under the coupling abnormal vibration can be represented in Table 6 and Figure 22. Facing the hardest abnormal vibration diagnosis, it can be noted that the proposed M^2AGCNs increased by 8.94% over the lowest-performance SAE-DBN method and by 1.346% over the highest-performance MSIDBN method. In summary, the coupling abnormal vibration diagnosis task, compared with the other vibration diagnosis tasks, can contain the combination of vertical and horizontal vibrations, and the proposed M^2AGCNs are able to perform the extraction of promising features from multi-sensor signals and achieve the maximum classification results.

Table 6. The detailed diagnosis results under the coupling abnormal vibration.

Dataset Name	Models	Min-Acc (%)	Max-Acc (%)	Avg-Acc (%)
Case study 3	SAE-DBNs	75.672	77.154	76.791
	MPCNNs	77.192	79.651	78.359
	1DLSTMs	79.231	81.672	80.584
	Mix-VAEs	80.681	82.571	81.730
	MSCNNs	81.530	83.336	82.415
	MSIDBNs	83.512	85.415	84.385
	M^2AGCNs	84.783	86.461	85.731

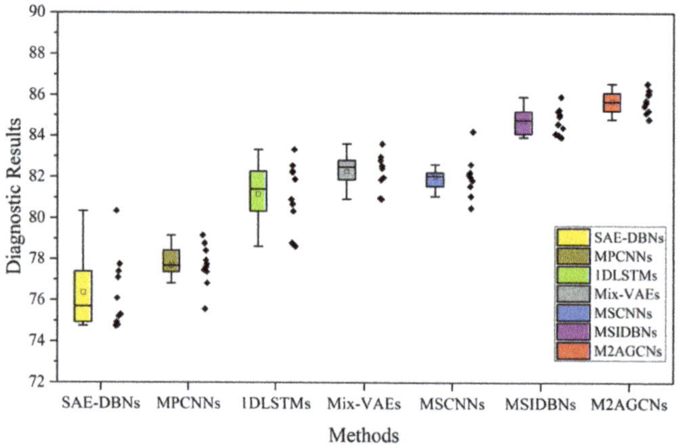

Figure 22. The avg-acc (%) box plot under the coupling abnormal vibration of the rolling mill.

The features extracted by the proposed M^{2}AGCNs per health status were visualized by the t-SNE technique, and the results are displayed in Figure 23. From the visualization results, the features extracted by the proposed M^{2}AGCNs have clearer class boundaries, except for a slight confusion between four health states, which can be well distinguished. This also confirms the superiority of the designed M^{2}AGCNs from the side.

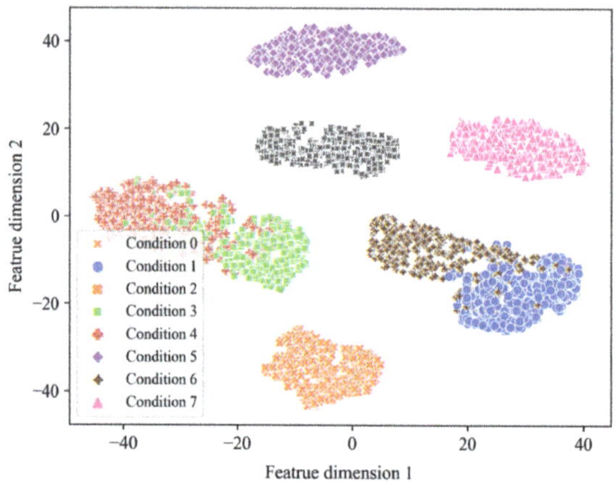

Figure 23. The visualized results of the proposed method under the coupling abnormal vibration.

The confusion matrix for the coupling abnormal vibration diagnosis results of the proposed M^{2}AGCNs is depicted in Figure 24. The confusion matrix illustrates the specific identification situation for each health status, from which one can observe that the proposed M^{2}AGCNs achieved a higher diagnostic accuracy overall than the other methods, with the identification accuracy for each abnormal vibration type reaching more than 80%. The confusion matrix results also indirectly demonstrate that the proposed M^{2}AGCNs can adequately extract and exploit the vibration information embedded in the multi-sensor signals and accomplish the abnormal vibration diagnosis task stably and accurately.

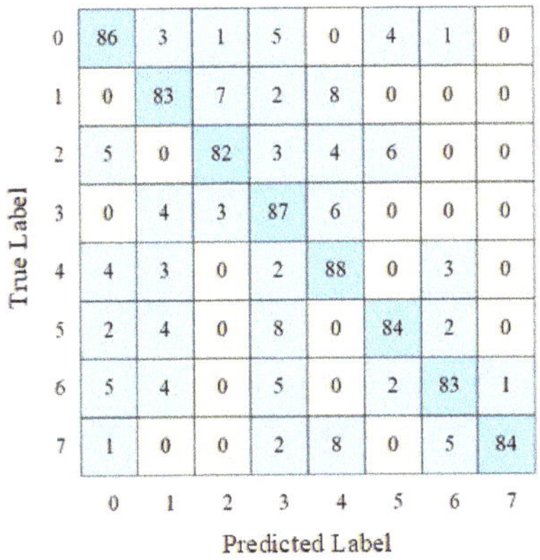

Figure 24. The confusion matrix of the proposed method under the coupling abnormal vibration.

6. Discussions

6.1. Influence of Rolling Speed

The rolling rpm plays a significant role in the abnormal vibration generation during the production process of the rolling mill, which identifies the absence and presence of the abnormal vibration: horizontal abnormal vibration, vertical abnormal vibration, and coupling abnormal vibration. Compared to the other two types of abnormal vibration, horizontal abnormal vibration is more prevalent and commonly observed in real plant environments. Hence, we should investigate the influence of the rolling rpm on different locations of horizontal abnormal vibration to propose future plans for subsequent vibration studies, as shown in Figure 25.

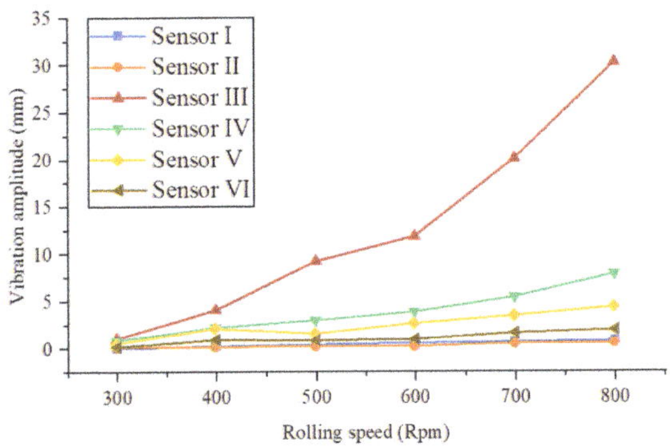

Figure 25. Time series vibration data collected by different sensors.

As can be seen in Figure 25, the amplitude of the abnormal vibration of Sensor I~VI increases with the increase in the rolling speed. In particular, Sensor III is much more susceptible to abnormal vibration magnitudes than many other sensors. As the lower working rolls of the mill are exposed under the combined influence of the rolling pressure from the mechanical system, the interactive stress amongst the roll system and the friction between the roll seams. Hence, as the rolling speed of the rolling mill increases, the amplitude of the abnormal vibration becomes greater and more sensitive. As a result, Sensor III contains the most meaningful information about abnormal vibrations compared to the other sensors. When the number of sensors is limited, we should focus on Sensor III. The use of multiple sensors enables comprehensive system monitoring, as each captures different aspects of vibration behavior. This is crucial in complex environments, where a single sensor might miss subtle or transient fault features. By integrating data from multiple sensors, the model can leverage a wider range of features, enhancing generalization and reducing the risk of undetected faults. When sensors are abundant, their installation becomes flexible, providing varied perspectives on the system's vibration and facilitating the detection of different fault types.

6.2. Ablation Experiment

To further demonstrate the superiority of the proposed M^{2}AGCNs, the ablation experiment is performed. The low-version methods come from the proposed method: (1) Method 1: The M^{2}AGCNs with the normal graph construction and (2) Method 2: The M^{2}AGCNs without the MSGCNs. The experiments on abnormal vibration diagnosis are performed to compare three methods, the results are shown in Table 7.

Table 7. Experimental results of ablation experiment.

Task	Method 1	Method 2	M^{2}AGCNs
Case 1	93.361%	92.562%	94.254%
Case 2	90.239%	89.663%	91.734%
Case 3	83.652%	82.681%	85.731%
Average	89.084%	88.302%	90.573%

It can be shown that the two essential components (the AFGC and MSGCNs) play a critical role in the proposed method, resulting in a 1.489% and 2.271% gain in accuracy, respectively. As a result, the proposed method outperforms the best.

7. Conclusions

In this study, a multi-source information fusion framework for the proposed M^{2}AGCNs is introduced to achieve the abnormal vibration diagnosis of rolling mills. First, the CNNs are applied to extract multiple features from the multi-sensor signals. Then, the AFGC is adapted to construct the graphs. After that, these MSGCNs mine complementary and enriched features embedded in the multi-sensor graphs through a multi-scale receptive field. Finally, a fusion layer is integrated into the network to trade off and fuse the multi-source features and identify and diagnose the various abnormal vibration types of the rolling mill. Finally, the cases of three types of abnormal vibration are introduced to show the effectiveness of the M^{2}AGCN method in this paper and that the proposed method on the basis of the multi-sensor information fusion can achieve the maximum abnormal vibration type diagnosis accuracy in comparison to the mainstream methods. In addition, detailed experiments on abnormal vibration based on the rolling mill and the specific locations of the sensors are analyzed.

In future work, to further enable real-time applications, we plan to explore model pruning and quantization techniques to reduce computational overhead while maintaining high diagnostic accuracy. Additionally, to validate its practicality, we will evaluate the proposed method in a real-time rolling mill scenario.

Author Contributions: Software, S.Z.; Data curation, C.G.; Writing—original draft, R.P.; Writing—review & editing, R.P.; Project administration, R.P.; Funding acquisition, R.P. All authors have read and agreed to the published version of the manuscript.

Funding: The studies were funded by the Science and Technology Research Project of Jiangxi Provincial Department of Education, grant numbers GJJ2402809 and GJJ2402814; the Initial Scientific Research Foundation for Talented Scholars of Nanchang Institute of Science and Technology, grant number NGRCZX-23-09; the Project of Nanchang Institute of Science and Technology, grant number NGKJ-23-01; and the Nonlinear Dynamics and Application Research Center, grant number NGYJZX-2021-04.

Data Availability Statement: Data are contained within the article.

Conflicts of Interest: The authors declare no conflict of interest.

References

1. Yu, Y.; He, Y.; Karimi, H.R.; Gelman, L.; Cetin, A.E. A two-stage importance-aware subgraph convolutional network based on multi-source sensors for cross-domain fault diagnosis. *Neural Netw.* **2024**, *179*, 106518. [CrossRef] [PubMed]
2. Han, D.; Zhang, Y.; Yu, Y.; Tian, J.; Shi, P. Multi-source heterogeneous information fusion fault diagnosis method based on deep neural networks under limited datasets. *Appl. Soft Comput.* **2024**, *154*, 111371. [CrossRef]
3. Rother, A.; Jelali, M.; Söffker, D. A brief review and a first application of time-frequency-based analysis methods for monitoring of strip rolling mills. *J. Process. Control.* **2015**, *35*, 65–79. [CrossRef]
4. Yu, Y.; Lan, L.; Ding, F.; Wang, L. Mechanical properties of hot-rolled and cold-formed steels after exposure to elevated tem-perature: A review. *Const Build Mater.* **2019**, *213*, 360–376. [CrossRef]
5. Zhang, X.; Gao, Y.; Guo, Z.; Zhang, W.; Yin, J.; Zhao, W. Physical model-based tool wear and breakage monitoring in milling process. *Mech. Syst. Signal Process.* **2022**, *184*, 109641. [CrossRef]
6. Du, Y.; Song, Q.; Liu, Z. Prediction of micro milling force and surface roughness considering size-dependent vibration of micro-end mill. *Int. J. Adv. Manuf. Tech.* **2022**, *119*, 5807–5820. [CrossRef]
7. Liu, Z.; Chai, T.; Tang, J.; Yu, W. Heterogeneous selective ensemble learning model for mill load parameters forecasting by using multiscale mechanical frequency spectrum. *Soft Comput.* **2022**, *26*, 13467–13484. [CrossRef]
8. Lu, X.; Sun, J.; Song, Z.; Li, G.; Wang, Z.; Hu, Y.; Wang, Q.; Zhang, D. Prediction and analysis of cold rolling mill vibration based on a data-driven method. *Appl. Soft Comput.* **2020**, *96*, 106706. [CrossRef]
9. Pan, J.; Chen, J.; Zi, Y.; Yuan, J.; Chen, B.; He, Z. Data-driven mono-component feature identification via modified nonlocal means and MEWT for mechanical drivetrain fault diagnosis. *Mech. Syst. Signal Process.* **2016**, *80*, 533–552. [CrossRef]
10. Li, K.; Qiu, C.; Zhou, X.; Chen, M.; Lin, Y.; Jia, X.; Li, B. Modeling and tagging of time sequence signals in the milling process based on an improved hidden semi-Markov model. *Expert Syst. Appl.* **2022**, *205*, 117758. [CrossRef]
11. Lei, Y.; Yang, B.; Jiang, X.; Jia, F.; Li, N.; Nandi, A.K. Applications of machine learning to machine fault diagnosis: A review and roadmap. *Mech. Syst. Signal. Pract.* **2020**, *138*, 106587. [CrossRef]
12. Tian, J.; Han, D.; Li, M.; Shi, P. A multi-source information transfer learning method with subdomain adaptation for cross-domain fault diagnosis. *Knowl.-Based Syst.* **2022**, *243*, 108466. [CrossRef]
13. Yu, J.; Liu, G. Knowledge extraction and insertion to deep belief network for gearbox fault diagnosis. *Knowl.-Based Syst.* **2020**, *197*, 105883. [CrossRef]
14. Lu, K.-D.; Zeng, G.-Q.; Luo, X.; Weng, J.; Luo, W.; Wu, Y. Evolutionary Deep Belief Network for Cyber-Attack Detection in Industrial Automation and Control System. *IEEE Trans. Ind. Inform.* **2021**, *17*, 7618–7627. [CrossRef]
15. Zhang, P.; Gao, D.; Lu, Y.; Ma, Z.; Wang, X.; Song, X. Cutting tool wear monitoring based on a smart toolholder with embedded force and vibration sensors and an improved residual network. *Measurement* **2022**, *199*, 111520. [CrossRef]
16. Shah, M.; Vakharia, V.; Chaudhari, R.; Vora, J.; Pimenov, D.Y.; Giasin, K. Tool wear prediction in face milling of stainless steel using singular generative adversarial network and LSTM deep learning models. *Int. J. Adv. Manuf. Technol.* **2022**, *121*, 723–736. [CrossRef]
17. Xu, G.; Liu, M.; Jiang, Z.; Shen, W.; Huang, C. Online Fault Diagnosis Method Based on Transfer Convolutional Neural Networks. *IEEE Trans. Instrum. Meas.* **2020**, *69*, 509–520. [CrossRef]
18. Zhao, C.; Sun, J.; Lin, S.; Peng, Y. Rolling mill bearings fault diagnosis based on improved multivariate variational mode de-composition and multivariate composite multiscale weighted permutation entropy. *Measurement* **2022**, *195*, 111190. [CrossRef]
19. Zhang, D.; Stewart, E.; Entezami, M.; Roberts, C.; Yu, D. Intelligent acoustic-based fault diagnosis of roller bearings using a deep graph convolutional network. *Measurement* **2020**, *156*, 107585. [CrossRef]
20. Yang, C.; Zhou, K.; Liu, J. SuperGraph: Spatial-Temporal Graph-Based Feature Extraction for Rotating Machinery Diagnosis. *IEEE Trans. Ind. Electron.* **2022**, *69*, 4167–4176. [CrossRef]
21. Liu, J.; Zhou, K.; Yang, C.; Lu, G. Imbalanced fault diagnosis of rotating machinery using autoencoder-based SuperGraph feature learning. *Front. Mech. Eng.* **2021**, *16*, 829–839. [CrossRef]
22. Liu, C.; Tong, J.; Zheng, J.; Pan, H.; Bao, J. Rolling bearing fault diagnosis method based on multi-sensor two-stage fusion. *Meas. Sci. Technol.* **2022**, *33*, 125105. [CrossRef]
23. Huo, Z.; Martinez-Garcia, M.; Zhang, Y.; Shu, L. A Multisensor Information Fusion Method for High-Reliability Fault Diagnosis of Rotating Machinery. *IEEE Trans. Instrum. Meas.* **2022**, *71*, 1–12. [CrossRef]

24. Huang, H.; Tang, B.; Luo, J.; Pu, H.; Zhang, K. Residual Gated Dynamic Sparse Network for Gearbox Fault Diagnosis Using Multisensor Data. *IEEE Trans. Ind. Inform.* **2022**, *18*, 2264–2273. [CrossRef]
25. Cai, B.; Liu, Y.; Fan, Q.; Zhang, Y.; Liu, Z.; Yu, S.; Ji, R. Multi-source information fusion based fault diagnosis of ground-source heat pump using Bayesian network. *Appl. Energy* **2014**, *114*, 1–9. [CrossRef]
26. Chen, F.; Yuan, Z.; Huang, Y. Multi-source data fusion for aspect-level sentiment classification. *Knowl.-Based Syst.* **2020**, *187*, 104831. [CrossRef]
27. Rezaeianjouybari, B.; Shang, Y. A novel deep multi-source domain adaptation framework for bearing fault diagnosis based on feature-level and task-specific distribution alignment. *Measurement* **2021**, *178*, 109359. [CrossRef]
28. Khan, M.R.; Blumenstock, J.E. Multi-GCN: Graph Convolutional Networks for Multi-View Networks, with Applications to Global Poverty. *Proc. AAAI Conf. Artif. Intell.* **2019**, *33*, 606–613. [CrossRef]
29. Kipf, T.; Fetaya, E.; Wang, C.; Welling, M.; Zemel, R. Neural relational inference for interacting systems. In Proceedings of the 35th International Conference on Machine Learning, Stockholm, Sweden, 10–15 July 2018; pp. 2688–2697.
30. Kipf, T.; Welling, M. Variational graph auto-encoders. *arXiv* **2016**, arXiv:1611.07308.
31. Li, T.; Zhao, Z.; Sun, C.; Yan, R.; Chen, X. Domain Adversarial Graph Convolutional Network for Fault Diagnosis Under Variable Working Conditions. *IEEE Trans. Instrum. Meas.* **2021**, *70*, 3515010. [CrossRef]
32. Li, J.; Li, X.; He, D.; Qu, Y. A Novel Method for Early Gear Pitting Fault Diagnosis Using Stacked SAE and GBRBM. *Sensors* **2019**, *19*, 758. [CrossRef] [PubMed]
33. Xia, P.; Huang, Y.; Li, P.; Liu, C.; Shi, L. Fault knowledge transfer assisted ensemble method for remaining useful life prediction. *IEEE Trans. Ind. Inform.* **2021**, *18*, 1758–1769. [CrossRef]
34. Hao, S.; Ge, F.-X.; Li, Y.; Jiang, J. Multisensor bearing fault diagnosis based on one-dimensional convolutional long short-term memory networks. *Measurement* **2020**, *159*, 107802. [CrossRef]
35. Wang, C.; Xin, C.; Xu, Z.; Qin, M.; He, M. Mix-VAEs: A Novel Multisensor Information Fusion Model for Intelligent Fault Diagnosis. *Neurocomputing* **2022**, *492*, 234–244. [CrossRef]
36. Shi, P.; Yu, Y.; Gao, H.; Hua, C. A novel multi-source sensing data fusion driven method for detecting rolling mill health states under imbalanced and limited datasets. *Mech. Syst. Signal Process.* **2022**, *171*, 108903. [CrossRef]
37. Yu, Y.; Shi, P.; Tian, J.; Xu, X.; Hua, C. Rolling mill health states diagnosing method based on multi-sensor information fusion and improved DBNs under limited datasets. *ISA Trans.* **2023**, *134*, 529–547. [CrossRef]

machines

MDPI

Article

A Dual-Level Intelligent Architecture-Based Method for Coupling Fault Diagnosis of Temperature Sensors in Traction Converters

Yunxiao Fu, Qiuyang Zhou * and Haichuan Tang

CRRC Industrial Academy Co., Ltd., Beijing 100071, China; yunxiaof2012@163.com (Y.F.); thc@crrc.tech (H.T.)
* Correspondence: qiuyangzhou@my.swjtu.edu.cn

Abstract: To address the coupled fault diagnosis challenge between temperature sensors and equipment in traction converter cooling systems, this paper proposes a dual-level intelligent diagnostic architecture. This method achieves online sensor fault isolation and early equipment anomaly warning by leveraging spatiotemporal correlation modeling of multimodal sensor data and ensemble learning-based prediction. At the first level, it integrates multi-source parameters such as outlet temperature and pressure to establish dynamic prediction models, which are combined with adaptive threshold mechanisms for detecting various sensor faults including offset, open-circuit, and noise interference. At the second level, it monitors the status of temperature sensors through time-series analysis of inlet temperature data. Verified on an edge computing platform, the proposed method effectively resolves the coupling misdiagnosis between sensor distortion and equipment faults while maintaining physical interpretability, thereby significantly enhancing diagnostic robustness under complex operating conditions.

Keywords: traction converter; temperature sensor; coupling fault diagnosis

1. Introduction

With the advancement of rail transit systems toward high-density and heavy-load operation, the reliability of traction converters has become a critical bottleneck affecting train operational safety and economic performance. As the core component for energy conversion, the power modules in converters generate thermal losses reaching megawatt levels under complex operating conditions [1]. The temperature monitoring network serves a dual role: acting as a real-time feedback channel for cooling system regulation and providing diagnostic features for equipment health assessment. However, the harsh operating conditions inherent to rail transit—such as high-frequency mechanical vibrations and strong electromagnetic interference—pose more severe reliability challenges for temperature sensors compared to industrial stationary scenarios. Sensor faults, such as drift or open-circuit faults, may cause the CCU (central control unit) to misinterpret the cooling system's status, potentially leading to abnormal shutdowns due to redundant protection mechanisms or, worse, masking actual equipment failures and causing catastrophic accidents. Notably, substantive faults within converters, such as insulated gate bipolar transistor (IGBT) module breakdowns, also manifest as abnormal temperature parameters. This creates a dual diagnostic dilemma: determining whether anomalous signals stem from sensor failures or reflect genuine equipment faults.

In the field of converter ontology fault diagnosis, physical modeling methods based on voltage/current signal analysis [2–4] and data-driven approaches [5–7] have reached a relatively mature stage. However, both types of methods are premised on the ideal assumption that the sensor system is entirely reliable, which may lead to the risk of misdiagnosis when sensor faults co-occur with equipment failures. Specifically, existing methods generally overlook two critical scientific challenges: First, sensor faults and equipment faults are strongly coupled in the time-frequency domain, making single-dimensional diagnostic models prone to generating "false-positive" alarms; second, the nonlinear heat transfer characteristics of the cooling system during start–stop transients and overload conditions complicate efforts to accurately decouple the real temperature field from sensor aberration signals using traditional static models.

Sensors, as critical sensing units in industrial control systems, directly influence operational safety through their reliability. In recent years, sensor fault diagnosis technologies have rapidly advanced for various applications, including motor drives, power electronic converters, and traction systems.

(1) Model-based methods

Model-based diagnostic methods achieve fault detection by constructing a dynamic system model and using an observer to generate residual signals [8]. These methods are widely applied in motor drives and power electronic systems. For instance, sliding mode observers [9–11] and Kalman filtering [12] enable current sensor fault detection by comparing the residuals of the feedback current with the reference value in matrix converter-permanent magnet synchronous motor (PMSM) drive systems. For induction motors and interior permanent magnet synchronous motors (IPMSMs), the strategy proposed by Dan et al. based on extended Kalman filtering (EKF) can simultaneously handle current and voltage sensor faults while reducing noise interference [13]. For the single-phase three-element current limiter (SPTL), Xu et al. designed a reduced-order observer [14], which resolves the coupling issue between sensor faults and IGBT open-circuit faults by reconstructing the system state and decoupling sensor faults from switching device faults via a state-space model. A mixed logic dynamic (MLD) model was established by Ge et al. to analyze the slope characteristics of residual change rates, combined with an adaptive threshold setting strategy, enabling the simultaneous differentiation of open-circuit faults in IGBTs and clamping diodes [15]. Additionally, a rapid sensor fault diagnosis method derived from parity space was proposed by Berriri et al., leveraging time-window redundancy and equivalent space analysis to ensure algorithm robustness against parameter variations while maintaining diagnostic effectiveness [16].

The model-driven approach offers advantages such as clear physical interpretability and suitability for systems requiring fast dynamic responses. However, its limitations include a reliance on precise mathematical models, sensitivity to parameter drift, and an accuracy that is heavily dependent on the precision of system models and parameter estimation.

(2) Data-driven methods

Data-driven methods leverage extensive historical data to extract deep knowledge characterizing implicit relationships among system variables through statistical analysis or artificial intelligence techniques, thereby training diagnostic models or algorithms. These approaches are particularly effective for complex nonlinear systems [17]. This category encompasses two primary technical pathways: the first involves purely data-driven multi-class fault identification, exemplified by the explainable artificial intelligence-low-cost sensors (XAI-LCS) framework proposed by Sinha and Das in [18], which directly employs the XGBoost algorithm to learn from raw data of low-cost sensors (LCS). By utilizing the SHAP (SHapley Additive exPlanations) framework to interpret model decisions, it detects various fault types, including offset, drift, and complete failure. While this approach adapts to complex nonlinear relationships without relying on physical models, it requires substantial labeled data and faces computational resource limitations in real-time applications. Dong et al. [19] presents a novel intelligent diagnostic method based on long short-term memory (LSTM) networks, which autonomously learns hidden fault characteristics from multi-sensor signals in traction converters, enabling comprehensive fault diagnosis for high-speed train traction converters. Haque et al. presents a lightweight framework named EnsembleXAI-Motor to achieve fault classification for electric vehicle traction motors through feature selection, ensemble learning, and explainable AI techniques, demonstrating enhanced classification performance while maintaining computational efficiency [20]. Wang et al. presents a novel bearing fault diagnosis method by integrating an improved convolutional neural network (CNN) with variational Bayesian inference to strengthen feature extraction capabilities, while adopting weighted voting strategies for the decision-level fusion of multi-sensor signals to improve diagnostic reliability [21]. Fast Fourier transform (FFT)-based feature extraction combined with Relieff algorithm-based feature selection was employed by Gou et al. for diagnosing and classifying sensor and IGBT faults in inverters [22]. These methods establish nonlinear mapping relationships between fault characteristics and fault types, demonstrating applicability for identifying and tracing multiple compound faults in complex systems.

The second pathway involves residual-based data-driven analysis, where residual signals are generated either through physical models or data-driven approaches, followed by threshold optimization for fault localization. For instance, Tao et al. establishes a DC-side voltage estimation model for three-level rectifiers, distinguishing between current and voltage sensor faults by analyzing voltage residual patterns and capacitor voltage differences using only single voltage residual signals, without requiring additional hardware [23]. Meanwhile, Tao et al. introduces a data-driven adaptive threshold learning mechanism in traction system inverters, achieving the joint diagnosis of multiple faults by integrating voltage residuals and current characteristics, effectively addressing misjudgment issues in multi-fault coupling scenarios that arise with conventional fixed thresholds [24]. A data-driven signal predictor was developed by Zhang et al. to monitor the sensor status of rectifiers, combining the strengths of both data-driven and model-based methods, enabling online sensor fault identification via residual evaluation [25]. For intermittent voltage sensor faults in traction inverters, regression prediction combined with sliding time windows was employed by Gou et al. to construct current residuals, and quantitative evaluation metrics were established based on the temporal characteristics of fault occurrence and disappearance to assess fault severity [26]. These methods integrate residual generation (physics-model-driven) with threshold decision-making (data-driven optimization), maintaining interpretability while enhancing computational efficiency, making them particularly suitable for embedded deployment.

The fundamental distinction lies in feature extraction: Pure data-driven methods process raw high-dimensional data directly, while residual analysis generates low-dimensional residual features through physical rules or simplified models. These two approaches exhibit complementary advantages in complex system diagnostics. The former excels at handling nonlinear high-dimensional problems, while the latter demonstrates superior performance in miniaturized deployment for systems with well-defined physical characteristics.

This study proposes a dual-level intelligent diagnostic architecture based on equipment–sensor coupling analysis to address these technical challenges. The framework adopts a hierarchical progressive decision-making mechanism: At the first level, it establishes an ensemble machine learning prediction model that enables the online detection and isolation of inlet water temperature sensor faults by leveraging spatiotemporal correlations among multimodal sensor data, including cooling system outlet water temperature and inlet/outlet pressures. At the second level, it develops a predictor-based approach that extracts dynamic evolution patterns from historical sensor data to provide early warning of temperature surge risks caused by converter abnormalities. The core advantage of this method lies in its dual-verification mechanism, which effectively resolves the coupled diagnostic challenge of sensor distortion masking equipment faults.

The technical innovations of this approach are highlighted in two key aspects:

(1) The ensemble learning-based predictor achieves a 12.5% reduction in root mean square error (RMSE) for temperature prediction compared to conventional single-model approaches.

(2) Breakthrough in edge computing real-time performance: The system demonstrates real-time inference capability when deployed on NVIDIA Jetson edge computing platforms, achieving significant improvements in computational efficiency.

2. Problem Description

The traction converter cooling system is a critical subsystem for ensuring the stable operation of electric multiple unit (EMU) core power equipment, utilizing closed-loop water circulation to achieve the precise thermal management of high-heat components such as IGBT power modules and transformers. The system consists of cooling pumps, radiators, dual-speed cooling fans, expansion tanks, piping networks, and distributed temperature/pressure sensor arrays. The cooling pump drives coolant circulation through the closed-loop system, where high-temperature coolant undergoes forced-air heat exchange via radiators. Temperature sensors, functioning as the system's sensory network, are embedded in distributed configurations along the coolant inlet/outlet pipelines to monitor real-time temperature gradient variations during heat transfer processes.

From a thermodynamic control perspective, as illustrated in Figure 1, the cooling system evaluates heat dissipation efficiency by measuring the differential temperature (ΔT) between inlet (T_{in}) and outlet (T_{out}) temperature sensors. When ΔT exceeds predefined thresholds, the traction control unit (TCU) activates cooling fan speed escalation or redundant heat dissipation mechanisms. Temperature data not only directly participate in cooling power regulation but also form a coordinated monitoring network with pressure sensors. Notably, if pipeline pressure abnormalities cause coolant flow reduction, the system may trigger pre-protection measures based on abnormal temperature rise rates even before reaching temperature thresholds. This temperature–pressure coupled closed-loop control mechanism makes temperature sensor measurement accuracy critical for maintaining the converter's thermal safety margin.

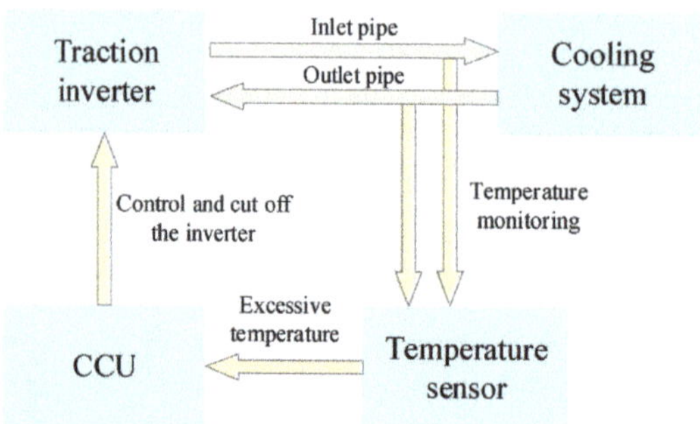

Figure 1. Diagram of the temperature safety protection mechanism for the traction inverter cooling system.

The main converter's coolant flows through outlet pipes to the cooling unit, where temperature sensors provide real-time monitoring. Upon detecting overtemperature conditions, these sensors transmit fault signals to the central control unit (CCU), which may, consequently, execute converter shutdown procedures. This establishes a defined fault propagation path from the cooling unit to the CCU and then to the main converter, as illustrated in the accompanying schematic. The path demonstrates how sensor-level anomalies can escalate to system-level protective actions through the control hierarchy.

The predominant fault modes affecting temperature sensors include the following:

(1) Bias Fault

The bias fault is mathematically characterized as

$$y(t) = x(t) + \Delta. \tag{1}$$

where Δ denotes the offset magnitude. This fault typically arises from electrolyte contamination or sensing element degradation, leading to persistent positive or negative deviations. A positive bias may trigger excessive cooling actions by the TCU, whereas a negative bias can conceal overheating risks, thereby compromising the reliability of thermal management.

(2) Open-Circuit Fault

The open-circuit fault is represented by

$$y(t) = 0. \tag{2}$$

This open-circuit fault ($y(t) = 0$) is defined as a complete interruption of signal transmission path leading to a zero-input state, fundamentally distinct from grounding short-circuit faults or valid zero-value measurements. Typically caused by vibration-induced cable fractures or connector disengagements, the interruption of the signal path induces TCU fail-safe activation. This safety mechanism can result in unintended train stoppages during operation due to critical control signal loss.

(3) Noise Fault

The noise fault is modeled as

$$y(t) = x(t) + n(t) \tag{3}$$

where $n(t)$ represents random noise components. Primarily caused by converter switching electromagnetic interference (EMI), this non-Gaussian noise leads to erroneous activation of TCU control logic.

Figure 2 illustrates the waveform variations of the temperature sensor at the cooling system outlet when it is subjected to offset faults, open-circuit faults, and noise faults during the same time interval.

It is particularly noteworthy that the failure consequences of temperature sensors exhibit significant location-dependent characteristics. Outlet temperature sensor malfunctions may cause the TCU to erroneously judge insufficient heat dissipation efficiency, leading to the excessive activation of cooling fans and subsequent energy consumption surges. Conversely, inlet sensor failures might potentially mask actual equipment overheating risks. Although the cooling system's redundant design (e.g., dual-fan configuration, intermittent pump operation) can alleviate the effects of single-point failures, coupled failures of multiple sensors may still trigger cascading protection mechanisms, resulting in unplanned operational interruptions.

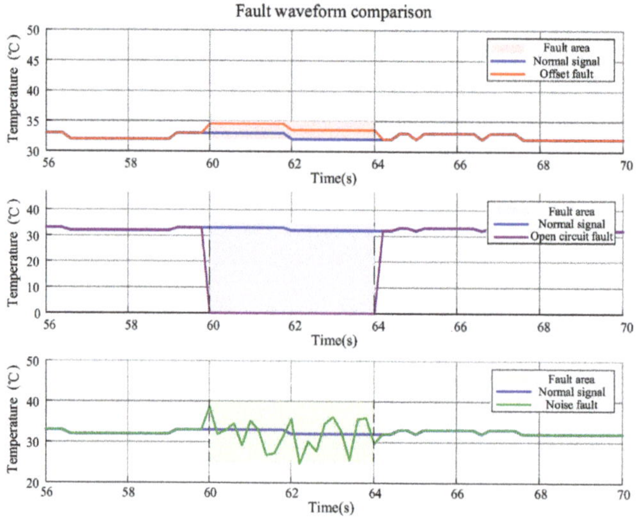

Figure 2. Fault waveform comparison.

This study proposes a hybrid-criteria cooperative diagnostic method for addressing fault diagnosis of inlet temperature sensors in traction converter cooling systems. The approach integrates multi-source data to construct time-series prediction models combined with dynamic threshold optimization algorithms, specifically targeting online identification of three independent fault types: bias, open-circuit, and noise interference. This effectively prevents maladjustment of thermal management systems caused by sensor faults. Furthermore, a dedicated time-series predictor trained on inlet temperature sensor data enables early warning of converter overheating risks.

3. Research Methodology

3.1. General Framework

The proposed dual-level intelligent diagnostic architecture employs a cooperative ensemble prediction model to achieve coupled fault analysis in cooling systems, with its core operational mechanism illustrated in Figure 3 (model training details provided in Figure 4).

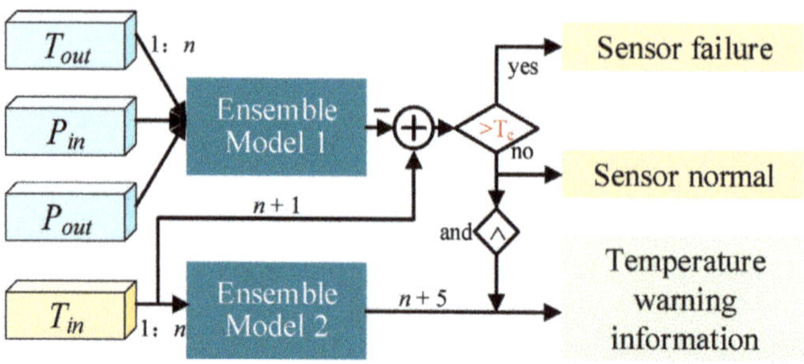

Figure 3. Main process diagram of dual-level diagnosis architecture.

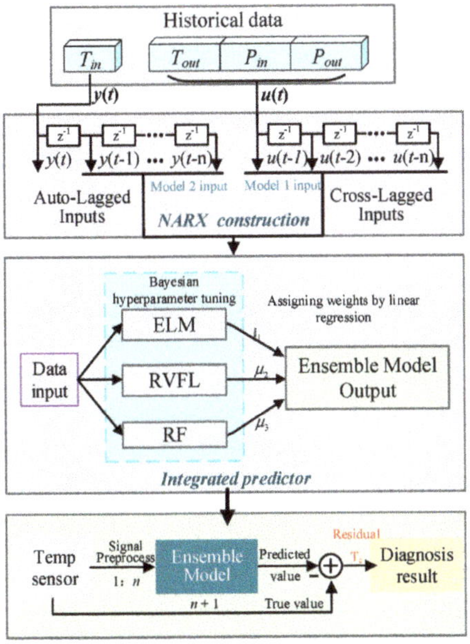

Figure 4. Schematic diagram of collaborative integration model training optimization.

At the first level, Ensemble Prediction Model 1 integrates spatiotemporal correlation features extracted from multimodal sensor data (including outlet temperature, inlet pressure, and outlet pressure) to construct a dynamic prediction model under normal operating conditions. This model generates real-time physical parameter predictions for the n + 1 timestep. By comparing deviations between predicted and measured data against predefined thresholds, it achieves online sensor fault detection and signal isolation, effectively preventing interference from abnormal sensor data in subsequent diagnostics.

The second level implements Ensemble Prediction Model 2, which develops a machine learning predictor for inlet temperature time-series data. By capturing dynamic evolution patterns and periodic characteristics in historical sensor data, this model provides advanced temperature signal prediction, enabling early warning of abnormal conditions such as overheating.

The connection point between Integrated Prediction Model 1 and Integrated Prediction Model 2 represents a critical component of the overall process. When Integrated Prediction Model 1 verifies that the sensor is functioning within normal parameters by comparing the deviation against the predefined threshold T_C, the validated coolant inlet temperature data are forwarded to Integrated Prediction Model 2 as input. This ensures that only sensor data confirmed to be reliable under operational conditions are utilized in Model 2 for advanced temperature prediction and anomaly detection, thus preventing inaccurate predictions and false alarms that could arise due to sensor malfunctions.

This hierarchical dual-model architecture establishes a closed-loop verification mechanism through decoupling analysis of equipment-sensor parameters: It simultaneously validates individual sensor reliability via multi-parameter cooperative prediction while inversely deriving temperature information through single-parameter prediction. This approach overcomes the diagnostic bottleneck of coupled sensor distortion and equipment faults present in conventional methods, significantly enhancing the robustness and accuracy of fault diagnosis under complex operating conditions.

The proposed integrated predictor construction and fault diagnosis process, as illustrated in Figure 4, consists of three core technical components: data-driven modeling, multi-model fusion optimization, and residual-based diagnosis. During the model construction phase, the system first performs spatiotemporal feature reconstruction on historical data of cooling system parameters, including inlet temperature, outlet temperature, and pressure. By adopting a Nonlinear AutoRegressive with eXogenous inputs (NARX) model architecture, automatic lag processing is applied to the target variable $y(t)$ to generate time-delayed sequences from $y(t-1)$ to $y(t-n)$. Simultaneously, cross-lag processing is performed on correlated variables $u(t)$ to create an input matrix spanning $u(t-1)$ to $u(t-n)$, thereby capturing the dynamic coupling characteristics and temporal dependencies among system parameters.

During the training phase of the integrated predictor, a heterogeneous ensemble of base models was constructed, comprising the extreme learning machine (ELM), random vector functional link (RVFL) network, and random forest (RF) algorithm. This design leverages their complementary strengths: The ELM enables rapid computation via its single-hidden-layer architecture, the RVFL handles linear-nonlinear hybrid features through direct input-output linkages, and RF enhances robustness by employing bootstrap aggregation. Bayesian optimization performs a global search across the hyperparameter space of each base model to fully exploit their respective advantages—specifically, fine-tuning the hidden nodes of the ELM for speed–accuracy balance, optimizing the dimensionality of the RVFL's enhanced layer, and calibrating the tree depth of RF to mitigate overfitting risks. A weighted fusion mechanism based on linear regression (with weighting coefficients μ_1, μ_2, μ_3) dynamically integrates the outputs of multiple models.

In the fault diagnosis stage, historical sequences from temperature sensors are input into the trained ensemble model to generate predicted values for the $n+1$ timestep. These predictions are then compared with actual measurements to compute residuals. If the residuals exceed the predefined threshold T_C the system triggers corresponding fault codes based on hierarchical diagnostic logic. This mechanism not only enables sensor anomaly detection through multi-parameter cooperative prediction but also facilitates inverter fault identification via single-parameter deep prediction inversion, thereby establishing a cross-verification closed-loop for equipment-sensor status monitoring.

3.2. Extreme Learning Machines

The ELM is an efficient machine learning algorithm based on the single-hidden layer feedforward neural network (SLFN), proposed by Professor Guang-Bin Huang [27]. This algorithm revolutionizes the traditional neural network training paradigm by randomly initializing hidden layer parameters and analytically determining output layer weights, significantly improving model training efficiency. The core concept of ELM lies in treating the hidden layer as a random feature mapping space and directly fitting outputs through linear regression, demonstrating broad application potential in classification, regression, and feature learning tasks.

The advantages of the ELM lie in its computational efficiency and theoretical completeness: The combination of randomly initialized hidden layer parameters and analytical solutions enables extremely fast training speeds, making it suitable for real-time applications; the global optimality of output weights avoids local optima issues inherent in gradient descent methods; and regularization and random feature mapping enhance generalization performance; while the algorithm remains compatible with diverse activation functions and task types. However, significant limitations are evident: Random initialization of hidden layer parameters may induce performance variability, necessitating multiple experimental trials to mitigate fluctuations; the hidden layer node count L relies on manual configuration, where excessive nodes risk overfitting and insufficient nodes lead to underfitting; the model exhibits sensitivity to data normalization; and the model's single-hidden-layer structure hinders extension to deep networks while offering weaker theoretical interpretability.

3.3. Random Vector Functional Link

The RVFL represents an efficient learning algorithm based on single-hidden-layer feedforward neural networks [28]. Similar to the extreme learning machine (ELM), the RVFL significantly enhances training efficiency by randomly generating and fixing the weight matrix and bias vectors between the input and hidden layers, thereby eliminating iterative optimization processes. The calculation of the hidden layer output matrix $H = g(W^T X + b)$ aligns completely with that of the ELM.

However, the fundamental improvement of the RVFL lies in its feature augmentation mechanism: By concatenating the original input X with the hidden layer output H to form an augmented matrix $A = [X|H] \in \mathbb{R}^{n\times(d+L)}$, the model preserves both the linear components of input features and the nonlinear components of hidden layer mappings, thereby constructing a more robust feature representation space. The output weight matrix $\beta \in \mathbb{R}^{(d+L)\times m}$ is derived through the closed-form solution of regularized least squares:

$$\beta = (A^T A + \lambda I)^{-1} A^T Y \quad (4)$$

This formulation directly contrasts with the ELM, reflecting the RVFL's innovative design that extends the regression basis from pure hidden-layer features H to a hybrid feature space A. This enhancement effectively mitigates the potential information loss in input features caused by the ELM's complete reliance on random projections, demonstrating improved performance in scenarios with significant linear correlations between inputs and outputs. However, the increased dimensionality of the augmented matrix (d + L) introduces a marginal increase in computational complexity. For high-dimensional input features, overfitting risks must be suppressed through careful regularization coefficient λ adjustment. Despite these considerations, the RVFL retains the ELM's core advantages of computational efficiency and global optimal solution characteristics, establishing itself as an ideal choice for small-to-medium-scale datasets requiring balanced linear and nonlinear modeling capabilities.

3.4. Random Forest

RF is an ensemble learning algorithm that enhances model generalization performance by constructing multiple decision trees and aggregating their prediction results [29]. The core mechanism of RF combines bootstrap aggregation (bagging) with random feature selection: First, multiple subsampled training sets are generated through with-replacement sampling, each used to train an individual decision tree. During node splitting, the algorithm randomly selects a subset of candidate features (typically the square root of the total feature count) from the complete feature set to increase inter-tree diversity and reduce overfitting risks. For regression tasks, optimal split points are determined by minimizing the mean squared error ($\text{MSE} = \frac{1}{n}\sum_{i=1}^{n} (y_i - \hat{y}_i)^2$). After generating all decision trees, classification results are integrated through majority voting, while regression outputs are averaged.

The advantages of RF include high robustness (insensitivity to noise and outliers), overfitting resistance (achieved through bagging and random feature selection to suppress model variance), and quantifiable feature importance. However, it exhibits higher computational costs and weaker interpretability compared to single decision trees. This algorithm is widely applicable to high-dimensional data and complex classification/regression tasks, though caution is required regarding potential prediction biases in extreme noise scenarios.

4. Algorithm Validation

4.1. Data Acquisition

In the state monitoring of the traction inverter cooling system, an ensemble learning-based method was employed to predict the inlet temperature of the coolant. This method established time series models via two distinct operating modes, with all output results corresponding to the inlet temperature of the cooling water. In the first mode, Prediction Model 1 performs multivariate collaborative prediction by selecting system parameters such as cooling water outlet temperature T_{out}, cooling water inlet pressure P_{in}, and cooling water outlet pressure P_{out} that exhibit strong correlations with the cooling system's inlet water temperature. Consequently, a multi-dimensional feature time series input is constructed. The feature dimensionality expands to N dimensions (N $\geq$ 2), enhancing the model's adaptability to complex operating conditions by integrating cross-parameter dynamic coupling information. In the second mode, Prediction Model 2 adopts self-input prediction and directly constructs a univariate time series model using the historical data of the cooling water inlet temperature T_{in}. This approach segments the original continuously monitored temperature data into fixed time windows to characterize the temporal evolution patterns of the single temperature parameter.

The operational data originate from CR400 AF (developed by CRRC) EMU (electric multiple unit) runtime datasets provided by CRRC Academy, sampled at 5 Hz. The temperature and pressure data from the cooling system of the second traction inverter were selected and continuously recorded over a 100 min period. The structure of the dataset aligns with the specifications outlined in Table 1.

Table 1. Cooling system data for CR400 AF trainsets.

Model	Data Types	Symbol	Unit
Model 2 input	Cooling water inlet temperature	T_{in}	°C
Model 1 input	Cooling water outlet temperature	T_{out}	°C
Model 1 input	Cooling water inlet pressure	P_{in}	kPa
Model 1 input	Cooling water outlet pressure	P_{out}	kPa

Following outlier treatment, the data undergo normalization and reconstruction using the NARX structure.

The specific parameter settings of Model 1 are as follows:

- Lagged exogenous input parameters: cooling water outlet temperature T_{out}, cooling water inlet pressure P_{in}, and cooling water outlet pressure P_{out}.
- Lag step size: $n = 10$, corresponding to a 2-s historical window (sampling frequency of 5 Hz).
- Prediction step size: $k = 1$, enabling real-time verification of sensor operational status.

For Model 2, the lag step size remains at 10, while the prediction step size $k = 5$ allows for a 1 s forward predictive horizon.

The selection of the lag step size $n = 10$ and the prediction horizon $k = 5$ is justified based on the following analyses:

- $n = 10$: Experimental evaluation of prediction performance across various lag step sizes ($n = 5, 10, 15$) demonstrated that $n = 10$—corresponding to a 2 s historical window at a sampling frequency of 5 Hz—effectively captures the dynamic behavior of the data while minimizing the risk of overfitting.
- $k = 5$: This setting satisfies the requirement for short-term prediction in real-world engineering applications, offering a forward-looking capability of 1 s, while maintaining acceptable prediction stability.

The processed dataset is partitioned into training, validation, and test sets in a 0.7:0.15:0.15 ratio.

4.2. Model Training

The proposed fault predictor employs a Bayesian-optimized heterogeneous ensemble architecture consisting of three core stages: base model optimization, dynamic weight integration, and residual-driven fault detection. Initially, three heterogeneous base models—the extreme learning machine (ELM), the random vector functional link (RVFL) network, and random forest (RF)—are constructed with customized hyperparameter search spaces. For the ELM, the optimization variables encompass hidden layer nodes (search range: 50–500) and activation functions ('hardlim', 'sig', and 'sin'). RVFL optimization focuses on the regularization coefficient (log-uniform sampling range $[10^{-2}, 10^{2}]$) and hidden layer nodes (50–500). RF hyperparameter tuning targets maximum tree depth (5–30) and tree quantity (50–500). The Bayesian optimization algorithm, leveraging a 5-fold cross-validation mean squared error objective function, conducts 200 optimization iterations on the training set to determine optimal hyperparameter configurations for each base model. The hyperparameter optimization process is visualized in Figure 5. The final optimal model training parameters are determined as follows:

- ELM (extreme learning machine):

  Hidden layer nodes: 158; activation function: 'sin'.

- RVFL (random vector functional link):

  Activation function: 'hardlim'; regularization coefficient: 2.53; hidden layer nodes: 103.

- RF (random forest):

  Maximum tree depth: 8; tree quantity: 466.

Subsequently, the prediction matrix $\hat{Y} \in \mathbb{R}^{n\times 3}$ (where n denotes the number of samples) is generated from the base models using validation set data. The actual signal Y is the regression target, and the closed-form solution to solve for the weight coefficients with L2 regularization is

$$w = \left(\hat{Y}^T\hat{Y} + \lambda I\right)^{-1}\hat{Y}^T y \tag{5}$$

The regularization coefficient λ is determined through the hold-out validation set to mitigate the negative impact of base model overfitting on ensemble performance. The final integrated prediction output is expressed as

$$\hat{y}_{ens} = \hat{Y}w \tag{6}$$

The regularization coefficient λ was determined via grid search to be 0.4, and the weight coefficients of the linear regression model were optimized as [0.507, 0.375, 0.117].

As shown in Table 2, this study compares the root mean square error (RMSE) between the ensemble model and individual sub-models and also includes a comparison with the LSTM network. The results indicate that the prediction accuracy of the integrated model is significantly enhanced after introducing weight coefficients via linear regression optimization. The core strength of the integration method lies in its capacity to consistently improve the performance of diverse model architectures, thereby significantly reducing prediction errors. The experimental results further confirm that the weighted fusion integration strategy reduces the average RMSE across key baseline models—including RF, the RVFL, and the ELM—by 28.64%, with a minimum RMSE reduction of 12.5% observed for individual sub-models.

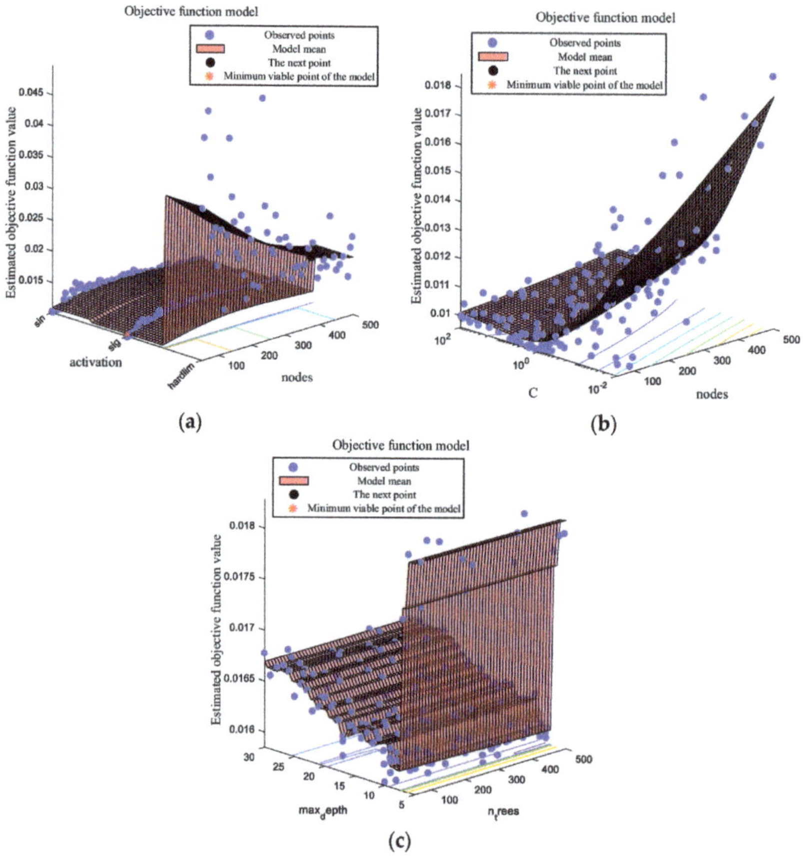

Figure 5. Bayesian objective function model. (**a**) ELM. (**b**) RVFL. (**c**) RF.

Table 2. Comparison of model prediction accuracy.

Data Model	RMSE	MAE	R2
Integrated Learning	0.0049	0.0027	0.8784
ELM	0.0056	0.0035	0.8415
RVFL	0.0062	0.0040	0.8056
RF	0.0088	0.0053	0.6086
LSTM	0.0094	0.0075	0.5653

4.3. Offline Prediction Results

Figure 6a illustrates the dynamic prediction model established under normal operating conditions using Ensemble Prediction Model 1. This model incorporates $n = 10$ lagged sequences of cooling water outlet temperature T_{out}, cooling water inlet pressure P_{in}, and cooling water outlet pressure P_{out} to predict the physical parameters of the inlet temperature sensor at the $(n + 1)$th time step in real-time. The figure provides a visual comparison between the predicted results and the actual values. The model exhibits an RMSE value of 0.0049, indicating high predictive accuracy.

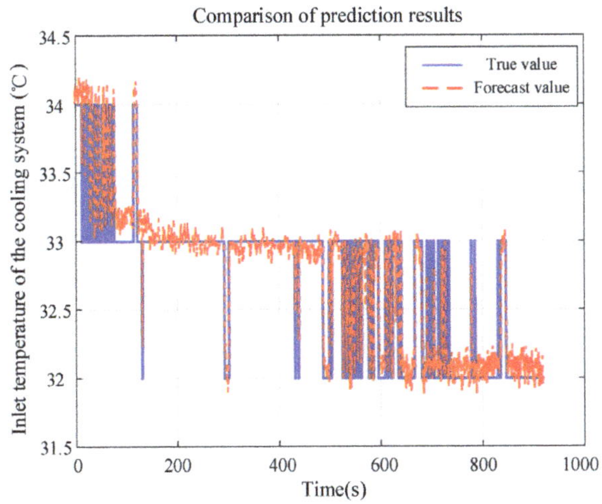

(**a**) Model 1 Test Set Results

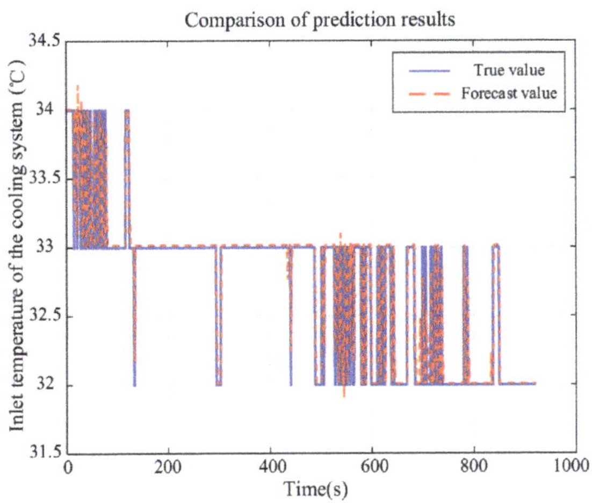

(**b**) Model 2 Test Set Results

Figure 6. Test set prediction results.

Figure 6b validates the time-series prediction capability of temperature sensor historical data by constructing a prediction model based on single historical temperature inputs. Sensor integrity is confirmed through first-level diagnostic verification. By utilizing $n = 10$ lagged sequences (equivalent to a 2 s historical window), the model achieves $k = 5$-step-ahead prediction (1 s) of temperature trends. Under normal operating conditions, the prediction curve shows high consistency with actual monitored values, and the model exhibits an RMSE value of 0.0056. The discrepancy between these parameters and the visual results can be primarily attributed to the root mean square error introduced by the time delay inherent in the predictor's multi-step-ahead predictions. This analysis confirms that intelligent prediction based on historical temperature data can uncover underlying patterns in sensor data and decouple the coupled characteristics between equipment anomalies and sensor faults, thereby providing proactive decision-making support for converter health management when sensor reliability is assured.

During the fault detection phase, real-time residual sequences are computed as follows:

$$e_t = |y_t - \hat{y}_{ens,t}| \quad (7)$$

Dynamic thresholds are set by fitting residual distributions based on historical normal data:

$$\tau = \mu_e + 3\sigma_e \quad (8)$$

where μ_e and σ_e represent the residual mean and standard deviation, respectively. If the residual difference between the predicted result and the true value exceeds the threshold, it indicates the occurrence of a fault. Furthermore, μ_e and σ_e are dynamically updated based on real-time operational conditions, ensuring that the system adapts to the current system state and minimizes false alarms while maintaining sensitivity to faults.

Figure 7 presents the MATLAB R2023b/Simulink simulation results of the cooling system's operational dynamics. The three subplots are configured with identical dual-panel structures: The upper panel contrasts Predictor 1 outputs with ground-truth measurements, while the lower panel visualizes residual sequences alongside dynamically determined threshold boundaries. An alarm protocol is automatically activated when residuals exceed established thresholds, thereby validating the detection capability for abnormalities in the inlet temperature sensor.

Diagnostic scenarios for offset and open-circuit faults are illustrated in Figure 7a and b. The verification process was further extended through diagnostic testing for intermittent open-circuit faults, as illustrated in subfigure c, to enhance model robustness and practical applicability. Three operational scenarios were systematically simulated at a 5 Hz sampling rate, featuring fault durations of 0.2 s, 0.4 s, and 1 s, respectively. This progressive fault duration scheme accurately replicates the transient characteristics of real-world intermittent faults. A significant discrepancy between predicted and actual signals emerges during cooling system inlet temperature sensor failures, attributable to the predictor's incorporation of correlated signal inputs. Subsequent threshold-based evaluation triggers immediate alarm signal activation, conclusively demonstrating the methodological effectiveness.

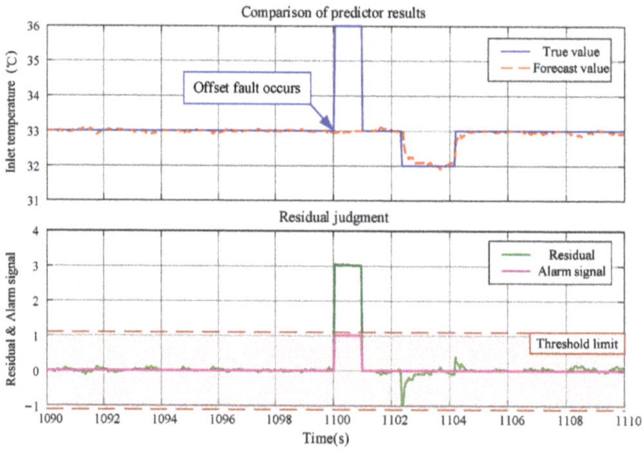

(a) Offset fault

Figure 7. *Cont.*

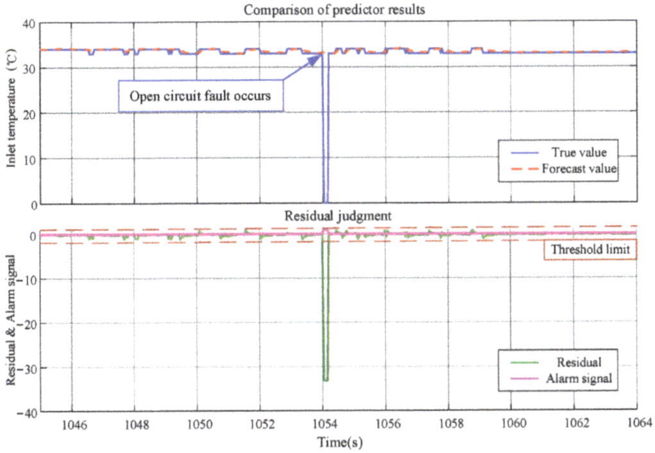

(**b**) Open-circuit fault

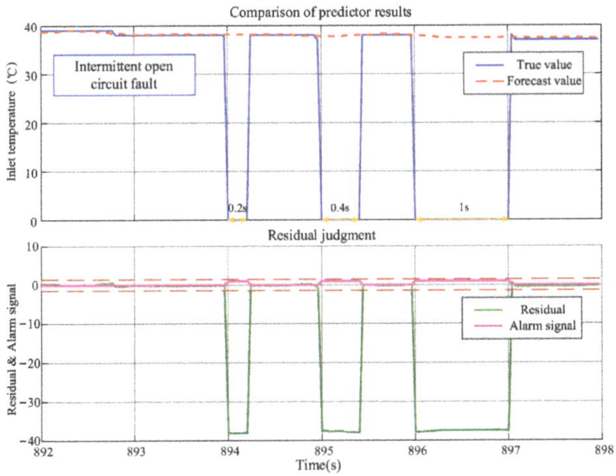

(**c**) Intermittent open-circuit fault

Figure 7. Online diagnosis result graph of Matlab/Simulink.

4.4. Online Validation Results

To validate the engineering deployment capability and real-time performance of intelligent algorithms on embedded edge devices, this study establishes an experimental platform for predicting the temperature of a traction converter cooling system based on the NVIDIA Jetson Orin NX. The edge computing system features a compact modular design and utilizes network interface communication to achieve low-latency data exchange. Benchmarking results confirm that under typical operating loads, the system achieves an average inference latency of 10 milliseconds for Ensemble Prediction Model 1 while the sampling period of the sensor is 200 milliseconds. These metrics demonstrate the system's ability to meet the millisecond-level real-time processing requirements of complex algorithms. The platform architecture consists of the following core modules:

- Edge Computing Unit: executes temperature prediction algorithms to process sensor time-series data in real time.

- Operating Condition Simulator: generates real-time sensor output data to simulate the dynamic behavior of a traction converter cooling system under both normal and predefined fault conditions (e.g., sensor bias, drift, or complete failure).
- Communication Relay: establishes deterministic, low-latency network transmission channels between devices.
- Verification Terminal: conducts closed-loop validation by comparing predicted values against the simulator-generated "ground truth" data through a visual interface.

The platform operates in three sequential phases: First, the operating condition simulator generates real-time cooling system data, which are transmitted via Ethernet to the edge computing unit. Next, the prediction model extracts spatiotemporal features and analyzes temporal correlations from the incoming temperature data stream to generate predictions for future time windows. Finally, the prediction results are returned to the simulator, and both the predicted curves and the simulator's output are displayed synchronously on the verification terminal for performance evaluation. Figure 8 illustrates the prediction model's output fidelity under normal conditions, while Figure 7 presents comprehensive fault detection results using residual analysis and thresholding.

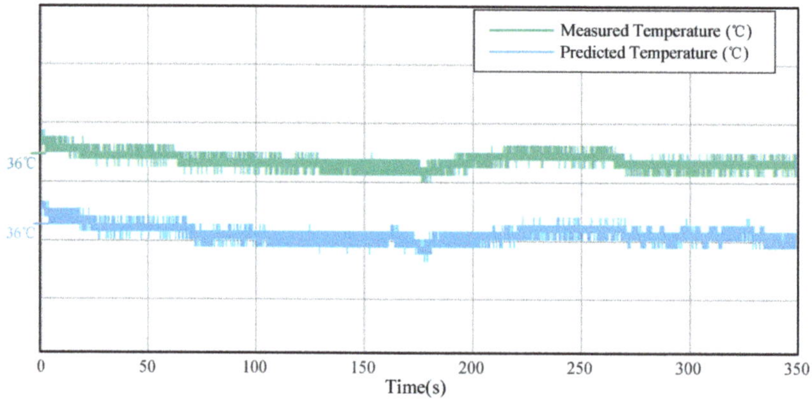

Figure 8. Temperature online prediction results.

Figure 8 demonstrates the output accuracy of the predictive model (Ensemble Prediction Model 2) running on the edge platform under simulated normal operating conditions. The predicted inlet temperature (prediction value) derived from the model and its corresponding "true value" (simulation value) are displayed in this figure. To improve waveform visibility, the two signals are vertically separated and annotated with their respective 36 °C reference points for amplitude comparison, with both waveforms scaled at 5 °C per grid division. The model's capability to accurately track the system dynamics in real time on the edge computing platform has been experimentally validated.

4.5. Robustness Verification

The prediction performance of Predictor 1 under 5% noise interference in the input signal is presented in Figure 9, with diagnostic results for offset, open-circuit, and noise-induced faults simultaneously demonstrated. It should be noted that due to the predictor's lag step size (n = 10), a 2 s temporal offset was introduced between the two subplots in Figure R5-2 to ensure the proper alignment of input and output sequences. A comparative analysis of input signals before and after noise injection is demonstrated in Subplot (a), quantitatively characterizing the noise contamination level. Subplot (b) exhibits the model's predictive outputs and fault diagnosis outcomes under noisy conditions. Although the presence of noise was observed to slightly degrade prediction accuracy, the majority of faults were effectively identified through the threshold-based diagnostic mechanism. These experimental results confirm that significant predictive and diagnostic capabilities are retained by the model under noisy input conditions, thereby validating its robustness.

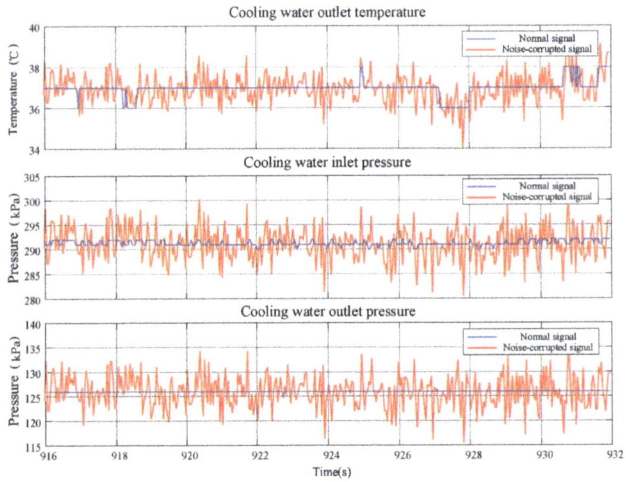

(**a**) The impact of noise on the input signal.

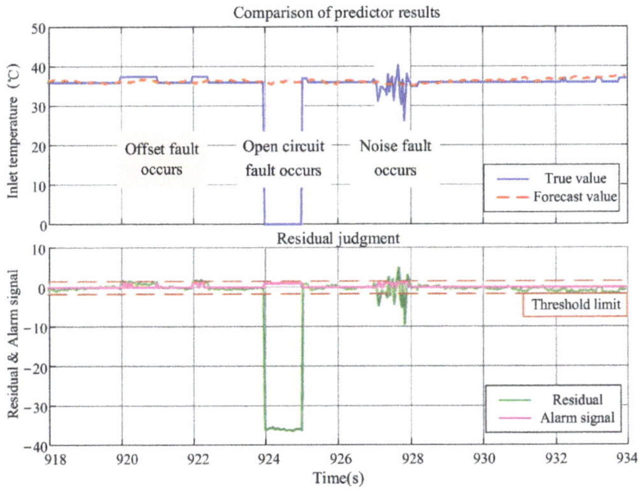

(**b**) Online diagnosis results under noise interference.

Figure 9. Noise interference experiment.

Operational data from the traction cooling system of a different train unit are employed for external validation of data source independence, as shown in Figure 10. As illustrated in Figure 10a, when the pre-trained Predictor 1 is directly applied to the independent dataset, the overall temperature trend is successfully captured; however, a discrepancy of approximately 2 °C between predicted and actual values is observed due to differences in data ranges across trains, resulting in an RMSE of 0.0491. To address this issue, 20% of the dataset is reserved as an independent validation set for weight re-calibration, yielding optimized weights of [0.087, 0.574, and 0.292]. The improved prediction outcomes, with an RMSE significantly reduced to 0.0116, are presented in Figure 10b. These findings confirm that the proposed ensemble learning model can effectively predict temperature trends; however, performance may be affected when test data fall outside the training range. Importantly, the improvement achieved through weight re-adjustment highlights the efficacy of the weighting mechanism in mitigating discrepancies caused by data range variations.

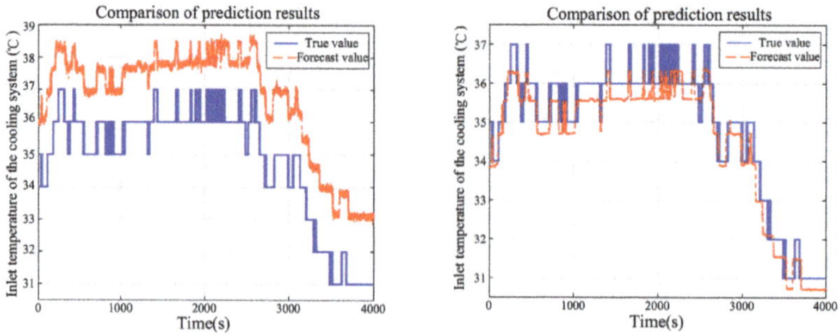

(a) (b)

Figure 10. Test using an independent data source. (**a**) The prediction results derived directly from Model 1. (**b**) The prediction results obtained after recalculating with adjusted weights.

5. Conclusions

This study presents a collaborative diagnostic framework based on a dual-layer intelligent architecture, designed to address the complex coupling between temperature sensor faults and equipment failures in traction inverter cooling systems. By integrating spatiotemporal features extracted from multimodal sensor data through an ensemble learning model, the proposed method enables the rapid identification of inlet temperature sensor faults and accurate prediction of the thermal behavior of the inverter cooling system. The robustness of the model has been systematically evaluated and validated within this study. The experimental results demonstrate that the model is capable of handling noisy environments; however, its predictive accuracy decreases when tested on data outside the training distribution. Future work will aim to improve the model's generalization across diverse data sources through algorithmic optimization and will emphasize real-time validation under practical operational conditions.

Author Contributions: Conceptualization, Y.F. and Q.Z.; methodology, Y.F.; software, Y.F.; validation, Y.F., H.T. and Q.Z.; formal analysis, Y.F.; investigation, Y.F.; resources, Q.Z.; data curation, Y.F.; writing—original draft preparation, Y.F.; writing—review and editing, Q.Z. and H.T.; visualization, Y.F.; supervision, Q.Z.; project administration, Q.Z.; funding acquisition, Q.Z. All authors have read and agreed to the published version of the manuscript.

Funding: The work was supported by the National Key Research Development Program of China (No. 2022YFB3207704).

Data Availability Statement: The data supporting the reported results in this study are not publicly available due to confidentiality restrictions.

Conflicts of Interest: Yunxiao Fu, Qiuyang Zhou, and Haichuan Tang are employed by CRRC (China Railway Rolling Stock Corporation). The research was funded by the National Key Research Development Program of China (No. 2022YFB3207704). The funding sponsor had no role in the design of the study; in the collection, analyses, or interpretation of data; in the writing of the manuscript; or in the decision to publish the results. No other commercial or financial relationships that could be construed as a potential conflict of interest exist among the authors.

References

1. International Electrotechnical Commission. *Railway Applications-Power Converters Installed on Board Rolling Stock-Part 1: Characteristics and Test Methods (IEC 61287-1:2014)*; International Electrotechnical Commission: Geneva, Switzerland, 2014.
2. Wu, X.; Chen, C.Y.; Chen, T.F.; Cheng, S.; Mao, Z.H.; Yu, T.J. A fast and robust diagnostic method for multiple open-circuit faults of voltage-source inverters through line voltage magnitudes analysis. *IEEE Trans. Power Electron.* **2019**, *35*, 5205–5220. [CrossRef]
3. Zhang, W.; He, Y.; Wang, C.; Chen, M.; Sui, C. A multiple open-circuit fault diagnosis method for two-level three-phase voltage source converters based on average-phase voltage model. *IEEE J. Emerg. Sel. Top. Power Electron.* **2021**, *10*, 1138–1152. [CrossRef]
4. Chen, M.; He, Y. Multiple open-circuit fault diagnosis method in NPC rectifiers using fault injection strategy. *IEEE Trans. Power Electron.* **2022**, *37*, 8554–8571. [CrossRef]
5. Hu, X.; Jia, H.; Zhang, Y.; Deng, Y. An open-circuit faults diagnosis method for MMC based on extreme gradient boosting. *IEEE Trans. Ind. Electron.* **2022**, *70*, 6239–6249. [CrossRef]
6. Ye, S.; Zhang, F.; Gao, F.; Zhou, Z.; Yang, Y. Fault diagnosis for multilevel converters based on an affine-invariant riemannian metric autoencoder. *IEEE Trans. Ind. Inform.* **2022**, *19*, 2619–2628. [CrossRef]
7. Wu, F.; Chen, K.; Qiu, G.; Zhou, W. Robust open circuit fault diagnosis method for converter using automatic feature extraction and random forests considering nonstationary influence. *IEEE Trans. Ind. Electron.* **2024**, *71*, 13263–13273. [CrossRef]
8. Gao, Z.; Cecati, C.; Ding, S.X. A survey of fault diagnosis and fault-tolerant techniques—Part I: Fault diagnosis with model-based and signal-based approaches. *IEEE Trans. Ind. Electron.* **2015**, *62*, 3757–3767. [CrossRef]
9. Zhang, G.; Zhou, H.; Wang, G.; Li, C.; Xu, D. Current sensor fault-tolerant control for encoderless IPMSM drives based on current space vector error reconstruction. *IEEE J. Emerg. Sel. Top. Power Electron.* **2019**, *8*, 3658–3668. [CrossRef]
10. Zhang, G.; Wang, G.; Zhou, H.; Li, C.; Wang, G.; Xu, D. Current sensor fault diagnosis and fault-tolerant control for encoderless PMSM drives based on dual sliding-mode observers. In Proceedings of the 2019 IEEE Applied Power Electronics Conference and Exposition (APEC), Anaheim, CA, USA, 17–21 March 2019; IEEE: New York, NY, USA; pp. 2549–2553.
11. Manohar, M.; Das, S. Current sensor fault-tolerant control for direct torque control of induction motor drive using flux-linkage observer. *IEEE Trans. Ind. Inform.* **2017**, *13*, 2824–2833. [CrossRef]
12. Foo, G.H.B.; Zhang, X.; Vilathgamuwa, D.M. A sensor fault detection and isolation method in interior permanent-magnet synchronous motor drives based on an extended Kalman filter. *IEEE Trans. Ind. Electron.* **2013**, *60*, 3485–3495. [CrossRef]
13. Dan, H.; Yue, W.; Xiong, W.; Liu, Y.; Su, M.; Sun, Y. Open-switch and current sensor fault diagnosis strategy for matrix converter-based PMSM drive system. *IEEE Trans. Transp. Electrif.* **2021**, *8*, 875–885. [CrossRef]
14. Xu, S.; Xu, X.; Du, H.; Wang, H.; Chai, Y.; Zheng, W. Comprehensive diagnosis strategy for power switch, grid-side current sensor, DC-link voltage sensor faults in single-phase three-level rectifiers. *IEEE Trans. Circuits Syst. I Regul. Papers* **2024**, *71*, 3343–3356. [CrossRef]
15. Ge, X.; Pu, J.; Gou, B.; Liu, Y. An open-circuit fault diagnosis approach for single-phase three-level neutral-point-clamped converters. *IEEE Trans. Power Electron.* **2017**, *33*, 2559–2570. [CrossRef]
16. Berriri, H.; Naouar, M.W.; Slama-Belkhodja, I. Easy and fast sensor fault detection and isolation algorithm for electrical drives. *IEEE Trans. Power Electron.* **2011**, *27*, 490–499. [CrossRef]
17. Fu, X.; Fang, Y.; Xu, Y.; Xu, H.; Ma, G.; Peng, N. Current Status of Research on Fault Diagnosis Using Machine Learning for Gear Transmission Systems. *Machines* **2024**, *12*, 679. [CrossRef]
18. Sinha, A.; Das, D. XAI-LCS: Explainable AI-Based Fault Diagnosis of Low-Cost Sensors. *IEEE Sens. Lett.* **2023**, *7*, 1–4. [CrossRef]
19. Dong, H.; Chen, F.; Wang, Z.; Jia, L.; Qin, Y.; Man, J. An adaptive multisensor fault diagnosis method for high-speed train traction converters. *IEEE Trans. Power Electron.* **2020**, *36*, 6288–6302. [CrossRef]
20. Haque, M.E.; Zabin, M.; Uddin, J. EnsembleXAI-Motor: A Lightweight Framework for Fault Classification in Electric Vehicle Drive Motors Using Feature Selection, Ensemble Learning, and Explainable AI. *Machines* **2025**, *13*, 314. [CrossRef]
21. Wang, Z.; Xu, X.; Song, D.; Zheng, Z.; Li, W. A Novel Bearing Fault Diagnosis Method Based on Improved Convolutional Neural Network and Multi-Sensor Fusion. *Machines* **2025**, *13*, 216. [CrossRef]
22. Gou, B.; Xu, Y.; Xia, Y.; Deng, Q.; Ge, X. An Online Data-Driven Method for Simultaneous Diagnosis of IGBT and Current Sensor Fault of Three-Phase PWM Inverter in Induction Motor Drives. *IEEE Trans. Power Electron.* **2020**, *35*, 13281–13294. [CrossRef]
23. Tao, H.; Yu, Y.; Song, Y.; Hu, J.; Xu, W.; Yu, C. A Sensor Fault Diagnosis Method for Single-Phase Three-Level Rectifier Used in Traction Systems. In Proceedings of the 2023 CAA Symposium on Fault Detection, Supervision and Safety for Technical Processes (SAFEPROCESS), Yibin, China, 22–24 September 2023; IEEE: New York, NY, USA; pp. 1–5.
24. Tao, H.; Peng, T.; Yang, C.; Yin, S.; Chen, Z.; Fan, X. A diagnosis method for IGBT and current sensor faults of two-level inverter used in traction systems. In Proceedings of the 2021 CAA Symposium on Fault Detection, Supervision, and Safety for Technical Processes (SAFEPROCESS), Chengdu, China, 6–8 August 2021; IEEE: New York, NY, USA; pp. 1–6.
25. Zhang, K.; Gou, B.; Xiong, W.; Feng, X. An online diagnosis method for sensor intermittent fault based on data-driven model. *IEEE Trans. Power Electron.* **2022**, *38*, 2861–2865. [CrossRef]

26. Gou, B.; Xiong, W.; Wang, Y.; Wu, Q.; Zhang, K.; Ge, X. An Intermittent Fault Diagnosis Metod for Multiple Sensors Based on Multi-AI Predictors in Induction Motor Drive System. *IEEE Trans. Ind. Electron.* **2024**, *72*, 3148–3160. [CrossRef]
27. Huang, G.B.; Zhu, Q.Y.; Siew, C.K. Extreme learning machine: Theory and applications. *Neurocomputing* **2006**, *70*, 489–501. [CrossRef]
28. Pao, Y.H.; Park, G.H.; Sobajic, D.J. Learning and generalization characteristics of the random vector functional-link net. *Neurocomputing* **1994**, *6*, 163–180. [CrossRef]
29. Breiman, L. Random forests. *Mach. Learn.* **2001**, *45*, 5–32. [CrossRef]

machines

MDPI

Article

A High-Speed Train Axle Box Bearing Fault Diagnosis Method Based on Dimension Reduction Fusion and the Optimal Bandpass Filtering Demodulation Spectrum of Multi-Dimensional Signals

Zhongyao Wang [1,2], Zejun Zheng [3,*], Dongli Song [3] and Xiao Xu [3]

1 School of Mechanical Engineering, Dalian Jiaotong University, Dalian 116028, China; wzy7889388@126.com
2 CRRC Changchun Railway Vehicles Co., Ltd., Changchun 130062, China
3 State Key Laboratory of Rail Transit Vehicle System, Southwest Jiaotong University, Chengdu 610031, China; sdlcds@swjtu.edu.cn (D.S.); xuxiao@my.swjtu.edu.cn (X.X.)
* Correspondence: zhengjuner@163.com

Abstract: The operating state of axle box bearings is crucial to the safety of high-speed trains, and the vibration acceleration signal is a commonly used bearing-health-state monitoring signal. In order to extract hidden characteristic frequency information from the vibration acceleration signal of axle box bearings for fault diagnosis, a method for extracting the fault characteristic frequency based on principal component analysis (PCA) fusion and the optimal bandpass filtered denoising signal analytic energy operator (AEO) demodulation spectrum is proposed in this paper. PCA is used to measure the dimension reduction and fusion of three-direction vibration acceleration, reducing the interference of irrelevant noise components. A new type of multi-channel bandpass filter bank is constructed to obtain filtering signals in different frequency intervals. A new, improved average kurtosis index is used to select the optimal filtering signals for different channel filters in a bandpass filter bank. A dimensionless characteristic index characteristic frequency energy concentration coefficient (CFECC) is proposed for the first time to describe the energy prominence ability of characteristic frequency in the spectrum and can be used to determine the bearing fault type. The effectiveness and applicability of the proposed method are verified using the simulation signals and experimental signals of four fault bearing test cases. The results demonstrate the effectiveness of the proposed method for fault diagnosis and its advantages over other methods.

Keywords: data fusion; axle box bearing; fault diagnosis; characteristic frequency; bandpass filtering

1. Introduction

The axle box bearing is one of the important components of high-speed train bogies [1], and its operating state is crucial to the safety of the train. The type of axle box bearing is mostly a rolling bearing. With an increase in operation time, pitting and peeling may form between the contact surfaces of various components inside the bearing. With the relative rotation between components, repetitive pulse impacts are generated. In an ideal situation, these pulse impacts are periodic, but in an actual situation, some additional impacts and disturbances are often pseudo-cyclostationary, such as the vibration impact between train wheels and rails and the randomness of relative sliding between bearing components, which increase the uncertainty of vibration changes, reduce the signal-to-noise ratio (SNR) of the signal [2], and lead the vibration acceleration change in the bearing to be pseudo-cyclostationary. It is often difficult to extract characteristic frequencies directly by Fourier transform [3]. Therefore, it is necessary to carry out research on the signal model and signal processing method of bearings and reduce the interference of irrelevant signals through various methods, improve the proportion of periodic impact components in the signal, and make it easier to extract the fault frequency of bearings.

The signal processing methods of bearings can be divided into time-domain, frequency-domain, and joint time–frequency-domain methods. Some scholars have established time-domain signal characteristic indicators, such as the sparsity impact measure index (SIMI) [4], multiscale symbolic dynamic entropy (MSDE) [5], and some representative statistical metrics in the time domain [6]. In the study of frequency-domain signal processing methods, there are also novel methods, such as the amplitude modulation bispectrum [7] and product envelope spectrum optimization-gram [8]. Some joint time–frequency-domain methods, such as the time–frequency spectral amplitude modulation method [9] and the novel time–frequency modulation bispectrum [10], have also achieved good results in bearing fault diagnosis. Some researchers have also studied signal denoising decomposition algorithms, such as adaptive mode decomposition, the improved morphological filtering algorithm, the blind deconvolution algorithm, etc., which provide novel and effective solutions for the research of the bearing fault diagnosis method, and also lay a solution foundation for the proposed fault diagnosis method in this paper.

With the development of bearing monitoring technology and the improvement of safety monitoring requirements for mechanical systems, the fault diagnosis method based on a single signal has gradually developed into multi-signal fault diagnosis methods and research on bearing fault diagnosis based on multi-signal fusion has also increased. During actual operation, the axle box bearing is subjected to vibration and impact from multiple directions, including lateral, longitudinal, and vertical directions between the wheel and rail, resulting in complex spatial curves for the motion trajectories of the internal components of the bearing. The vibration acceleration of the bearing is not a single-direction signal, which needs to be decomposed in a three-dimensional coordinate system to completely represent the vibration condition of the bearing. Therefore, the bearing signal collected by the triaxial accelerometer is more comprehensive, and how to reasonably use the fusion of three-dimensional vibration acceleration signals and other multivariate signals to improve the accuracy of bearing fault diagnosis has become a research hotspot [11]. Multi-signal data fusion can be divided into data-level, feature-level, and decision-level fusion [12]. The fusion based on the data level focuses more on the physical attributes of the signal itself and the external interference to the signal can be reduced by the weighting, linear combination, the Kalman filter [13] and other methods. The fusion of the feature level pays more attention to the extraction of signal features [14], which can be obtained from different physical data through continuous dimension reduction and fusion. The fusion of the decision level focuses on the role of probability in multi-decision fusion, such as the use of the Dempster–Shafer (DS) information fusion theory [15] and generalized methods [16] in fusion.

In the processing of multi-dimensional signals, such as three-dimensional vibration acceleration [17], various multi-dimensional signal analysis methods have been widely used in bearing fault diagnosis, such as bivariate empirical mode decomposition (BEMD) [18], trivariate empirical mode decomposition (TEMD) [19], multivariate empirical mode decomposition (MEMD) [20], noise assisted multivariate empirical mode decomposition (NA-MEMD) [21], multivariate variational mode decomposition (MVMD) [22], multivariate complex modulation model decomposition (MCMMD) [23], multivariate local characteristic-scale decomposition (MLCD) [24], and other methods. These types of decomposition methods can achieve synchronous correlation analysis of multi-signals, map the signal to the multi-dimensional space hypersphere direction vector, obtain the envelope lines in various directions, and then perform feature extraction. They integrate the fusion of the data level and the feature level and play a role in promoting the research into using multivariate signals for fault diagnosis. However, this type of method often takes a long time in obtaining ideal computational results. Principal component analysis (PCA) is a commonly used method for fast data dimension reduction, and it is also used in signal fusion [25]. Shahdoosti et al. [26] proposed an optimal filter using the PCA fusion method for image fusion, which improved the quality of the image. Mochammad et al. [27] used PCA to fuse the characteristics of bearings in multiple time windows and established a standard for detecting the degradation performance of bearings. Bashir et al. [28] used stationary wavelet transform (SWT) and PCA to fuse image input by multiple sensors, and the results showed that PCA had a better effect on image inputs with different brightness levels.

PCA is mostly used for image fusion and feature dimensionality reduction in neural networks and is rarely used in data-level fusion algorithms. Therefore, PCA is used to fuse multi-dimensional data in the data level in this paper, the filter bank is used to filter the fused signal to highlight the periodic pulse impact component, and the fault frequency of the signal is extracted by demodulation.

The remaining sections are as follows. Section 2 introduces the signal processing methods used in this paper and proposes improved average kurtosis and a dimensionless characteristic index for selecting the optimal signal. Section 3 is the simulation signal analysis of the simulation signal to verify the effectiveness of PCA fusion and the effectiveness of the proposed method process for the simulation signal with the inclusion of prior knowledge. Section 4 is the experimental signal analysis, and the fault diagnosis method is verified by four fault-bearing cases experimental signals collected by an axle box bearing test rig. Section 5 presents the conclusions of this paper.

2. Theoretical Background and the Proposed Method

Bearings often produce vibrations in multiple directions during operation. The vibration data are highly nonlinear and often contain strong noise [29]. In order to comprehensively and accurately evaluate the operating state of bearings, a method for bearing fault diagnosis by comprehensively using a three-directional vibration acceleration signal is proposed in this paper.

2.1. PCA Data Fusion Denoising Method

In statistics, PCA is a technique to simplify data sets by using linear transformations to represent the original high-dimensional data with new coordinate coefficients, arranged in order of data variance size and dimensionality reduction [30]. The purpose of PCA is to find a new set of orthogonal bases that maximize the distance between data after projecting multi-dimensional signals onto the plane formed by the orthogonal bases; the similarity between the multi-dimensional signals is maintained as much as possible, and the components with weak correlation are gradually removed. The maximum amount of original information is preserved with the least number of dimensions.

In signal processing and analysis, the signal containing periodic impulses often has a relatively large variance, while interference noise has a small variance. In order to make the signal obtain a higher SNR and reduce the interference of noise, the PCA dimension reduction method can be used to extract a new signal with a large variance as the optimized signal to achieve the purpose of noise reduction.

The PCA algorithm is illustrated with the example of reducing a two-dimensional array to a one-dimensional array.

The two-dimensional array $\mathbf{X}$ is as follows:

$$\mathbf{X} = \begin{bmatrix} x_{1,1} & x_{1,2} & x_{1,3} & \cdots & x_{1,n} \\ x_{2,1} & x_{2,2} & x_{2,3} & \cdots & x_{2,n} \end{bmatrix} \quad (1)$$

To simplify the calculation of variance in subsequent processes, each dimension of $\mathbf{X}$ can be removed from the mean value to make the mean of each dimension equal to 0.

It is hard for a two-dimensional array to contain all the data points on a straight line, and the two-dimensional coordinate system always has its own orthonormal basis. The data points can be described as linear combinations of orthonormal bases in the coordinate system. The variance of the projection of matrix $\mathbf{X}$ onto the orthogonal basis $\mathbf{u}_1$ is as follows:

$$Var_1 = \frac{1}{n}\sum_{i=1}^{n}\left(\mathbf{x}_i{}^T\mathbf{u}_1\right)^2 = \mathbf{u}_1{}^T \cdot \frac{1}{n}\sum_{i=1}^{n}\left(\mathbf{x}_i\mathbf{x}_i{}^T\right) \cdot \mathbf{u}_1 = \frac{1}{n}\mathbf{u}_1{}^T\mathbf{X}\mathbf{X}^T\mathbf{u}_1 \quad (2)$$

where $\mathbf{x}_i$ is the i-th column vector in matrix $\mathbf{X}$, $\mathbf{x}_i = [x_{1,i}\ x_{2,i}]^T$. Since n and $\mathbf{X}\mathbf{X}^T$ are constants, the above equation can be simplified as follows:

$$Var_1 = \mathbf{u}_1{}^T \cdot \mathbf{Q} \cdot \mathbf{u}_1 \tag{3}$$

According to the concept of PCA, the $\mathbf{u}_1$ corresponding to the maximum Var_1 needs to be solved. The constrained problem can be solved according to the Lagrange operator [31]:

$$Var_1 = \mathbf{u}_1{}^T \cdot \mathbf{Q} \cdot \mathbf{u}_1 \quad s.t.\mathbf{u}_1{}^T\mathbf{u}_1 = 1 \tag{4}$$

The extremal function is constructed as follows:

$$F(\mathbf{u}_1) = \mathbf{u}_1{}^T \cdot \mathbf{Q} \cdot \mathbf{u}_1 + \lambda\left(1 - \mathbf{u}_1{}^T\mathbf{u}_1\right) \tag{5}$$

When $\frac{\partial F}{\partial \mathbf{u}_1} = 0$, the above equation can be simplified as follows:

$$\mathbf{Q} \cdot \mathbf{u}_1 = \lambda \mathbf{u}_1 \tag{6}$$

Therefore, the key step in the process of solving PCA dimensionality reduction fusion is to solve the eigenvalues and eigenvectors of $\mathbf{Q}$ [32], where $\mathbf{Q}$ is the covariance matrix of matrix $\mathbf{X}$. The eigenvalues of the covariance matrix are sorted from large to small, and the corresponding number of eigenvalues can be selected for dimensionality reduction and fusion of the corresponding dimensions, and then the signal with higher SNR is obtained.

PCA, as a powerful data dimensionality reduction method, is able to transform multiple variables in the original data into a few principal components by linear transformations that are able to retain most of the information of the original data while removing noise. PCA achieves dimensionality reduction in the data by finding the directions of maximum variance in the data, using these directions as principal components. Therefore, the benchmark of the maximum variance direction will also be used as the benchmark of the subsequent signal dimension reduction fusion in this paper, so that the most relevant part of the signals can be retained.

2.2. Bandpass Filter Bank Signal Denoising Method

The collected signals are often accompanied by strong random pulse interference. The signal processing method, based on spectral kurtosis [33], is a relatively effective method for filtering and noise reduction, which is the process of signal filtering from different bandpass filters and selecting the signal corresponding to the optimal kurtosis value. In the case of complex signals, due to the sensitivity of kurtosis to single pulse impulses, the optimal bandwidth selected may likely lose information due to the extremely short bandwidth. Consequently, based on the processing method of spectrum kurtosis, an improved kurtosis index is adopted to avoid the interference of the single pulse impact component on the results in this section. In parallel, a more reasonable filter bank is established through the subsection processing of the whole frequency band, and the bandpass filtering of the signal is measured adaptively to extract the characteristic information.

2.2.1. FIR Bandpass Filter Bank

The filters of the filter bank are Finite Impulse Response (FIR) filters [34]. In order to balance the time resolution and frequency resolution of the signal after spectrum segmentation, the number of filters in the filter bank is determined according to the length of the original signal as shown in the following equation:

$$M = \lfloor \log_2(N) \rfloor - 6 \tag{7}$$

where M is the number of frequency domain segments, N is the number of original discrete signal sample points, and the calculation result is rounded down. If there is a discrete sequence of 18,000 sample points and the data sampling frequency is 12,000 Hz, the signal frequency spectrum can be divided into 8 components within the frequency range of 0–6000 Hz, and the frequency corresponding to the maximum amplitude in each component is taken as the center frequency of the bandpass filter. In order to cover the entire frequency range, the bandwidth length is twice the length of the frequency component, and the filter window function is selected as the Blackman window. The decomposition of the frequency bandwidth occupied by the individual bandpass filters of the signal in the frequency domain is shown in Figure 1.

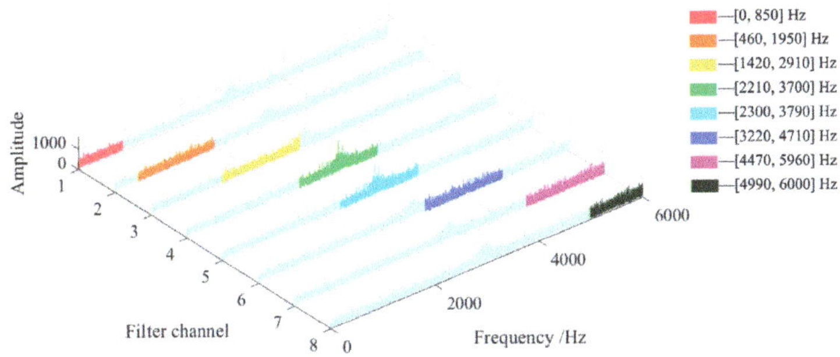

Figure 1. Decomposition of the frequency bandwidth.

In Figure 1, it can be seen that the lower-cut-off frequency and upper-cut-off frequency of each filtering channel are different, and the upper-cut-off frequency of the previous channel is larger than the upper-cut-off frequency of the next channel. After determining the center frequency, bandwidth and window function type of each filter, the filtered signal corresponding to different order filter coefficients can be calculated, and the optimal filtered signal under the filter can be selected as the output result. The different bandpass filters are arranged in order of increasing center frequency to form a multi-channel bandpass filter bank of the signal. The amplitude of the filtered signal of each bandpass filter is linearly scaled to the interval of $[-1, 1]$, and the results are combined to form the final output of the FIR bandpass filter bank.

2.2.2. Improved Average Kurtosis

In order to avoid the interference of monopulse impact in the signal, a characteristic index named improved average kurtosis is proposed, as shown in Equation (8):

$$\mathrm{KT}_{final} = \frac{1}{z}\sum_{i=1}^{z} \mathbf{K}(i) \cdot \mathbf{K}_{\min}/\mathbf{K}_{\max} \tag{8}$$

where KT_{final} is the improved average kurtosis, $\mathbf{K}$ is the vector composed of the kurtosis of each component after dividing the original time-domain signal into z consecutive components, $\mathbf{K}_{\min}$ is the minimum value in $\mathbf{K}$, and $\mathbf{K}_{\max}$ is the maximum value in $\mathbf{K}$.

If there are a few single pulse impact components with large amplitude in the signal, these will have a significant impact on the calculation of the signal kurtosis, causing an abnormal increase in the kurtosis and affecting the judgment of the overall change in the signal. During the iterative filtering process of the signal, it is easy to cause an abnormal increase in the amplitude near the endpoint due to the continuous accumulation of endpoint values. If the kurtosis is used to select the optimal filtered signal, it is easy to produce a judgment with large errors. Two sets of signals are shown in Figure 2.

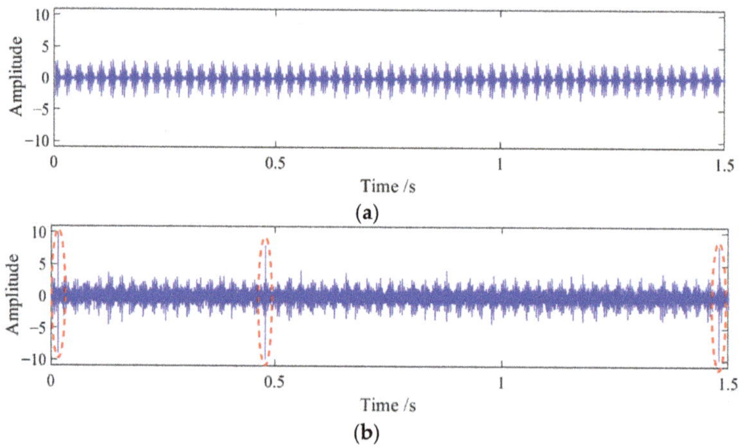

Figure 2. Decomposition of the frequency bandwidth. (**a**) Periodic amplitude modulated signal without noise; (**b**) periodic amplitude modulated signal with −5 dB noise and a large amplitude monopulse.

The signal in Figure 2a is a periodic amplitude-modulated signal without noise, while the signal in Figure 2b is a combination of the signal in Figure 2a with −5 dB of white Gaussian noise, and also contains three large amplitude monopulse signals. The kurtosis of the signals in (a) and (b) are calculated, and signals are divided into 10 components to calculate the improved average kurtosis KT_{final}. The kurtosis of the signal in (a) is 6.5997, and the KT_{final} is 5.2971. The kurtosis of the signal in (b) is 7.6692, and the KT_{final} is 1.8780. If the maximum kurtosis selection method is used, it will be judged that the quality of the signal in (b) is better than that in (a), which is inconsistent with the actual situation. Using the characteristic index proposed in this section can exclude the influence of monopulse impact and improve the reliability of judgment results.

By iterating the filters of different orders continuously, the filtered signal with the largest KT_{final} among the filtered results of different orders is selected as the optimal filtered result of the bandpass filter, and the final output result of the FIR bandpass filter bank in Section 2.2.1 is obtained.

2.3. Analytic Energy Operator Demodulation Method

When the signal components are complex, there are often resonance frequency bands of the mechanical signal acquisition system with large energy levels in practice; it is difficult to find the characteristic frequency of the signal in the spectrum directly using Fourier transform. Consequently, the signal demodulation method is often used to remove the interference of the system itself and the influence of high-frequency resonance bands, and improve the accuracy of fault frequency extraction [35].

The concept of AEO is developed from the analytic signal, which combines the time domain representation of the signal with the corresponding Hilbert transform representation to form the AEO of the signal [36,37].

For a real-valued modulated signal $x(t) = A(t)\cos[\omega(t)]$, the signal obtained by Hilbert transformation is $h(t)$, and they can constitute an analytic signal, as shown in Equation (9):

$$q(t) = x(t) + jh(t) \tag{9}$$

The instantaneous amplitude and phase of the analytic signal $q(t)$ are as follows:

$$A(t) = \pm\sqrt{x^2(t) + h^2(t)} \tag{10}$$

$$\omega(t) = \arctan(h(t)/x(t)) \tag{11}$$

The instantaneous frequency can be obtained by differentiating the instantaneous phase as follows:

$$\dot{\omega}(t) = \left[\dot{h}(t)x(t) - h(t)\dot{x}(t)\right]/A^2(t) \tag{12}$$

If $\left[\dot{h}(t)x(t) - h(t)\dot{x}(t)\right]$ is taken as the quantity to be calculated, it can represent the signal amplitude demodulation and frequency demodulation components, and so it is defined as follows:

$$\Lambda(t) = \dot{h}(t)x(t) - h(t)\dot{x}(t) \tag{13}$$

where $\Lambda(t)$ is the AEO, which can be used as the demodulation form of the original time domain signal to further extract the characteristic components of the signal.

2.4. Optimal Demodulation Component Selection Method

AEO demodulation is performed on the final output result of the FIR bandpass filter bank. The characteristic information in different demodulation spectrums is also different, and it is necessary to find the optimal result from multiple spectrums to extract more obvious fault characteristic frequency components as far as possible. Some characteristic indices can be used to describe the obviousness of the characteristic frequency in the spectrum, such as the characteristic frequency intensity coefficient (CFIC) [38,39]. In order to describe the energy prominence ability of characteristic frequencies in the spectrum within a small local frequency band, a dimensionless characteristic index called the characteristic frequency energy concentration coefficient (CFECC) is proposed for the optimal demodulation spectrum selection in this paper, as shown in Equation (14).

$$\begin{cases} \mathrm{ECC}_{mid} = \sum\limits_{i=1}^{P} \dfrac{E_{\max}([(1-r)f_c,(1+r)f_c])\cdot(L-1)}{\sum E([(1-r)f_c,(1+r)f_c]) - E_{\max}([(1-r)f_c,(1+r)f_c])} \\ \mathrm{ECC}_{lef} = \sum\limits_{i=1}^{P} \dfrac{E_{\max}([(1-2r)f_c,(1-r)f_c])\cdot(L-1)}{\sum E([(1-2r)f_c,(1-r)f_c]) - E_{\max}([(1-2r)f_c,(1-r)f_c])} \\ \mathrm{ECC}_{rig} = \sum\limits_{i=1}^{P} \dfrac{E_{\max}([(1+r)f_c,(1+2r)f_c])\cdot(L-1)}{\sum E([(1+r)f_c,(1+2r)f_c]) - E_{\max}([(1+r)f_c,(1+2r)f_c])} \\ \mathrm{CFECC} = \dfrac{2\mathrm{ECC}_{mid}}{P\left(\mathrm{ECC}_{lef} + \mathrm{ECC}_{rig}\right)} \end{cases} \tag{14}$$

where P is the selectable harmonic order of the fault characteristic frequency, $E_{\max}([(1-r)f_c,(1+r)f_c])$ is the maximum amplitude energy in the demodulated spectrum interval $[(1-r)f_c,(1+r)f_c]$, $\sum E([(1-r)f_c,(1+r)f_c])$ is the sum of each frequency amplitude energy in the corresponding interval, L is the number of discretized frequency points of the spectrum in this interval, f_c is a fault characteristic frequency, r is the characteristic frequency relative error range, and CFECC is the dimensionless characteristic index. By calculating the CFECC, the ratio of the amplitude energy of the fault characteristic frequency f_c and the harmonics at the first P order in the relative error interval to the amplitude energy of the left and right adjacent intervals of the same length can be determined. The larger the CFECC is, the more concentrated the energy is in the error interval containing the characteristic frequency, and the spectrum characteristic frequency has a better representation.

2.5. Calculation of Bearing Fault Characteristic Frequency

When there is local damage in a component of the axle box bearing, the contact of other components at the damage point will produce the corresponding periodic impulse impact excitation. The excitation is different from the vibration of the normal bearing, and its duration is very short, but it has a very wide spectrum range. Different components will produce different impact cycles due to their different motion modes. Therefore, fault defects at different locations will generate different impact frequencies. These frequencies can reflect most of the faults of the bearing, so they can be regarded as the fault characteristic frequencies of the bearing [40]. Considering the geometrical structure and rotational velocity of the bearing, the fault characteristic frequency of each bearing component can be calculated.

When the outer ring is fixed, the fault characteristic frequency of the bearing outer ring f_o can be calculated by Equation (15).

$$f_o = \frac{rn}{120}\left(1 - \frac{d}{D}\cos\alpha\right) \tag{15}$$

The fault characteristic frequency of the bearing inner ring f_i can be calculated by Equation (16).

$$f_i = \frac{rn}{120}\left(1 + \frac{d}{D}\cos\alpha\right) \tag{16}$$

The fault characteristic frequency of the bearing rolling element f_b can be calculated by Equation (17).

$$f_b = \frac{rD}{120d}\left(1 - \left(\frac{d}{D}\right)^2\cos^2\alpha\right) \tag{17}$$

The fault characteristic frequency of the bearing cage f_c can be calculated by Equation (18).

$$f_c = \frac{r}{120}\left(1 - \frac{d}{D}\cos\alpha\right) \tag{18}$$

In Equations (15)–(18), r is the rotational velocity of the rolling bearing, n is the number of rolling elements of the bearing; d is the diameter of the rolling element, D is the diameter of the pitch circle of the rolling bearing, and α is the contact angle of the rolling element.

By substituting the geometric parameters of the axle bearing into the fault characteristic frequency calculation equation, the fault frequency of each part of the bearing can be calculated [41].

2.6. The Framework of the Proposed Fault Diagnosis Method

According to the above analysis and introduction, a fault diagnosis method based on PCA data fusion and bandpass filter bank signal decomposition is proposed to extract the fault characteristic frequency from bearing three-dimensional vibration acceleration signals in this paper. The framework of the method is shown in Figure 3.

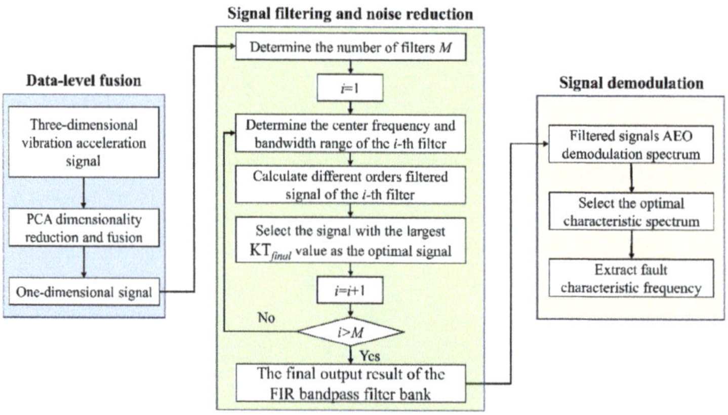

Figure 3. The framework of the fault diagnosis method.

The specific process of the fault diagnosis method is as follows.

Step 1: Firstly, according to the PCA dimension reduction rule, the three-dimensional vibration acceleration signals of the bearing fuse into a one-dimensional signal.

Step 2: According to the length of the signal, determine the number of filters, the center frequency, and the bandwidth, filter the signal by the multi-channel bandpass filter bank, and solve the optimal filtering result under each filter iteratively.

Step 3: Linearly scale the amplitude of the signal from each bandpass filter to the interval $[-1, 1]$ and form the filter bank final output result.

Step 4: Perform AEO demodulation on each filter component, solve the corresponding demodulated spectrum, calculate the CFECC of different characteristic frequencies in each spectrum, and select the characteristic frequency corresponding to the maximum value as the fault frequency and the corresponding demodulation spectrum as the final optimal demodulation spectrum.

3. Results Simulation Signal Analysis

In order to prove the effectiveness of the fault diagnosis method proposed in this paper, a numerical simulation signal was analyzed and verified.

3.1. Numerical Simulation Signal Model

Ideally, when the rolling bearing is faulty, a series of periodic impacts with cyclostationary characteristics will be generated. The rotating speeds of the rolling elements in the raceway are not always the same. The rotating speeds of the rolling elements in the bearing area are faster than those of the rolling elements in the non-bearing area, so there will be a certain deviation in the periodicity of the fault impact, which leads to the fault signal of the bearing being quasi-cyclostationary. When modeling the bearing fault impact signal, random variables can be introduced to cancel the complete cyclostationary characteristics of the signal. Due to the signal attenuation of the sensor during signal acquisition, the attenuation function can be introduced to simulate the attenuation process in the process of signal acquisition [38]. The final established bearing fault impact signal expression is shown in Equation (19).

$$y(t) = \sum_{i=1}^{N} [1 + A_0 \sin(2\pi f_r t)] \cdot e^{-C(t-iT-\tau_i)} \cdot \sin[2\pi f_s(t - iT - \tau_i)] + n(t) \tag{19}$$

where N is the number of fault impacts; A_0 is the initial value of amplitude; f_r is the bearing rotation frequency; C is the impact attenuation coefficient; T is the time interval between adjacent impacts; τ_i is the time error of mutual movement of internal components of bearing; f_s is the resonance frequency of the bearing system; and $n(t)$ is the complex strong noise generated by the bearing operating environment.

According to the simulation model parameters in references [38,41], the simulation signal of the bearing was established according to the parameters in Table 1.

Table 1. Simulation parameters of bearing simulation signal.

Simulation Variables	Parameter Value
N	435
A_0	0.95
f_r	30 Hz
C	450
T	6.289×10^{-3} s
τ_i	(1~2%) T
f_s	2500 Hz

3.2. Three-Dimensional Simulation Signals Analysis

In order to prove the effectiveness of the PCA data-level fusion method for signal denoising, the Hilbert envelope spectrum of the signal was extracted, and the effectiveness of data noise reduction was illustrated by the CFECC. According to the parameters in Table 1, the three-dimensional signals were constructed by using 3.2 dB, 2.9 dB, and 3.1 dB noises. The corresponding time domain diagram and Hilbert demodulation spectrum of each dimension signal and the PCA fusion signal are shown in Figures 4 and 5.

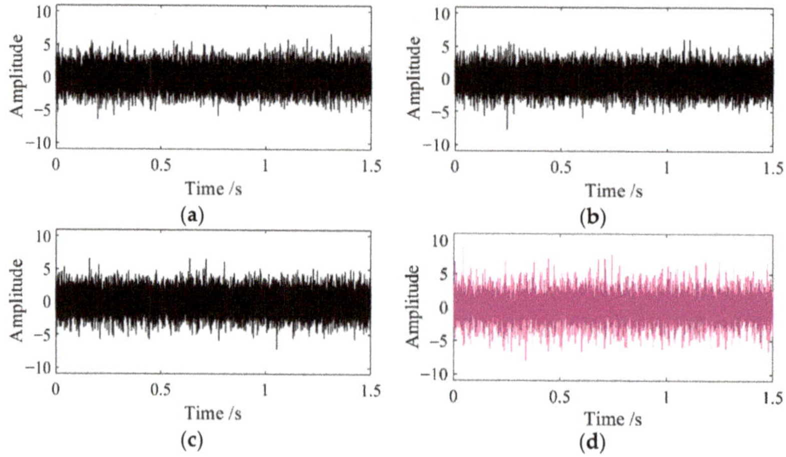

Figure 4. A time domain diagram of simulation signals. (**a**) The first-dimensional simulation signal with 3.2 dB noise; (**b**) the second-dimensional simulation signal with 2.9 dB noise; (**c**) the third-dimensional simulation signal with 3.1 dB noise; and (**d**) the PCA fusion signal.

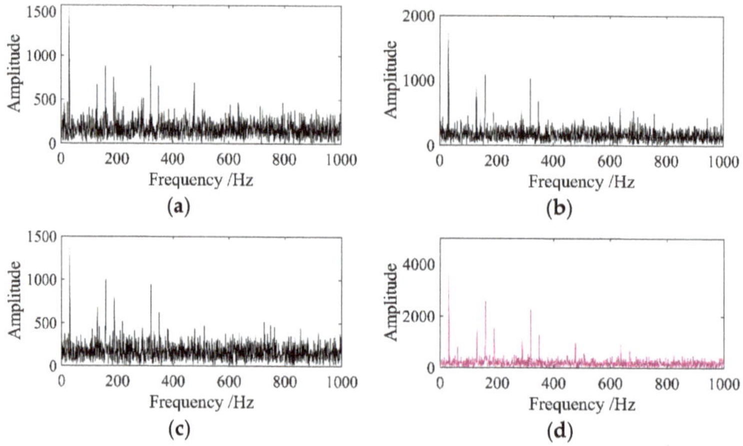

Figure 5. The Hilbert demodulation spectrum of simulation signals. (**a**) The first-dimensional simulation signal with 3.2 dB noise; (**b**) the second-dimensional simulation signal with 2.9 dB noise; (**c**) the third-dimensional simulation signal with 3.1 dB noise; and (**d**) the PCA fusion signal.

It can be seen from Figure 4 that after adding noise to the simulation signal, the periodic impact component of the signal is masked by noise, but this phenomenon can be weakened by the PCA fusion signal. In Figure 5, the noise content of the PCA fusion signal spectrum is significantly reduced. The CFECC value of each spectrum in Figure 5 is calculated with $f_f = 159$ Hz, $P = 4$, and $r = 0.01$, as shown in Table 2. From the calculation results, it can be seen that compared to the original three-dimensional signals, the CFECC of the PCA fusion signal is larger, it is easier to extract the characteristic frequency of the signal, and the characteristic frequency and its harmonic spectrum lines in the spectrum are more prominent.

Table 2. The CFECC values table.

Characteristic Parameter	The First-Dimensional Simulation Signal	The Second-Dimensional Simulation Signal	The Third-Dimensional Simulation Signal	PCA Fusion Signal
CFECC	2.9846	4.8208	3.0092	17.5054

In order to further verify the effectiveness of the proposed fault diagnosis method, the three-dimensional signals were constructed using 7.8 dB, 8 dB, and 8.3 dB noise. The corresponding time domain diagram is shown in Figure 6. The periodic impact component of the signal is masked by noise in Figure 6b–d.

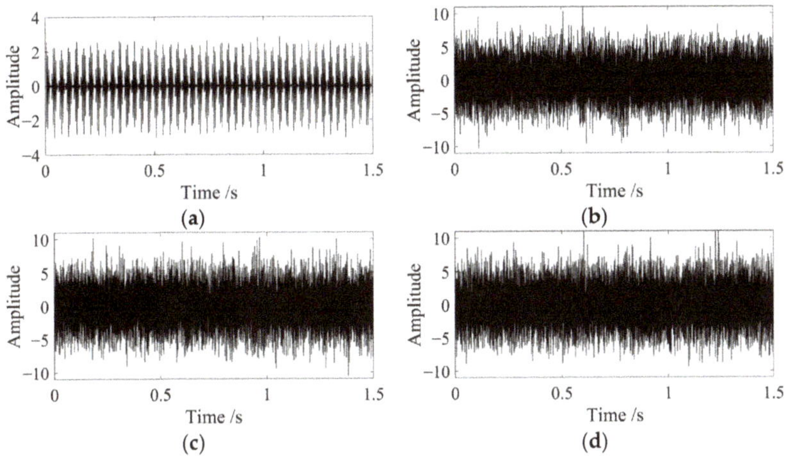

Figure 6. Time domain diagram. (**a**) Simulation signal without noise; (**b**) simulation signal with 7.8 dB noise; (**c**) simulation signal with 8 dB noise; and (**d**) simulation signal with 8.3 dB noise.

The signals in Figure 6b–d are fused by PCA. The PCA fusion signal was decomposed by a multi-channel bandpass filter bank, and the optimal signal obtained for each filtering channel is shown in Figure 7. The legends in Figure 7 illustrate the lower-cut-off frequency and upper-cut-off frequency of each filtering channel.

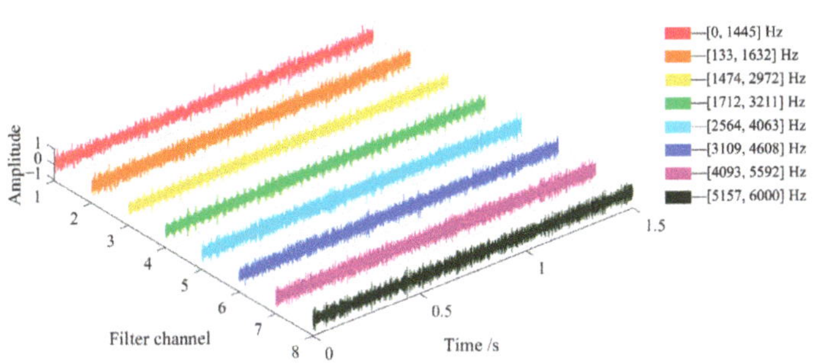

Figure 7. Time domain diagram of multi-filter channel signals.

The CFECC value of the AEO spectrum corresponding to each channel signal in Figure 7 was calculated with $f_f = 159$ Hz, $P = 4$, and $r = 0.025$, as shown in Table 3.

Table 3. The CFECC values table.

Characteristic Parameter	Channel 1	Channel 2	Channel 3	Channel 4	Channel 5	Channel 6	Channel 7	Channel 8
CFECC	0.8716	1.0435	4.7648	5.23178	2.62878	1.30738	1.08715	1.58524

The AEO demodulation spectrum of Channel 4 signal was selected as the optimal demodulation spectrum. To demonstrate the superiority of the proposed method compared to existing fault diagnosis methods, the commonly used multi-dimensional signal fault diagnosis methods NAMEMD–Hilbert envelope and MVMD–Hilbert envelope were adopted to extract the fault characteristics of the above three-dimensional signals. The demodulation spectrums of the Channel 4 signal and the optimal intrinsic mode function (IMF) component of the NAMEMD and MVMD methods are shown in Figure 8.

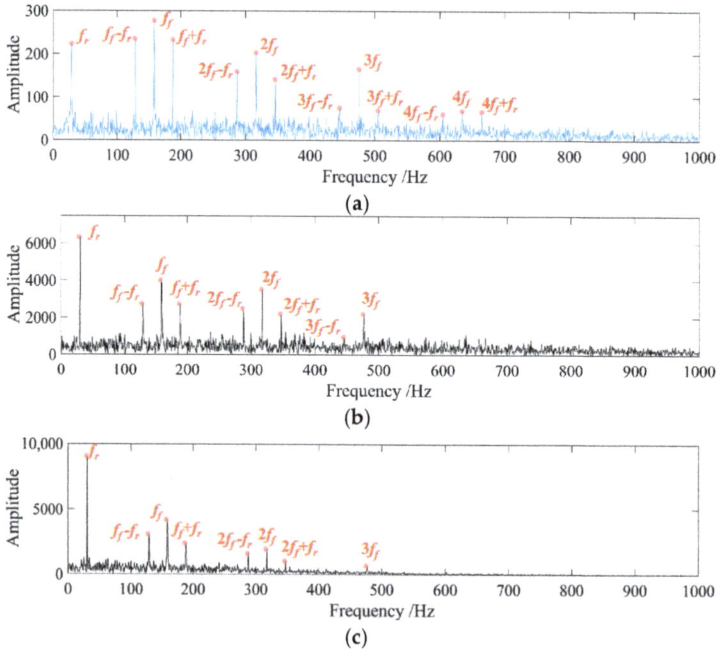

Figure 8. The demodulation spectrums. (**a**) The AEO spectrum of the optimal filtered signal. (**b**) The NAMEMD–Hilbert demodulation spectrum (the number of NAMEMD IMFs is 8); (**c**) the MVMD–Hilbert demodulation spectrum (the number of MVMD IMFs is 8).

It can be seen from Figure 8a that through the process of the fault diagnosis method proposed in this paper, the bearing rotation frequency f_r and fault frequency f_f contained in the original signal can be extracted, and the first four order harmonics of the fault frequency can be extracted, which indicates that the proposed method can extract the bearing fault frequency under complex background noise conditions.

In Figure 8b,c, the NAMEMD–Hilbert method and MVMD–Hilbert can also extract the rotation frequency and the preset fault characteristic frequency of the simulation signal, but the numbers of harmonics of the extracted fault characteristic spectrum are both less than the number of harmonics extracted by the proposed method. The CFECC values of the spectrum in Figure 8b,c are 3.1310 and 3.7779, respectively, which are smaller than the CFECC of the spectrum in Figure 8a. Therefore, the relative superiority of the proposed method can be illustrated, and the computation time of the proposed method is far less than that of the NAMEMD–Hilbert method and MVMD–Hilbert method.

4. Experimental Signal Acquisition and Analysis

4.1. Experimental Signals Acquisition

In order to further verify the effectiveness of the method proposed in this paper for fault diagnosis of actual vibration acceleration signals, the three-dimensional vibration acceleration signal of an axle box bearing collected by the rail vehicle bogie axle box bearing test rig of the State Key Laboratory of Rail Transit Vehicle System in Southwest Jiaotong University [39,41] were used for analysis. The test rig is shown in Figure 9.

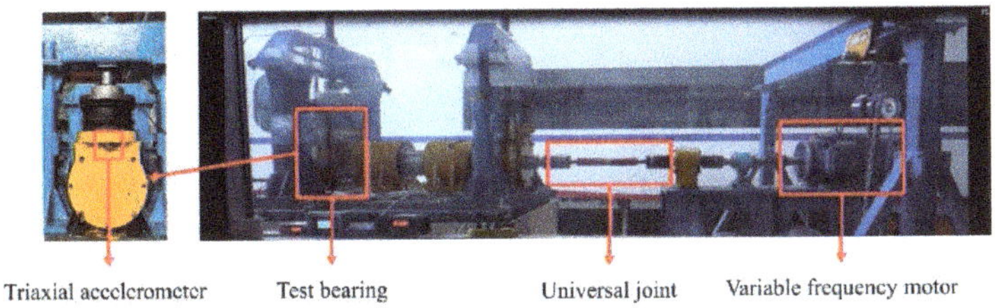

Figure 9. The rail vehicle bogie axle box bearing test rig.

Double-row tapered roller bearings (TRBs) and double-row cylindrical roller bearings (CRBs) commonly used in high-speed train were used in tests, and the three-directional vibration acceleration signals were collected under multi-speed conditions. The fault types of the bearings used in the test are shown in Figure 10.

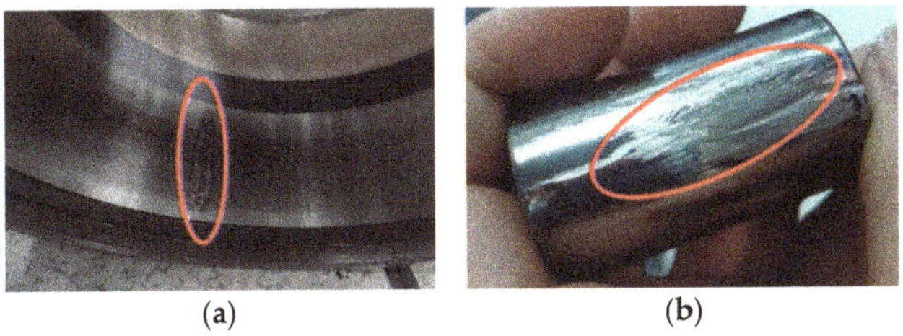

Figure 10. Physical diagram of axle box bearing fault. (**a**) Outer ring fault of TRB; (**b**) rolling element fault of CRB.

Four vibration acceleration signals acquisition tests were conducted according to the conditions in Table 4. The sampling frequency was set to 25.6 kHz.

Table 4. Test conditions table.

	Tested Bearing	Type of Bearing Failure	Bearing Rotation Speed	Sampling Frequency
Case 1	Double-row tapered roller bearing	Outer ring fault	1000 r/min	25.6 kHz
Case 2	Double-row tapered roller bearing	Outer ring fault	1150 r/min	25.6 kHz
Case 3	Double-row cylindrical roller bearing	Rolling element fault	500 r/min	25.6 kHz
Case 4	Double-row cylindrical roller bearing	Rolling element fault	1400 r/min	25.6 kHz

4.2. Experimental Signals Analysis

4.2.1. Case 1: TRB with Outer Ring Fault at 1000 r/min

The three-directional vibration acceleration signals of the bearing in Case 1 were collected as shown in Figure 11.

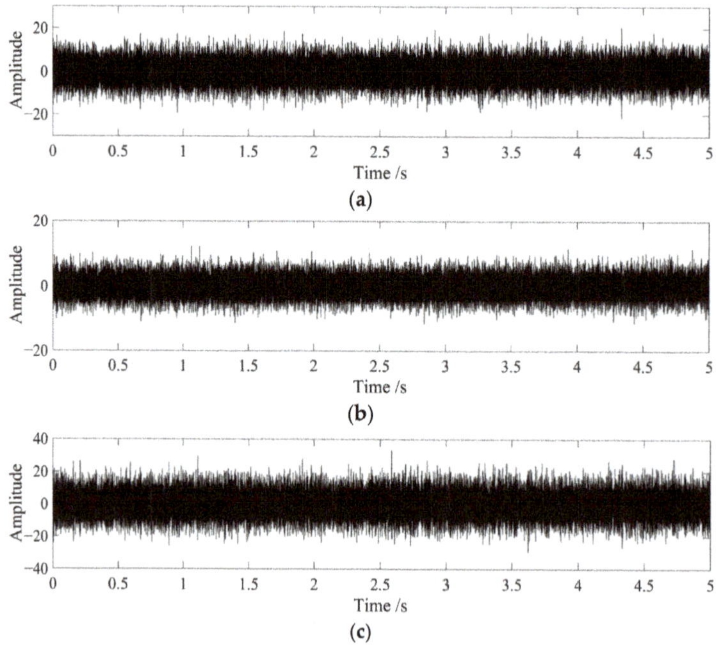

Figure 11. Three-directional vibration acceleration signals in Case 1. (a) Vibration acceleration signal in x direction; (b) vibration acceleration signal in y direction; and (c) vibration acceleration signal in z direction.

There is no obvious periodic fault impact in the time domain of the three-direction vibration acceleration signals in Figure 11. The proposed fault diagnosis method was applied to extract characteristics from the PCA fusion signal of the three-direction vibration acceleration signals in Figure 11, and the optimal signals obtained after filtering with the filter bank are shown in Figure 12.

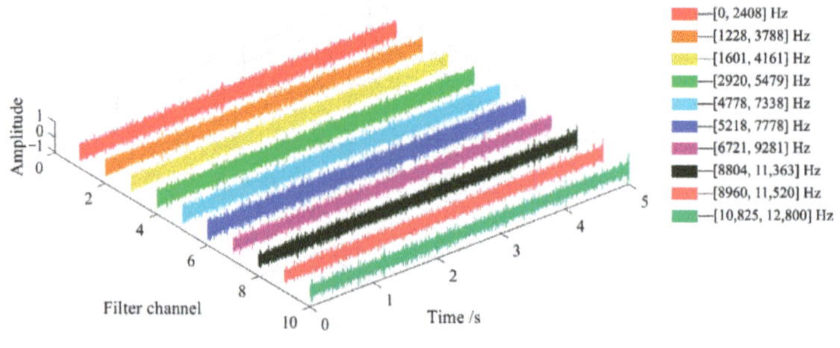

Figure 12. A time domain diagram of the multi-filter channel signals in Case 1.

The signals of different filtering channels in Figure 12 all have noise signals, and the optimal filtering signal cannot be directly selected. The legends illustrate the lower-cut-off frequency and upper-cut-off frequency of each filtering channel. Therefore, CFECC values were calculated for the AEO demodulation spectrums corresponding to each filter channel signal in Figure 12. If there is no prior knowledge of bearing fault types, it is necessary to calculate the CFECC value corresponding to each type of bearing fault characteristic frequency separately. If the CFECC for a certain frequency in the calculation result is significantly greater than the CFECC for the remaining fault characteristic frequencies, it can be preliminarily determined that this frequency is the fault frequency of the bearing, and the spectrum containing the maximum value is the optimal demodulation spectrum. The CFECC values of the AEO spectrum corresponding to each channel signal in Figure 12 were calculated with an outer ring fault frequency f_o = 122.85 Hz, an inner ring fault frequency f_i = 160.49 Hz, a rolling element fault frequency f_b = 60.26 Hz, P = 5, and r = 0.025, as shown in Figure 13.

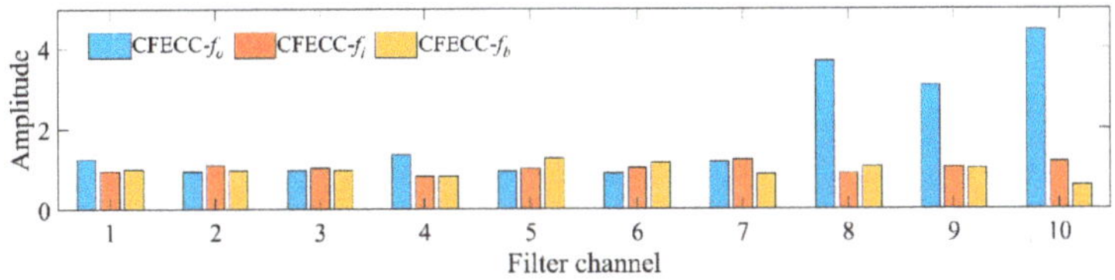

Figure 13. CFECC values of multiple fault characteristic frequencies in Case 1.

The AEO demodulation spectrum of the signal of Channel 10 was selected as the optimal demodulation spectrum and the fault type was determined to be the outer ring fault, which is consistent with the actual situation. The obtained characteristic spectrum is shown in Figure 14a. The demodulation spectrums of the optimal IMF component of the NAMEMD–Hilbert method and MVMD–Hilbert method are shown in Figure 14b,c.

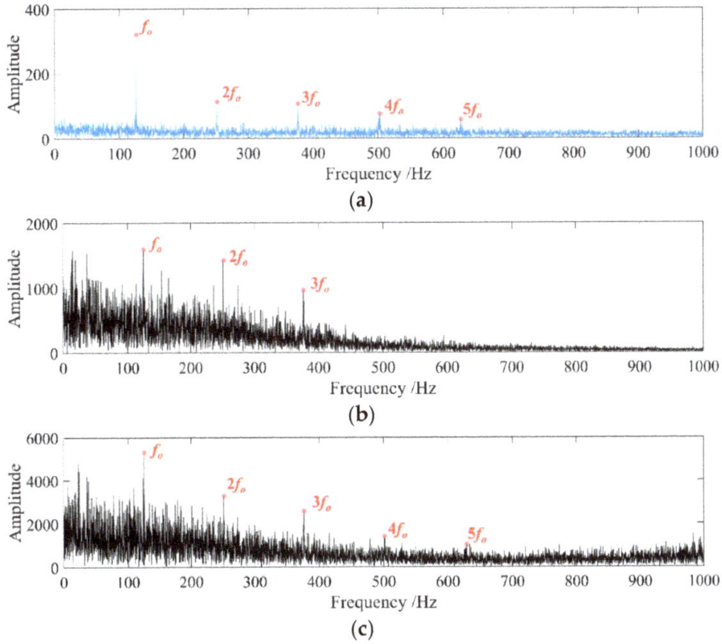

Figure 14. The demodulation spectrums in Case 1. (**a**) The optimal demodulation spectrum of the PCA fusion signal. (**b**) The NAMEMD–Hilbert demodulation spectrum; (**c**) the MVMD–Hilbert demodulation spectrum.

As can be seen from Figure 14a, after the processing of the proposed method, the final demodulation spectrum shows significant high-amplitude spectral lines at the theoretical outer ring fault frequency and its first five order harmonics, and the characteristic frequency information is prominently distributed in the spectrum. It can be seen that the proposed method extracts the characteristic frequency of the outer ring fault of the bearing well. The CFECC values of bearing outer ring fault signals demodulation spectrums in Figure 14b,c are 1.6043 and 1.8062, respectively, which are smaller than the CFECC value of the demodulation spectrum in Figure 14a.

4.2.2. Case 2: TRB with Outer Ring Fault at 1150 r/min

The three-directional vibration acceleration signals of the bearing in Case 2 were collected as shown in Figure 15.

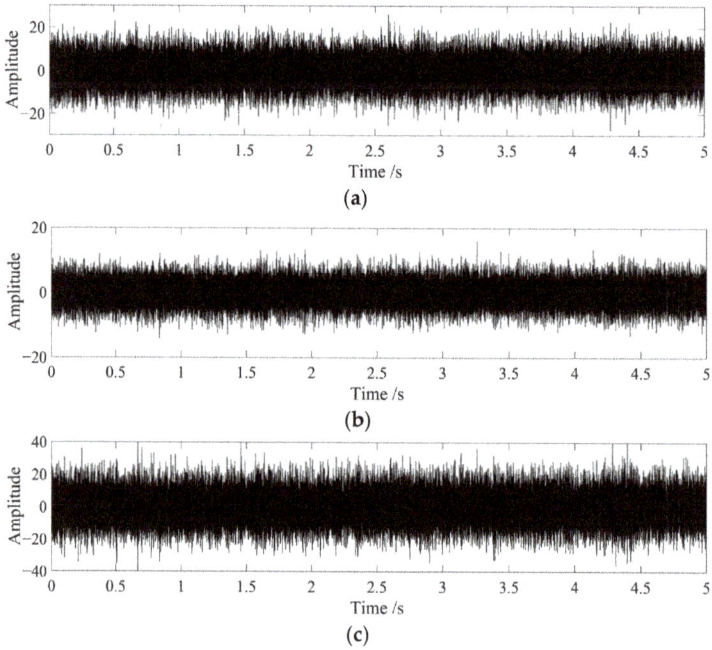

Figure 15. Three-directional vibration acceleration signals in Case 2. (**a**) Vibration acceleration signal in x direction; (**b**) vibration acceleration signal in y direction; (**c**) vibration acceleration signal in z direction.

There is no obvious periodic fault impact in the time domain of the three-directional vibration acceleration signals in Figure 15. The proposed fault diagnosis method was applied to extract characteristics from the PCA fusion signal of the three-directional vibration acceleration signals in Figure 15, and the optimal signals obtained after filtering with the filter bank are shown in Figure 16.

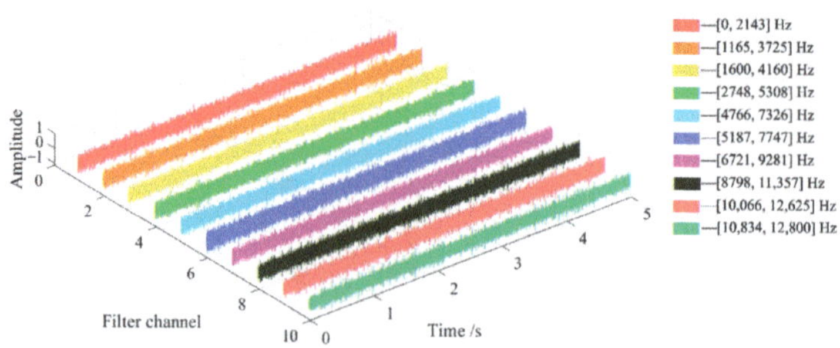

Figure 16. A time domain diagram of the multi-filter channel signals in Case 2.

The signals of different filtering channels in Figure 16 all have noise signals, and the optimal filtering signal cannot be directly selected. Therefore, The CFECC values of the AEO spectrum corresponding to each channel signal in Figure 16 were calculated with an outer ring fault frequency f_o = 141.27 Hz, an inner ring fault frequency f_i = 184.56 Hz, a rolling element fault frequency f_b = 69.29 Hz, P = 5, and r = 0.025, as shown in Figure 17.

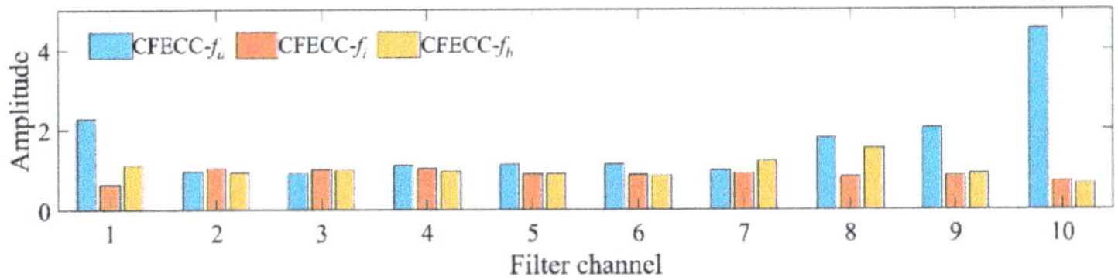

Figure 17. CFECC values of multiple fault characteristic frequencies in Case 2.

The AEO demodulation spectrum of the signal of Channel 10 was selected as the optimal demodulation spectrum and the fault type was determined to be the bearing outer ring fault, which is consistent with the actual situation. The final obtained characteristic spectrum is shown in Figure 18a. The demodulation spectrums of the optimal IMF components of the NAMEMD–Hilbert method and MVMD–Hilbert method are shown in Figure 18b,c.

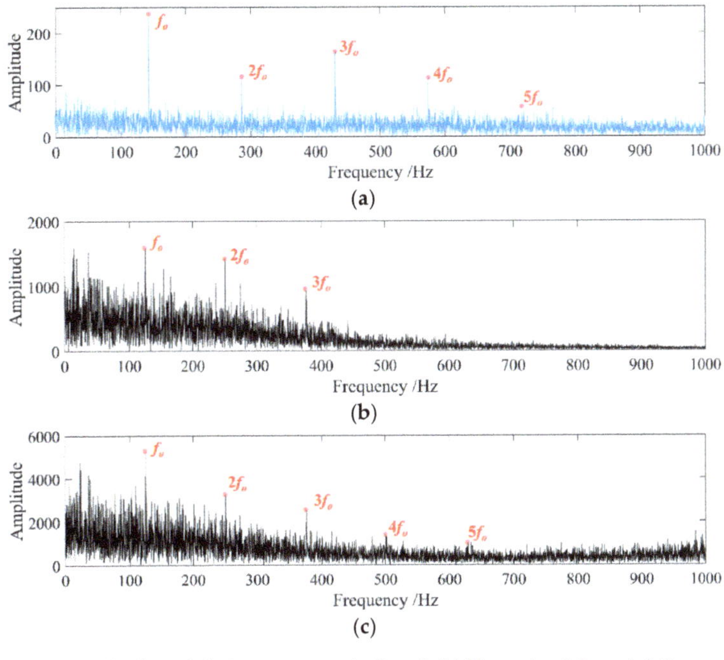

Figure 18. The demodulation spectrums in Case 2. (**a**) The optimal demodulation spectrum of the PCA fusion signal. (**b**) The NAMEMD–Hilbert demodulation spectrum; (**c**) the MVMD–Hilbert demodulation spectrum.

The obvious bearing outer ring fault frequency and its first five order harmonics are extracted in Figure 18a, which has a good characteristic extraction effect. The CFECC values of bearing outer ring fault signals demodulation spectrums in Figure 18b,c are 2.7174 and 2.3736, respectively, which are smaller than the CFECC value of the demodulation spectrum in Figure 18a.

4.2.3. Case 3: CRB with Rolling Element Fault at 500 r/min

The three-directional vibration acceleration signals of the bearing in Case 3 were collected as shown in Figure 19.

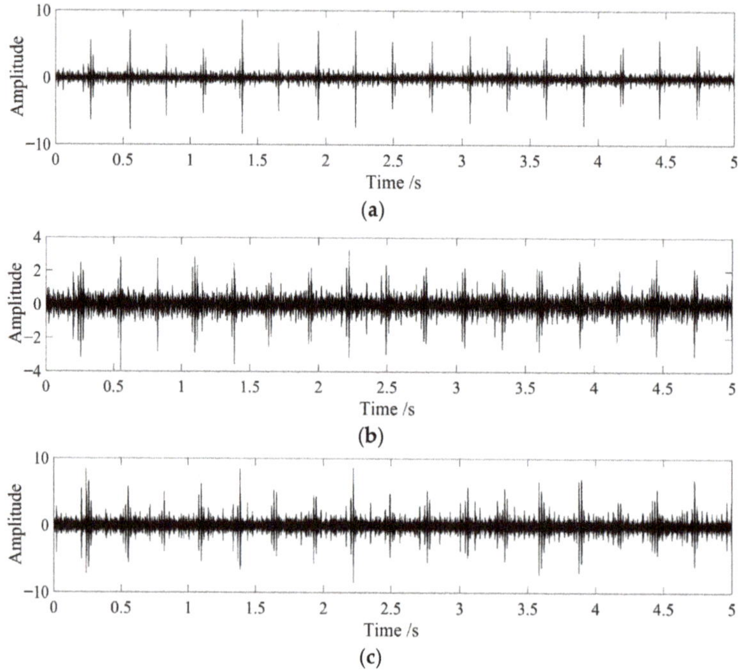

Figure 19. Three-directional vibration acceleration signals in Case 3. (**a**) Vibration acceleration signal in x direction; (**b**) vibration acceleration signal in y direction; (**c**) vibration acceleration signal in z direction.

The proposed fault diagnosis method was applied to extract characteristics from the PCA fusion signal of the three-directional vibration acceleration signals in Figure 19, and the optimal signals obtained after filtering with the filter bank are shown in Figure 20.

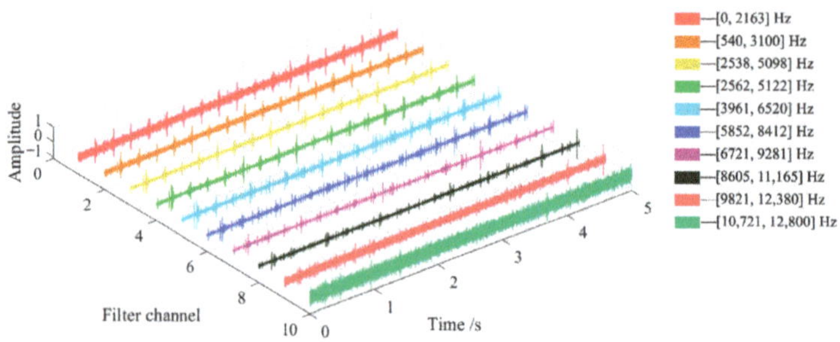

Figure 20. Time domain diagram of the multi-filter channel signals in Case 3.

The CFECC values of the AEO spectrum corresponding to each channel signal in Figure 20 were calculated with a bearing rolling element fault frequency f_o = 60.44 Hz, an inner ring fault frequency f_i = 81.23 Hz, a rolling element fault frequency f_b = 27.78 Hz, P = 5, and r = 0.025, as shown in Figure 21.

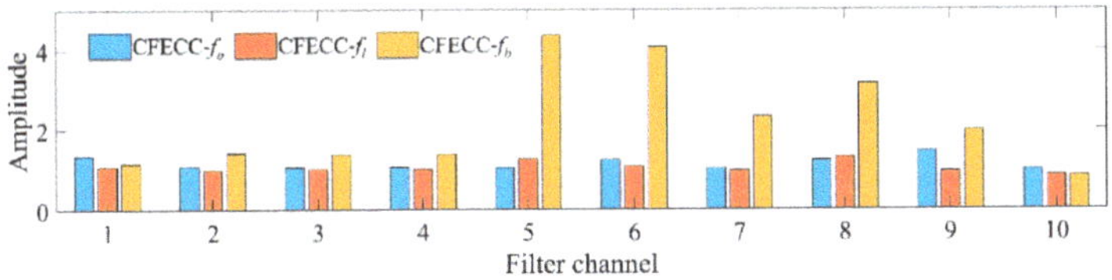

Figure 21. CFECC values of multiple fault characteristic frequencies in Case 3.

The AEO demodulation spectrum of the signal of Channel 5 was selected as the optimal demodulation spectrum, and the fault type was determined to be the bearing rolling element fault, which is consistent with the actual situation. The final obtained characteristic spectrum is shown in Figure 22a. The demodulation spectrums of the optimal IMF component of the NAMEMD–Hilbert method and MVMD–Hilbert method are shown in Figure 22b,c.

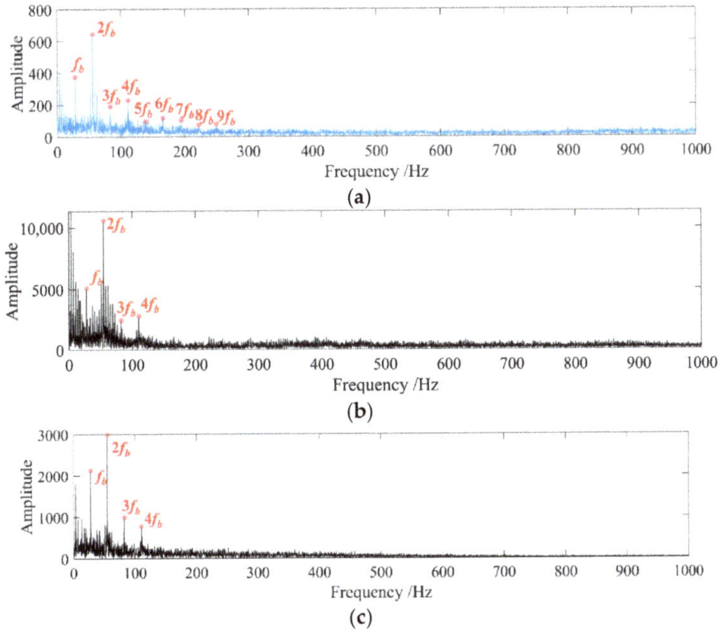

Figure 22. The demodulation spectrums in Case 3. (**a**) The optimal demodulation spectrum of the PCA fusion signal. (**b**) The NAMEMD–Hilbert demodulation spectrum; (**c**) the MVMD–Hilbert demodulation spectrum.

The obvious bearing rolling element fault frequency f_b and its first nine order harmonics are extracted in Figure 22a, which has a good characteristic extraction effect. The CFECC values of bearing rolling element fault signal demodulation spectrums in Figure 22b,c are 1.7146 and 3.1951, respectively, which are smaller than the CFECC value of the demodulation spectrum in Figure 22a.

4.2.4. Case 4: CRB with Rolling Element Fault at 1400 r/min

The three-directional vibration acceleration signals of the bearing in Case 4 were collected as shown in Figure 23.

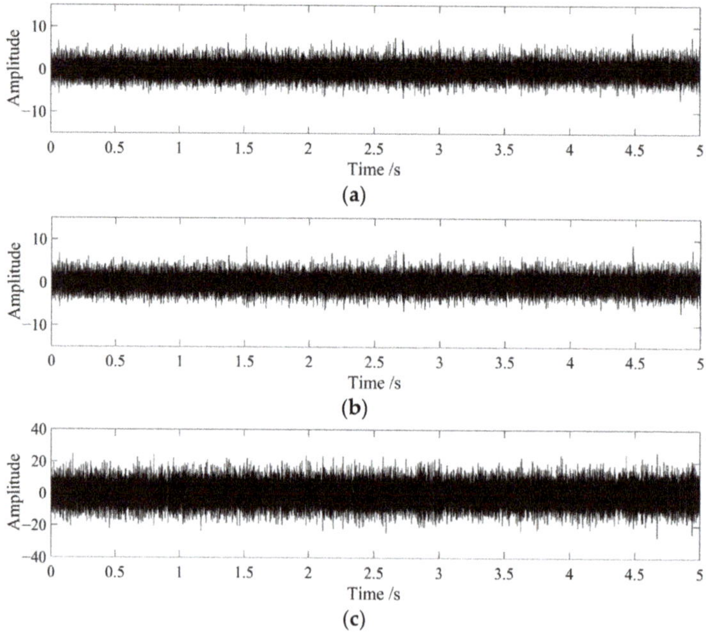

Figure 23. Three-directional vibration acceleration signals in Case 4. (**a**) Vibration acceleration signal in x direction; (**b**) vibration acceleration signal in y direction; (**c**) vibration acceleration signal in z direction.

There is no obvious periodic fault impact in the time domain of the three-direction vibration acceleration signals in Figure 23. The proposed fault diagnosis method was applied to extract characteristics from the PCA fusion signal of the three-directional vibration acceleration signals in Figure 23, and the optimal signals obtained after filtering with the filter bank are shown in Figure 24.

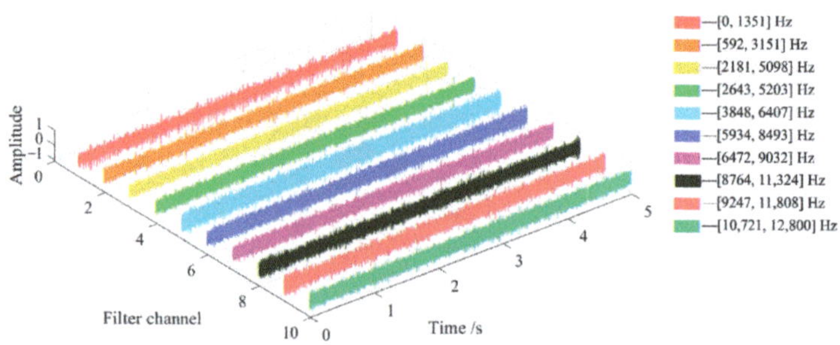

Figure 24. Time domain diagram of the multi-filter channel signals in Case 4.

There is no obvious periodic fault impact in the time domain of the three-direction vibration acceleration signals in Figure 24. The CFECC values of the AEO spectrum corresponding to each channel signal in Figure 24 were calculated with a bearing rolling element fault frequency $f_o = 169.23$ Hz, an inner ring fault frequency $f_i = 227.44$ Hz, a rolling element fault frequency $f_b = 77.79$ Hz, $P = 5$, and $r = 0.025$, as shown in Figure 25.

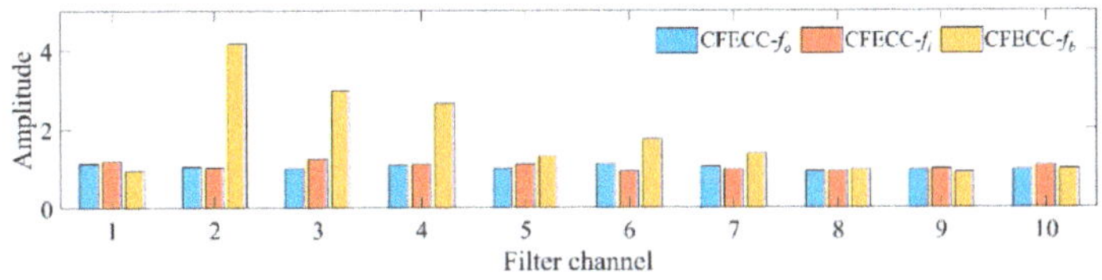

Figure 25. A time domain diagram of the multi-filter channel signals in Case 4.

The AEO demodulation spectrum of the signal of Channel 2 was selected as the optimal demodulation spectrum, and the fault type was determined to be the bearing rolling element fault, which is consistent with the actual situation. The final obtained characteristic spectrum is shown in Figure 26a. The optimal IMF demodulation spectrums extracted by the NAMEMD–Hilbert method and MVMD–Hilbert method for the signals in Case 4 are shown in Figure 26b,c.

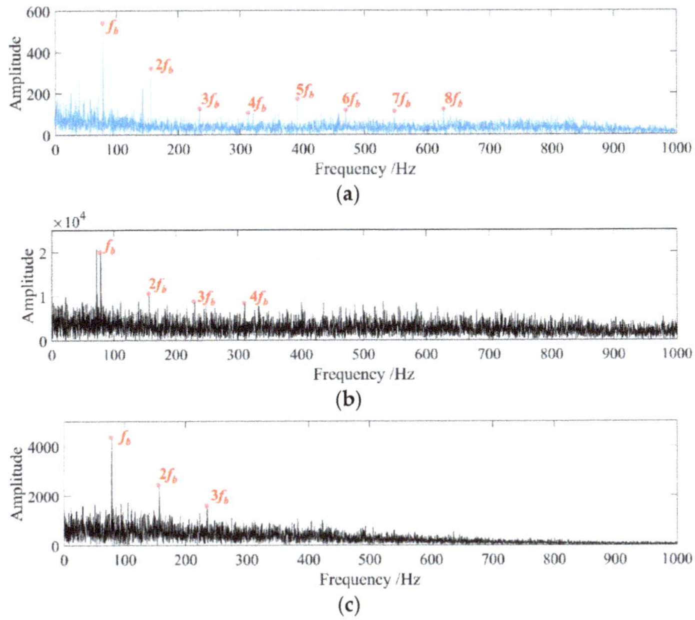

Figure 26. The demodulation spectrums in Case 4. (**a**) The optimal demodulation spectrum of the PCA fusion signal. (**b**) The NAMEMD–Hilbert demodulation spectrum; (**c**) the MVMD–Hilbert demodulation spectrum.

The obvious bearing rolling element fault frequency f_b and its first five order harmonics are extracted in Figure 26a, which has a good characteristic extraction effect. The CFECC values of bearing rolling element fault signals demodulation spectrums in Figure 26b,c are 2.3486 and 3.0779, respectively, which are smaller than the CFECC value of the demodulation spectrum in Figure 26a.

5. Discussions

The core of the fault diagnosis method proposed in this paper is bandpass filtering, and the optimal filtering signal is determined by the selection of the AEO demodulation spectrum of the filtering signal. In the simulation signal analysis, a periodic impact signal with a definite frequency was established, and the proportion of the periodic impact component in the signal was weakened by adding noise, which could make the simulation signal closer to the actual signal, and also could verify the superiority of the fault diagnosis method. The noise interference could be clearly seen in Figure 6b–d, while the periodic shocks were masked. After using the proposed fault diagnosis method to process the signal step by step, a more obvious characteristic frequency appeared in the spectrum as shown in Figure 8, which was the same as the preset 159 Hz. It could be considered that the proposed method accurately extracted the periodic impact characteristics in the signal. Compared with the spectrum obtained by the NAMEMD–Hilbert and MVMD–Hilbert method, the final spectrum obtained in this paper had more obvious impact spectrum peaks.

The vibration response characteristics after bearing fault impact were considered in the establishment of the simulation signal, and the frequency of 159 Hz was also within the main fault characteristic frequency range in the bearing experiment. If there is strong noise interference in the experimental environment, the characteristic information can still be extracted by the proposed method. The accuracy of the simulation signal analysis results lays a good foundation for subsequent experimental signal analysis. Through the verification of the feasibility of the proposed method, it is proven that the proposed method can be used to identify the bearing fault impact in the actual operation.

In the experimental signal analysis, the bearing signals under different working conditions were also inconsistent, such as the signals under four different working conditions in Case1–Case4. The periodic impact in the fault signal of the bearing rolling element at a low rotational velocity was more obvious, but there was also more interference from the bearing rotational frequency in the spectrum, which may affect the judgment of the bearing health state. Although the method proposed in this paper can extract the fault characteristic frequency of the bearing, the result will also be interfered with by the bearing rotational frequency. In subsequent research, how to avoid the interference of rotational frequency conversion on the diagnosis results must be considered.

6. Conclusions

In this paper, a bearing fault diagnosis method based on PCA data dimension reduction fusion, multi-channel bandpass filter bank filtering, and AEO demodulation is proposed. Through the proposed CFECC value, the fault characteristic frequency of a signal can be determined and the spectrum with the most obvious harmonic performance is selected as the optimal demodulation spectrum. The effectiveness of the proposed method is proved by simulation signals and experimental signals, and it has good extraction effects on the characteristic frequencies of multi-fault type bearings at different speed levels. The main innovations and conclusions of this paper are as follows.

(1) The PCA is used for data dimension reduction and fusion. The effect of PCA on data noise reduction in data dimensionality reduction and fusion is proven by simulation signals, and it is easier to extract the characteristic frequency of the fusion signal.

(2) An improved average kurtosis index KT_{final} is proposed, which can effectively avoid the interference of monopulse impact on signal characteristics. Based on the KT_{final}, the selection criteria of optimal filtered signals for each channel under a multi-channel bandpass filter bank are determined.

(3) The AEO is used to demodulate the filtered signal to find the fault characteristic frequency, and the spectrum characteristic index CFECC parameter is proposed. The CFECC can be used to calculate the local energy prominence ability of different characteristic frequencies and their harmonics in the spectrum, which is beneficial for determining the type of fault and selecting the demodulation spectrum containing the characteristic frequency and its harmonic component prominence to achieve a better fault diagnosis effect.

The method proposed in this paper provides a new solution for the comprehensive use of bearing three-direction vibration acceleration signals, which can improve the accuracy of bearing fault diagnosis. Meanwhile, it also has good application value for the research of multi-signal fault diagnosis methods for other rotating machinery components. More research is still needed in signal noise reduction.

Author Contributions: Conceptualization, Z.W. and Z.Z.; methodology, Z.W. and Z.Z.; software, Z.W. and Z.Z.; validation, Z.Z. and X.X.; formal analysis, Z.W. and Z.Z.; investigation, Z.W. and X.X.; resources, D.S.; data curation, D.S.; writing—original draft preparation, Z.W.; writing—review and editing, Z.Z.; visualization, Z.W. and Z.Z.; supervision, D.S.; project administration, D.S.; funding acquisition, D.S. and Z.W. All authors have read and agreed to the published version of the manuscript.

Funding: This research was funded by Science and Technology Research and development program of China State Railway Group Co., Ltd., grant number N2023J011.

Data Availability Statement: The data presented in this study are available on request from the corresponding author.

Acknowledgments: The authors want to express their gratitude to Ping He at Southwest Jiaotong University, Jianxiong Dong at Chongqing University, Zhiheng Chen and Cong Deng at CRRC Zhuzhou Locomotive Co., Ltd (Zhuzhou, China) for their support with the rail vehicle bogie axle box bearing test rig.

Conflicts of Interest: Author Zhongyao Wang was employed by the company CRRC Changchun Railway Vehicles Co., Ltd. The remaining authors declare that the research was conducted in the absence of any commercial or financial relationships that could be construed as a potential conflict of interest.

References

1. Lu, Z.; Wang, X.; Yue, K.; Wei, J.; Yang, Z. Coupling model and vibration simulations of railway vehicles and running gear bearings with multitype defects. *Mech. Mach. Theory* **2021**, *157*, 104215.
2. Liu, Y.; Kang, J.; Bai, Y.; Guo, C. Research on the health status evaluation method of rolling bearing based on EMD-GA-BP. *Qual. Reliab. Eng. Int.* **2023**, *39*, 2069–2080.
3. Chen, J.; Hua, C.; Dong, D.; Ouyang, H. Adaptive scale decomposition and weighted multikernel correntropy for wheelset axle box bearing diagnosis under impact interference. *Mech. Mach. Theory* **2023**, *181*, 105220.
4. Chauhan, S.; Vashishtha, G.; Kumar, R.; Zimroz, R.; Gupta, M.; Kundu, P. An adaptive feature mode decomposition based on a novel health indicator for bearing fault diagnosis. *Measurement* **2024**, *226*, 114191.
5. Tan, H.; Xie, S.; Zhou, H.; Ma, W.; Yang, C.; Zhang, J. Sensible multiscale symbol dynamic entropy for fault diagnosis of bearing. *Int. J. Mech. Sci.* **2023**, *256*, 108509.
6. Zhou, K.; Tang, J. A wavelet neural network informed by time-domain signal preprocessing for bearing remaining useful life prediction. *Appl. Math. Model.* **2023**, *122*, 220–241.
7. Yang, M.; Zhang, K.; Sheng, Z.; Zhang, X.; Xu, Y. The amplitude modulation bispectrum: A weak modulation features extracting method for bearing fault diagnosis. *Reliab. Eng. Syst. Saf.* **2024**, *250*, 110241.
8. Chen, B.; Zhang, W.; Gu, J.; Song, D.; Cheng, Y.; Zhou, Z.; Gu, F.; Ball, A. Product envelope spectrum optimization-gram: An enhanced envelope analysis for rolling bearing fault diagnosis. *Mech. Syst. Signal Process.* **2023**, *193*, 110270.
9. Jiang, Z.; Zhang, K.; Xiang, L.; Yu, G.; Xu, Y. A time-frequency spectral amplitude modulation method and its applications in rolling bearing fault diagnosis. *Mech. Syst. Signal Process.* **2023**, *185*, 109832.
10. Zou, X.; Zhang, H.; Jiang, Z.; Zhang, K.; Xu, Y. Toward accurate extraction of bearing fault modulation characteristics with novel time–frequency modulation bispectrum and modulation Gini index analysis. *Mech. Syst. Signal Process.* **2024**, *219*, 111629.
11. Yang, J.; Bai, Y.; Wang, J.; Zhao, Y. Tri-axial vibration information fusion model and its application to gear fault diagnosis in variable working conditions. *Meas. Sci. Technol.* **2019**, *30*, 095009.
12. Chahinez, O.; Sadok, B. Data fusion for ITS: A systematic literature review. *Inf. Fusion* **2023**, *89*, 267–291.
13. Demirci, M.; Gozde, H.; Taplamacioglu, M. Improvement of power transformer fault diagnosis by using sequential Kalman filter sensor fusion. *Int. J. Electr. Power Energy Syst.* **2023**, *149*, 109038.
14. He, D.; Lao, Z.; Jin, Z.; He, C.; Shan, S.; Miao, J. Train bearing fault diagnosis based on multi-sensor data fusion and dual-scale residual network. *Nonlinear Dyn.* **2023**, *111*, 14901–14924.
15. Dubois, D.; Prade, H. Upper and lower possibilities induced by a multivalued mapping. *IFAC Proc.* **1983**, *16*, 325–339.
16. Tang, X.; Gu, X.; Rao, L.; Lu, J. A single fault detection method of gearbox based on random forest hybrid classifier and improved Dempster-Shafer information fusion. *Comput. Electr. Eng.* **2021**, *92*, 107101.
17. Panigrahy, P.; Chattopadhyay, P. Tri-axial vibration based collective feature analysis for decent fault classification of VFD fed induction motor. *Measurement* **2021**, *168*, 108460.
18. Rilling, G.; Flandrin, P.; Goncalves, P.; Lilly, J. Bivariate empirical mode decomposition. *IEEE Signal Process. Lett.* **2007**, *14*, 936–939.
19. Rehman, N.; Mandic, D. Empirical mode decomposition for trivariate signals. *IEEE Trans. Signal Process.* **2010**, *58*, 1059–1068.
20. Lv, Y.; Yuan, R.; Song, G. Multivariate empirical mode decomposition and its application to fault diagnosis of rolling bearing. *Mech. Syst. Signal Process.* **2016**, *81*, 219–234.

21. An, X.; Yang, J. A method of eliminating the vibration signal noise of hydropower unit based on NA-MEMD and approximate entropy. *Proc. Inst. Mech. Eng. Part E J. Process Mech. Eng.* **2014**, *231*, 317–328.
22. Rehman, N.; Aftab, H. Multivariate variational mode decomposition. *IEEE Trans. Signal Process.* **2019**, *67*, 6039–6052.
23. Wu, H.; Cheng, J.; Nie, Y.; Wang, J.; Yang, Y. Multivariate complex modulation model decomposition and its application to gear fault diagnosis. *Digit. Signal Process.* **2023**, *135*, 103940.
24. Zhou, J.; Yang, Y.; Li, X.; Shao, H.; Cheng, J. Multivariate local characteristic-scale decomposition and 1.5-dimensional empirical envelope spectrum based gear fault diagnosis. *Mech. Mach. Theory* **2022**, *172*, 104772.
25. Wang, H.; Ni, G.; Chen, J.; Qu, J. Research on rolling bearing state health monitoring and life prediction based on PCA and Internet of things with multi-sensor. *Measurement* **2020**, *157*, 107657.
26. Shahdoosti, H.; Ghassemian, H. Combining the spectral PCA and spatial PCA fusion methods by an optimal filter. *Inf. Fusion* **2016**, *27*, 150–160.
27. Mochammad, S.; Noh, Y.; Kang, Y. Bearing fault degradation modeling based on multitime windows fusion unsupervised health indicator. *IEEE Sens. J.* **2023**, *23*, 19623–19634.
28. Bashir, R.; Junejo, R.; Qadri, N.; Fleury, M.; Qadri, M. SWT and PCA image fusion methods for multi-modal imagery. *Multimed. Tools Appl.* **2019**, *78*, 1235–1263.
29. Soleimani, M.; Campean, F.; Neagu, D. Diagnostics and prognostics for complex systems: A review of methods and challenges. *Qual. Reliab. Eng. Int.* **2021**, *37*, 3746–3778.
30. Fairley, N.; Bargiela, P.; Huang, W.; Baltrusaitis, J. Principal Component Analysis (PCA) unravels spectral components present in XPS spectra of complex oxide films on iron foil. *Appl. Surf. Sci. Adv.* **2023**, *17*, 100447.
31. Lei, X.; Yu, J.; Aini, H.; Wu, W. Data-driven alternating current optimal power flow: A Lagrange multiplier based approach. *Energy Rep.* **2022**, *8*, 748–755.
32. Ding, H.; Wan, G.; Zhou, Y.; Tang, J.; Zhou, Z. Nonlinearity analysis based algorithm for indentifying machine settings in the tooth flank topography correction for hypoid gears. *Mech. Mach. Theory* **2017**, *113*, 1–21.
33. Antoni, J.; Randall, R. The spectral kurtosis: Application to the vibratory surveillance and diagnostics of rotating machines. *Mech. Syst. Signal Process.* **2006**, *20*, 308–331.
34. Mintzer, L. *Handbook of Digital Signal Processing*; Academic Press: Cambridge, MA, USA, 1987; pp. 941–973.
35. Chen, X.; Wang, S.; Qiao, B.; Chen, Q. Basic research on machinery fault diagnostics: Past, present, and future trends. *Front. Mech. Eng.* **2018**, *13*, 264–291.
36. Cheng, J.; Xiang, T.; Yang, X.; Yang, Y. Application of analytic energy operator demodulation method to fault diagnosis of rolling bearings. *Noise Vib. Control* **2017**, *37*, 151–155.
37. Xu, Y.; Wang, Y.; Wang, L.; Qu, J. Bearing fault detection using an alternative analytic energy operator: A fast and non-filtering method. *Meas. Sci. Technol.* **2021**, *32*, 105101.
38. Chen, B.; Song, D.; Zhang, W.; Cheng, Y.; Wang, Z. A performance enhanced time-varying morphological filtering method for bearing fault diagnosis. *Measurement* **2021**, *176*, 109163.
39. Li, Q.; Chen, B.; Zhang, W.; Song, D. Research on mathematical morphological operators for fault diagnosis of rolling element bearings. *Measurement* **2022**, *203*, 111964.
40. Liu, J.; Zhao, Z.; Ren, G. An Intelligent Fault Diagnosis Method for Bogie Bearings of Train Based on Wavelet Paceket Decomposition and EEMD. *J. China Railw. Soc.* **2015**, *37*, 40–45.
41. Zheng, Z.; Song, D.; Xu, X.; Lei, L. A fault diagnosis method of bogie axle box bearing based on spectrum whitening demodulation. *Sensors* **2020**, *20*, 7155. [CrossRef]

machines

MDPI

Article

A Resonance-Identification-Guided Autogram for the Fault Diagnosis of Rolling Element Bearings

Mingxuan Liu [1], Yiping Shen [1,2] and Yuandong Xu [1,2,*]

[1] The School of Mechanical Engineering, Hunan University of Science and Technology, Xiangtan 411100, China; 22010301024@mail.hnust.edu.cn (M.L.); ypsh@hnust.edu.cn (Y.S.)
[2] Hunan Provincial Key Laboratory of Health Maintenance for Mechanical Equipment, Hunan University of Science and Technology, Xiangtan 411100, China
* Correspondence: yuandong.xu@hnust.edu.cn

Abstract: Rolling element bearings are key components for reducing friction and supporting rotors. Harsh working conditions contribute to the wear of bearings and consequent breakdown of machines, which leads to economic losses and even catastrophic accidents. Faulty impulses from bearings can excite resonance behavior in a system and produce modulation phenomena. Fault characteristics in modulated signals can be extracted using demodulation analysis methods, significantly improving the reliability and effectiveness of the fault diagnosis of rolling bearings. Optimal demodulation frequency band selection is a primary step for the demodulation-analysis-based fault diagnosis of bearing faults. To exploit the resonant modulation mechanism in the fault diagnosis of rolling element bearings, resonant frequencies identified through stochastic subspace identification are employed to guide the impulsive sparsity measures of an Autogram for bearing fault diagnosis, which combines physical modulation dynamics and data characteristics. The frequency band that not only matches the natural frequencies but also shows highly sparse impulsive characteristics is selected as the optimal demodulation frequency band for bearing fault diagnosis. The results of simulations and experimental data validate the advantages of the proposed method, which exploits physics-guided data processing for optimal demodulation frequency band determination.

Keywords: resonant frequencies; autogram; optimal demodulation frequency band; bearing fault diagnosis

1. Introduction

Rolling element bearings play a critical role in rotating machines, and their failure can lead to serious consequences [1]. The diagnosis of bearing faults early on is highly desired to reduce accident probability [2,3]. Demodulation analysis is popular in early fault diagnosis for bearings, for which the prior step is to determine the optimal central frequency and the frequency bandwidth for the purpose of maximizing the ratio of fault features to noise. In the beginning, the optimal demodulation frequency band (ODFB) is determined artificially using expert knowledge, a process that is highly contingent in practice [4].

Spectral Kurtosis (SK) can be used to extract the impulsive transient responses of modulated signals [5]. Antoni utilized the properties of SK to develop a method named the Kurtogram for ODFB selection, which can identify a frequency band containing rich impulse components related to bearing faults [6]. The recognition accuracy of the Kurtogram based on short-time Fourier transform or finite impulse response filters is not robust enough in machine fault diagnosis. Some researchers improved the Kurtogram in terms of filter design, including via wavelet packets [7] and dual-tree wavelets [8]. A fast-computing algorithm for the Kurtogram called Fast Kurtogram (FK) was developed by Antoni. It enables SK to be used in online industrial condition monitoring [9]. The FK method is widely used in academic and industrial applications, and its weakness is that it is not reliable against strong background noise and irregular impulses. To address the limitations of kurtosis-based methods for bearing fault feature extraction, researchers have explored two main approaches. Some have combined kurtosis analysis with other techniques, such as Empirical Wavelet Transform [10], Ensemble Localized Average Decomposition [11], and Average Local Kurtosis [12]. Alternatively, other scholars have focused on replacing kurtosis entirely with other sparsity measures like the Gini index [13] and the Hoyer measure [14]. Both approaches have demonstrated promising results in enhancing fault detection. To increase the accuracy of the ODFB selection, the Protrugram method was proposed by Barszcz and validated by means of aero-engine vibration signals [15]. The kurtosis is calculated from the envelope spectrum instead of raw signals. Antoni proposed the Infogram method by introducing the negentropy of the squared envelope and its corresponding spectrum to increase the reliability of FK [16]. The Infogram was then improved by separating cyclostationary bearing signals from gear meshing vibrations [17]. The separation is susceptible to irrelevant cyclostationary sources. An Autogram method based on maximum overlapping discrete wavelet packet transform (MODWPT) and unbiased auto-correlation (AC) was proposed by Moshrefzadeh. It reduces the amplitude of random noise and demodulates the fault characteristics [18]. Wang proposed a maximum-envelope-based-Autogram (MEAutogram), which changes the spectral segmentation structure of an Autogram via the maximum value envelope method, enabling it to adaptively decide the segmentation position based on signal characteristics [19]. The Kurtogram and the other similar methods mentioned above focus on characterizing bearing vibration data by measuring the impulse sparsity of first-order or second-order cyclostationarity. These methods utilize the data's properties and do not take into account the physical resonant modulation of faulty bearing systems. Consequently, the dynamic mechanism is exploited in this paper to increase the effectiveness of bearing fault diagnosis from a new perspective.

The stochastic subspace identification (SSI) method obtains the resonant frequencies from the output time domain signals only. Its applications have been widely recognized in the structural dynamics of large buildings and machines [20]. An approach to recognizing modal patterns online called Correlation-Subset-Based Stochastic Subspace Identification was developed by Liu, and it was experimentally verified on a scaled bogie rig system [21]. A covariance-driven stochastic subspace identification method (SSI-COV) was proposed by Magalhaes and experimentally validated in bridge modal parameter estimation experiments [22]. Cheynet compared the SSI-COV method with automated Frequency Domain Decomposition (FDD) in the identification of modal damping ratios for large-span suspension bridges and found that the SSI-COV method has less bias [23]. However, the SSI-COV method cannot accurately identify excited resonant behavior due to the presence of other modes and fake modes under periodic excitations.

As the physical modulation mechanism is neglected in bearing fault diagnosis, a resonance-identification-guided Autogram (RIAutogram) was developed in this study to increase the accuracy and performance of ODFB-based envelope analysis for bearing fault diagnosis. The rest of this paper is organized as follows: Section 2 highlights the theoretical background of the proposed RIAutogram. In Section 3, the validity of the RIAutogram method in a low-signal-to-noise-ratio (SNR) environment is demonstrated through simulated bearing signals and compared with Autogram and FK. The improved RIAutogram method is compared to other methods using the experimental bearing datasets from Paderborn University Dataset (PUD) and a DC Generator fault bearing vibration signal in Section 4. Section 5 presents the conclusions.

2. Theoretical Background

The SSI-COV method is an operational modal analysis method that is well suited for resonance identification. The resonance can be used to reweight the impulsive sparsity measures to determine the ODFB under low SNR.

2.1. Covariance Driven Stochastic Subspace Identification

A linear dynamic model of mechanical systems can be transformed into discrete-time state spaces.

$$\begin{cases} x_{k+1} = Ax_k + w_k \\ y_k = Cx_k + v_k \end{cases} \tag{1}$$

where $x_k \in R^n$ and $y_k \in R^l$ are, respectively, the state variables and output of the system at the discrete time k; $A \in R^{n\times n}$ is the system matrix; $C \in R^{l\times n}$ is the output matrix of the system; and $w_k \in R^n$ and $v_k \in R^l$ are the system noise and measurement noise, respectively.

The SSI method consists of three key steps, which are orthogonal decomposition, singular value decomposition (SVD), and system parameter estimation. The method first constructs the Hankel matrix H from the following signals:

$$H = \left[\begin{array}{cccc} y_1 & y_2 & \cdots & y_n \\ y_2 & y_3 & \cdots & y_{n+1} \\ \vdots & \vdots & \ddots & \vdots \\ y_i & y_{i+1} & \cdots & y_{i+n-1} \\ y_{i+1} & y_{i+2} & \cdots & y_{i+n} \\ \hline y_{i+2} & y_{i+3} & \cdots & y_{i+n+1} \\ \vdots & \vdots & \ddots & \vdots \\ y_{i+j+1} & y_{i+j+2} & \cdots & y_{i+j+n} \end{array}\right] = \begin{bmatrix} Y_p \\ Y_f \end{bmatrix} \tag{2}$$

where Y_p is the past portion of the matrix H, and Y_f is the future portion of the matrix H. Unlike SSI, the SSI-COV method starts from a covariance matrix of a structural response-based time series, which is formulated as follows:

$$\Lambda_i = \frac{1}{n-i}\sum_{k=0}^{n-i-1} y_{k+i}\cdot y_k^T \tag{3}$$

where n indicates the point number in the time series, and $(\cdot)^T$ denotes the transpose of the matrix. These are constructed into a Toeplitz matrix T, and a singular value is decomposed:

$$\begin{aligned} T &= \begin{bmatrix} \Lambda_i & \Lambda_{i-1} & \Lambda_{i-2} & \cdots & \Lambda_1 \\ \Lambda_{i+1} & \Lambda_i & \Lambda_{i-1} & \cdots & \Lambda_2 \\ \Lambda_{i+2} & \Lambda_{i+1} & \Lambda_i & \cdots & \Lambda_3 \\ \vdots & \vdots & \vdots & \ddots & \vdots \\ \Lambda_{2i-1} & \Lambda_{2i-2} & \Lambda_{2i-3} & \cdots & \Lambda_i \end{bmatrix} = Y_f Y_p^T \\ &= U\begin{bmatrix} \Sigma & 0 \\ 0 & 0 \end{bmatrix} V^T = USV^T \end{aligned} \tag{4}$$

where U, V are m, n-dimensional orthogonal matrices, respectively. Σ is a diagonal matrix and can be expressed as

$$\sum = diag\{\sigma_1, \sigma_2, \ldots, \sigma_i, \ldots, \sigma_n\} \tag{5}$$

where the element σ_i on the diagonal is the ith nonzero singular value of the matrix S. Also, the Toeplitz matrix T can be denoted as the multiplication of the observable matrix Γ_i of the system and the Kalman filtered estimate $\hat{X}_i$:

$$T = \Gamma_i \hat{X}_i = \begin{bmatrix} C \\ C \\ CA^i \end{bmatrix} \hat{X}_i = (US^{\frac{1}{2}})(S^{\frac{1}{2}}V^T) \tag{6}$$

From (4), (6) and the least squares method, the system matrix A and the output matrix C can be estimated as

$$\begin{bmatrix} \hat{A} \\ \hat{C} \end{bmatrix} = \begin{bmatrix} \hat{X}_{k+1} \\ Y_k \end{bmatrix} (\hat{X}_k)^{\dagger} \tag{7}$$

where $(\cdot)^{\dagger}$ denotes the generalized inverse of the matrix, and $\hat{A}$ and $\hat{C}$ represent the estimates of A and C. Via eigenvalue decomposition of A, the eigenvalue matrix Δ can be obtained:

$$\Delta = diag[\lambda_{a,1}, \lambda_{a,2}, \ldots, \lambda_{a,n}] \tag{8}$$

The modal frequency (MF) f_i and damping ratio r_i can be derived from eigenvalue $\lambda_{a,i}$:

$$\begin{aligned} \lambda_{b,i} &= f_s \ln \lambda_{a,i} \\ f_i &= \frac{|\lambda_{b,i}|}{2\pi} \\ r_i &= -\frac{R(\lambda_{b,i})}{|\lambda_{b,i}|} \end{aligned} \tag{9}$$

where f_s is the sampling frequency, $\lambda_{b,i}$ is the ith (i = 1, 2,. . . . . ., nth) pole of the continuous state space mode, and $R(\lambda_{b,i})$ denotes the real part of $\lambda_{b,i}$.

The resonant frequencies of the dynamic system can be obtained sequentially with the increase in the system orders. The normalized ratios of stable modes can be accurately calculated in the iteration of system orders, which contributes to the strength of the resonance.

2.2. Autogram

An Autogram based on unbiased AC is a new approach for fault diagnosis in rotating machinery. On account of the downsampling operation, the discrete wavelet packet transform (DWPT) filter limits the ability of the wavelet transform coefficients to decompose the signal. The first step of Autogram employs a MODWPT filter instead of a DWPT filter because it eliminates the downsampling step in the DWPT and preserves the full-time resolution. The second step calculates the unbiased AC value of the squared envelope (SE) of the MODWPT-filtered signal for each frequency band.

$$\hat{R}_{XX}(\theta) = \frac{1}{N-q} \sum_{i=1}^{N-q} E(t_i)E(t_i + \theta) \tag{10}$$

where E is the SE of the filtered signal, $\theta = q/f_s$ is the delay factor, and $q = 0, 1, \ldots, N-1$.

The advantage of this process is that it removes interfering parts of the signal, such as periodic impulses. Finally, the kurtosis of unbiased AC is calculated to generate an Autogram, and the band with the highest kurtosis is located. Figure 1 illustrates the basic steps of Autogram.

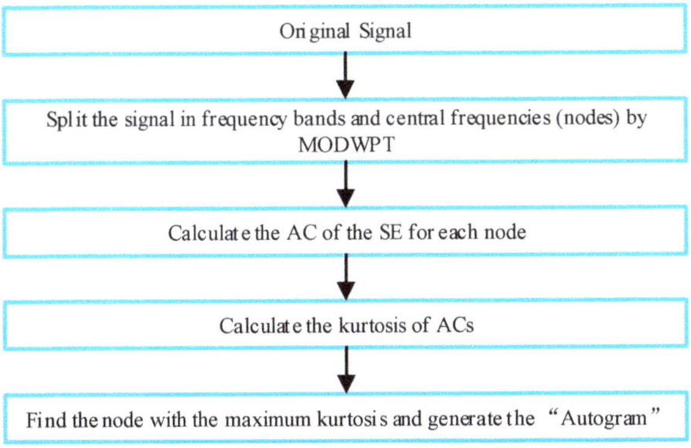

Figure 1. Basic steps of Autogram.

2.3. Resonance Identification Guided Autogram

Resonant frequencies excited in the vibration responses can be identified using SSI-COV. However, not all resonant frequencies are ideal for demodulation analysis. It is necessary to select the resonance behavior with the richest bearing fault information. The SSI method is not capable of figuring out the proper resonance that is most helpful for bearing fault diagnosis. Thereafter, Autogram, which is sensitive to the impulsive features of the data, is incorporated into the procces for accurately identifying ODFB. Autogram itself only selects the frequency band with maximum kurtosis, which is not always reliable for bearing fault diagnosis due to strong background noise or other interference components from surrounding environments.

In order to improve the accuracy and robustness of ODFB determination in challenging conditions, this study introduces the complementary physical modulation mechanism to the conventional data impulsiveness detecting method by developing a physics-guided processing route for ODFB selection.

The impulsive measures obtained by Autogram are reweighted according to the resonant frequencies identified by the SSI-COV using the following equation:

$$Kurtosis_{new} = Kurtosis_{raw} * (NR + 1) \tag{11}$$

where $Kurtosis_{new}$ denotes the weighted kurtosis values of the nodes matching the resonant frequencies. $Kurtosis_{raw}$ is the original kurtosis values of nodes from different levels, and NR represents the normalized ratio of the resonant frequencies in the identification process.

Finally, a new resonance-identification-guided Autogram is generated by combining the physical properties of the dynamic system with the impulse characteristics of the data sets, leading to robust and accurate determination of the optimal demodulation frequency band. The key steps of the proposed approach are illustrated in Figure 2.

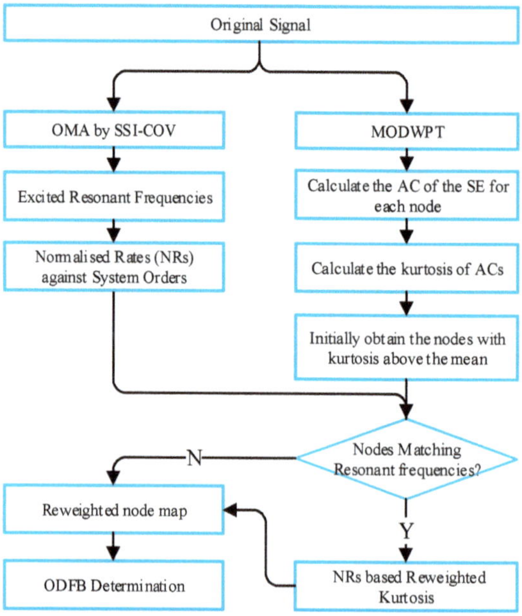

Figure 2. Flowchart of the RIAutogram.

3. Comparison on Simulated Signals

In engineering practice, the probability of bearing damage occurring in the outer and inner rings is higher. The simulated signals under outer and inner ring failures will be used in this section for the comparison and evaluation of the methods.

3.1. Vibration Signal Model of Bearing Faults

Considering that a rolling bearing is a linear system, the bearing vibration signal is represented by impulse response convolution with continuous shocks caused by a localized defect [24]. It can be expressed as

$$x(t) = h(t) * a(t) + n(t) \tag{12}$$

where $h(t) = e^{-\eta 2\pi f_m t_h} \times \cos(2\pi f_m t_h)$ is the system impulse response function, η is the damping parameter, f_m is the natural frequency (NF) of the bearing system, and t_h is a short decay period of damped vibrations. The pulse response of the bearing system is presumed as a single resonant mechanical system. $a(t)$ denotes the periodic shock generated by the roller passing through the fault, and $n(t)$ is the noise.

It is assumed that the bearing operates with a fixed outer race (i.e., stationary) and an inner race that follows the shaft in synchronized rotation. The position is fixed when there is a fault in the outer race of the bearing, and the amplitude of impact at the fault is constant. On the contrary, the location of the fault in the bearing inner race varies with the rotation, and the amplitude of impact varies in and out loading zone with the rotation of the shaft. Therefore, in the simulated signal, the amplitude of the signal is constant under the outer race fault, and the signal varies with the rotational frequency under the inner race fault. Thus, $a(t)$ in (12) can be expressed as

$$a(t) = \begin{cases} \delta(iT^f) & , T^f = \frac{1}{f_{outer}} \\ \left|\cos(2\pi \frac{f_r}{2} \delta(iT^f))\right| & , T^f = \frac{1}{f_{inner}} \end{cases} \tag{13}$$

where f_r is the rotational frequency; i denotes the ith pulse; T^f is the cycle of fault events (the reciprocal of the fault frequency); $\delta(\cdot)$ is the Dirac delta function; and f_{outer} and f_{inner} represent the ball pass frequency on the outer race (BPFO) and ball pass frequency on the inner race (BPFI), respectively.

Table 1 lists the variables for the simulated signal. The time and frequency domains are illustrated in Figure 3.

Table 1. Simulated signal variable settings.

Parameters	Value
Sampling Frequency	100,000 Hz
Sampling Time	60 s
BPFO	89.8 Hz
BPFI	135.2 Hz
NF	7094 Hz

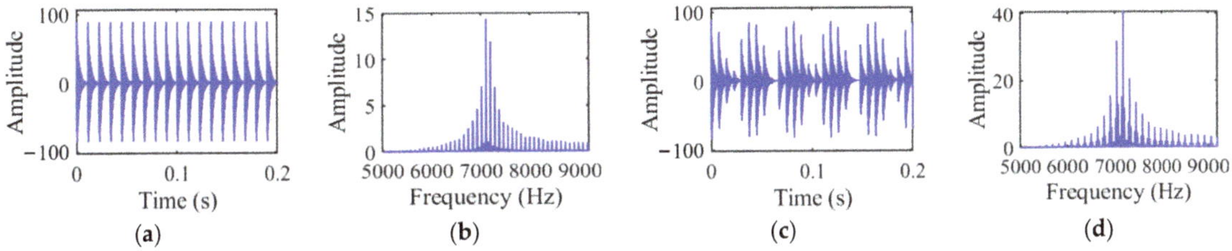

Figure 3. Time and frequency domain of simulated signals: (**a**,**b**) outer race fault and (**c**,**d**) inner race fault.

Ideally, the motion between the rollers and the bearing races is considered to be purely rolling. There is minor relative slippage between the rollers and the bearing raceways due to the surface roughness of the contact area and other factors. This slippage tends to cause random variations in the timing of the fault pulse. The pulse signal of the faulty bearing can only be considered as approximately periodic, and stochastic sliding of bearings can be simulated by the impact timing:

$$t_i = t_{i-1} + T^f + \Delta t_i^{SP} = iT^f + \sum_{j=1}^{i} \Delta t_j^{SP} \tag{14}$$

where Δt_i^{sp} is the stochastic sliding of the ith impact.

The smearing effect can be caused by bearing vibration due to random slippage. For example, Figure 4 shows a bearing vibration signal for an outer race fault that contains 2% of the slippage. The signals collected from the bearing system in the operation process inevitably suffer from various noise interference. The effectiveness of the RIAutogram is verified by adding Gaussian noise to the simulated signals. Figure 5a shows simulated signals for the outer race fault under an SNR of −25 dB. The impulse and harmonic components in the signal are indistinguishable. Detecting bearing faults directly from the raw signal is impossible.

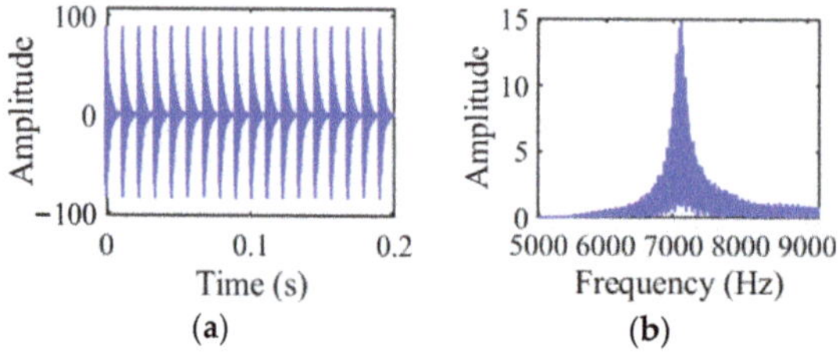

Figure 4. Impacts from random slippage: (**a**) time domain and (**b**) frequency domain.

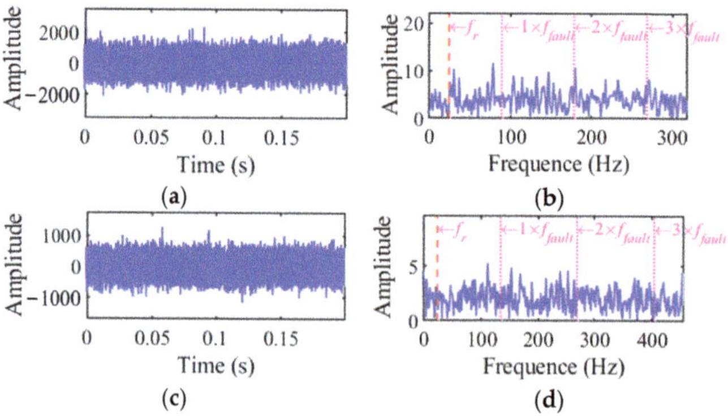

Figure 5. Simulated signals and frequency spectrum: (**a**,**b**) outer race fault and (**c**,**d**) inner race fault.

3.2. Diagnostic Results of Simulated Signals

A simulated case study is conducted to highlight the diagnostic performance of the resonance-identification-guided Autogram, benchmarked with the FK and Autogram. To simulate an early bearing fault, the slippage is set to 2%, and the SNR is −25 dB. Figure 5 shows the simulated signal.

The demodulation results of the three ODFB selection methods are illustrated in Figure 6. The center frequencies of the inner race fault are selected as 12,500 Hz using FK, 11,718.8 Hz using Autogram, and 7031.3 Hz using RIAutogram. It can be seen that only RIAutogram obtains the simulated resonant frequency, and the fault characteristic frequencies are easily distinguished.

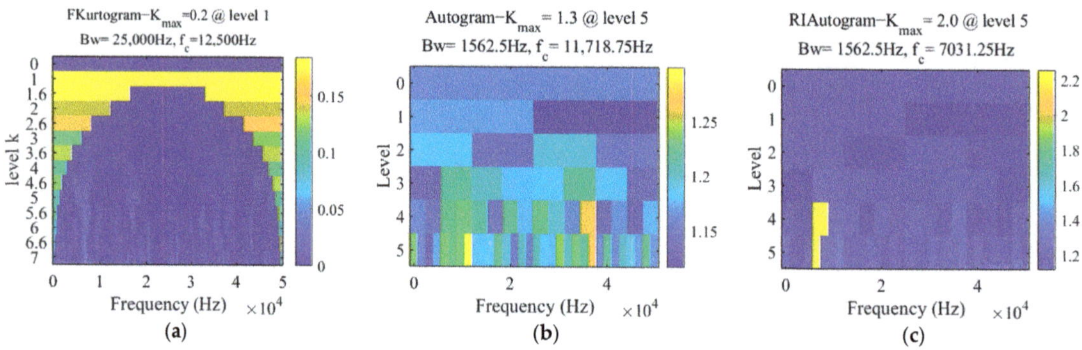

Figure 6. *Cont.*

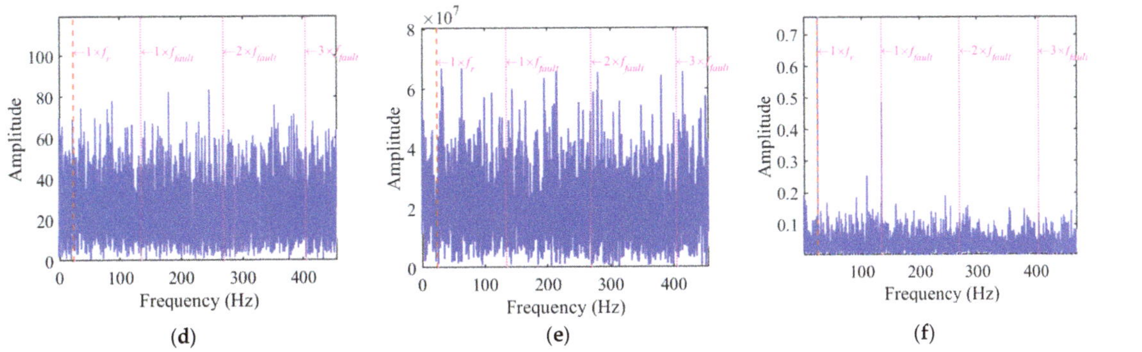

Figure 6. Processing results of inner ring fault simulated signal: (**a**,**d**) FK, (**b**,**e**) Autogram, and (**c**,**f**) RIAutogram.

The ODFB and diagnostic results for outer race fault signals are displayed in Figure 7. Similarly, only RIAutogram can obtain the proper frequency band for bearing fault diagnosis. The bandwidth determined by FK is wider compared to that of the other two approaches because FK is more sensitive to the impacting property than to the periodicity of the signal. Autogram is able to identify narrow bands containing impulsive information, but the determined frequencies are not coincided with the resonant frequencies, resulting in failure of bearing fault detection. The resonance identification makes an excellent complement to Autogram using the dynamic resonant modulation of the bearing systems.

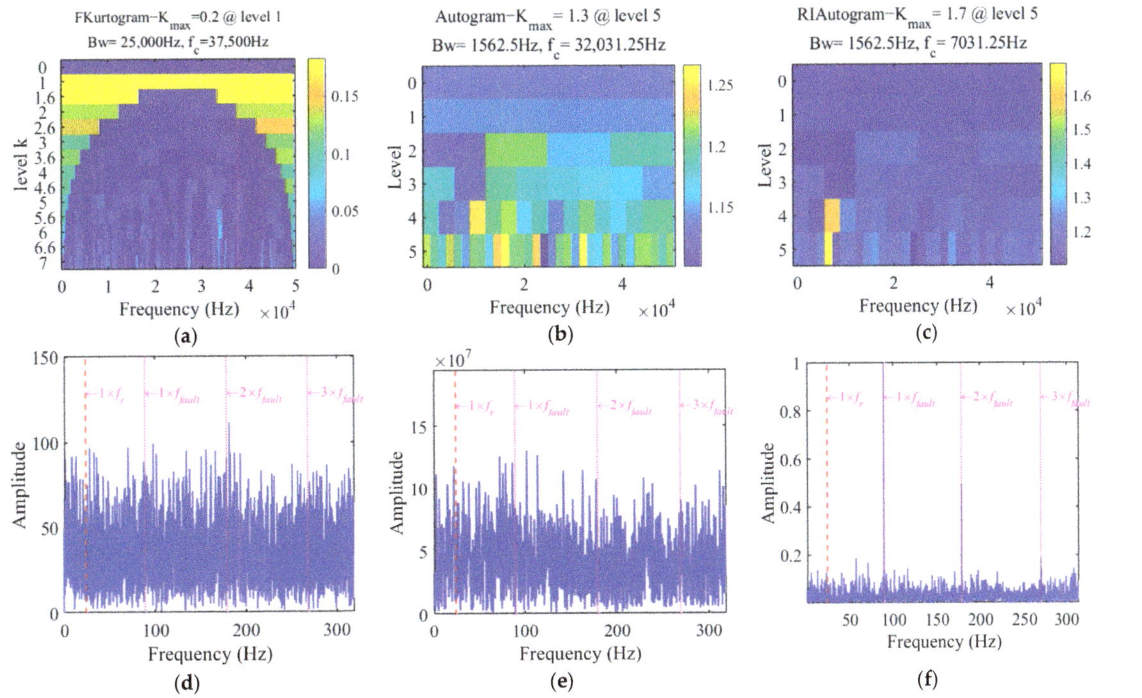

Figure 7. Processing results of outer ring fault simulated signal: (**a**,**d**) FK, (**b**,**e**) Autogram, and (**c**,**f**) RIAutogram.

RIAutogram shows pronounced advantages over FK and Autogram at various conditions.

3.3. Impact of Noise on RIAutogram Diagnostic Performance

To evaluate noise robustness, Gaussian white noise is added to the simulated vibration signals with SNR levels ranging from −25 dB to 0 dB. The length of the signal is taken as 1 s. A simulated signal of inner race fault is used, and the number of tests is 30 at each SNR value. The diagnostic results are shown in Table 2. When SNR > −15, all three methods provide an accurate diagnosis for the fault. Therefore, the table only counts the effect of SNR values below −15 on the diagnostic effectiveness. The table indicates that the RIAutogram maintains a precision above 90% for SNR ≥ −21 dB, outperforming Autogram (83.3% at SNR = −21 dB). However, at SNR = −23 dB, the precision drops to 30%, indicating a limitation in extreme noise conditions.

Table 2. Comparison of the effect of noise on diagnostic performance.

Method	Precision at Different SNR Values (%)						
	0	−15	−20	−21	−22	−23	−25
RIAutogram	100	100	93.3	90	70	30	3.3
Fast Kurtogram	100	96.7	67.7	50	36.7	6.7	0
Autogram	100	100	93.3	83.3	50	26.7	0

4. Comparison of Methods on Experimental Signals

In this section, two test cases are utilized to validate the RIAutogram method. The first case study uses an open data sets of the PUD bearing signals [25], and the second one is based on a motor bearing test rig.

4.1. Case 1 Fault Diagnosis of Open PUD Bearing Data

The PUD test rig has a 6203 bearing with 8 rollers that have a diameter of 6.75 mm. The contact angle is 0, and the pitch diameter is 29.05 mm. Inner and outer race fault vibration signals are acquired in the same operating conditions. The sampling frequency is configured as 64 kHz, and the time duration is 1 s. The shaft rotation frequency is 25 Hz, and the BPFI and BPFO of the 6203 bearing are 123.2 Hz and 76.8 Hz, respectively.

Figure 8 illustrates the ODFB selection and demodulation outcomes of three methods to process the PUD inner race fault signals. Comparative ODFB selections across the three methods highlight critical differences: FK yields a frequency band of 16,000–32,000 Hz, and Autogram picks the frequency band at 21,000–22,000 Hz. The proposed RIAutogram indicates that the frequency band 9000–10,000 Hz is the best for bearing fault diagnosis. The demodulation results of both Autogram and FK failed to reveal fault frequencies due to strong interference. In contrast, RIAutogram successfully identified fault characteristic frequencies of the inner race defects.

For outer race fault analysis, three approaches are also used to select distinct ODFBs. The frequency band of 31,333.3–32,000 Hz is determined by FK, and that of Autogram is 5000–6000 Hz, while RIAutogram recommends the frequency band of 0–4000 Hz. It is consistent with Figure 9 that the demodulation results from Autogram and FK are ineffective for outer race fault detection, while RIAutogram clearly reveals the BPFO. The high-order harmonics of the fault frequency demonstrate its superiority in bearing fault diagnosis in low-SNR environments.

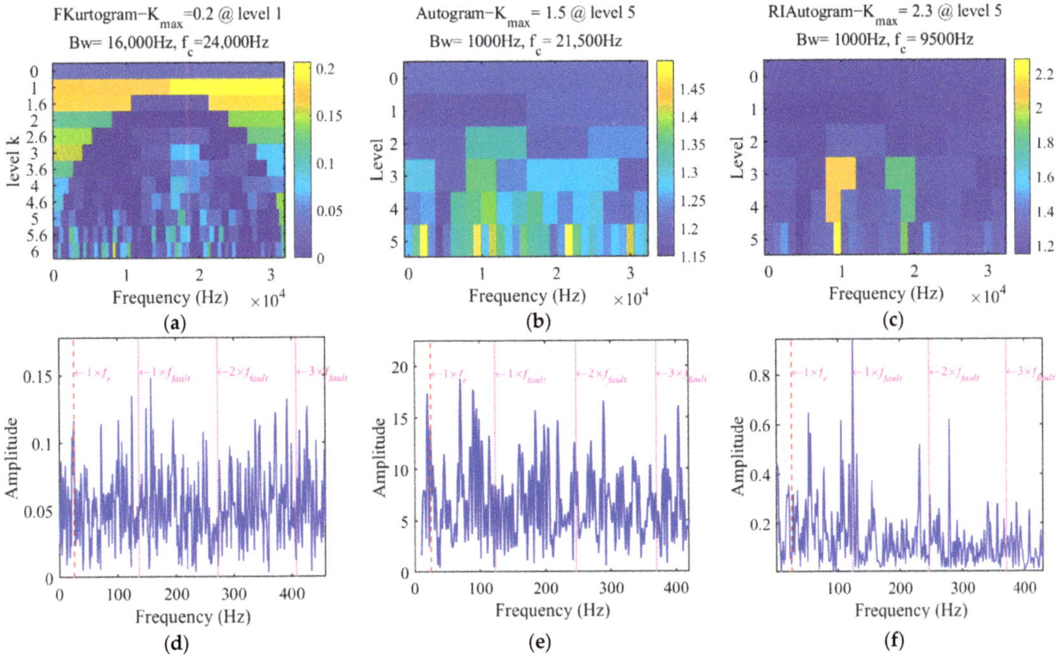

Figure 8. Processing results of PUD inner signal under −18 dB SNR: (**a**,**d**) FK, (**b**,**e**) Autogram, and (**c**,**f**) RIAutogram.

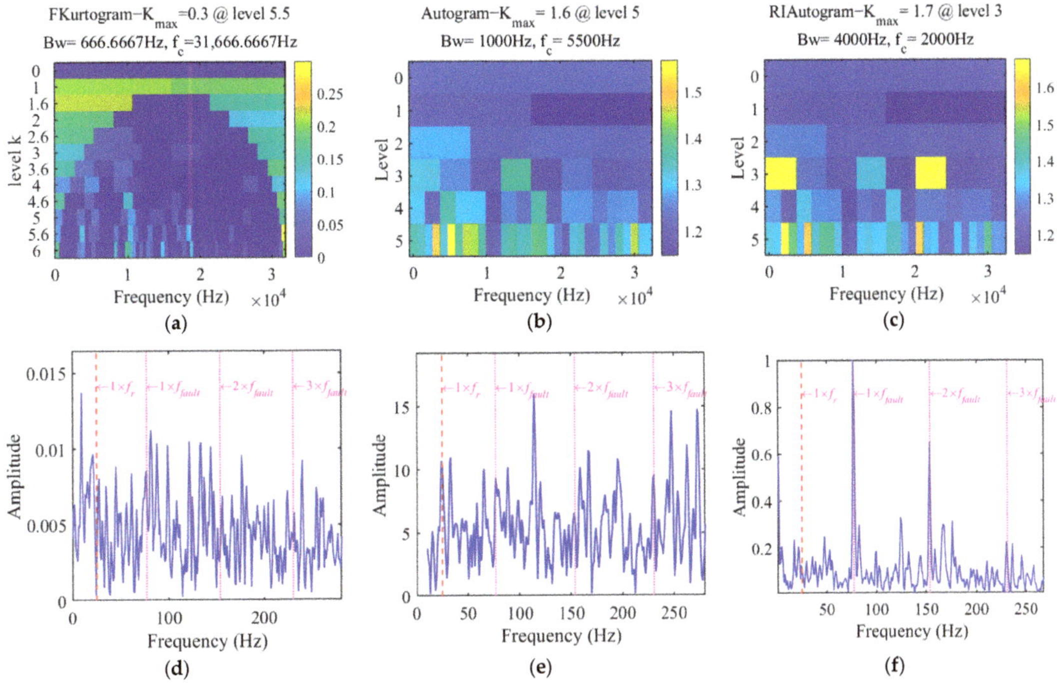

Figure 9. Processing results of PUD outer signal under −18 dB SNR: (**a**,**d**) FK, (**b**,**e**) Autogram, and (**c**,**f**) RIAutogram.

4.2. Case 2 Fault Diagnosis of Motor Bearings

In this section, the effectiveness of the RIAutogram is further verified using experimental data from induction motors. The vibration data are generated from the induction motor bearing test rig shown in Figure 10. The localized bearing faults are seeded as displayed in Figure 11. The test bench has a 6206 bearing with 9 rollers that have a diameter of 9.53 mm. The inner and outer race fault signals are acquired at a sampling rate of 96,000 Hz with a duration of 1 s. The rotating frequency is about 25 Hz.

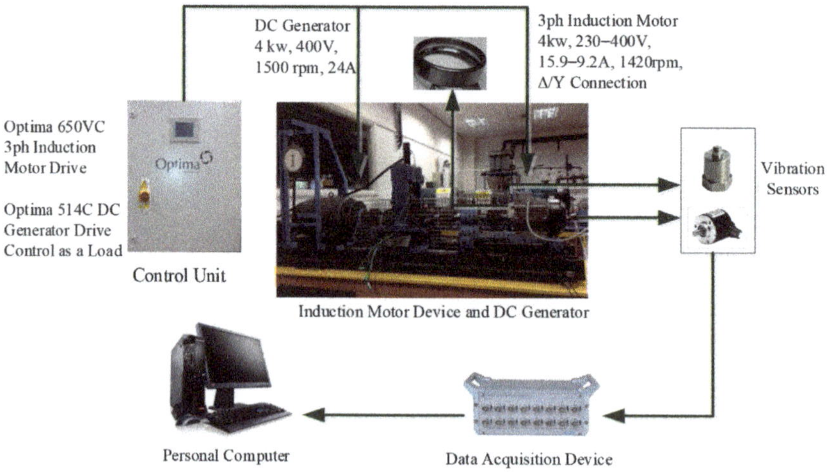

Figure 10. Schematic diagram of the bearing test system.

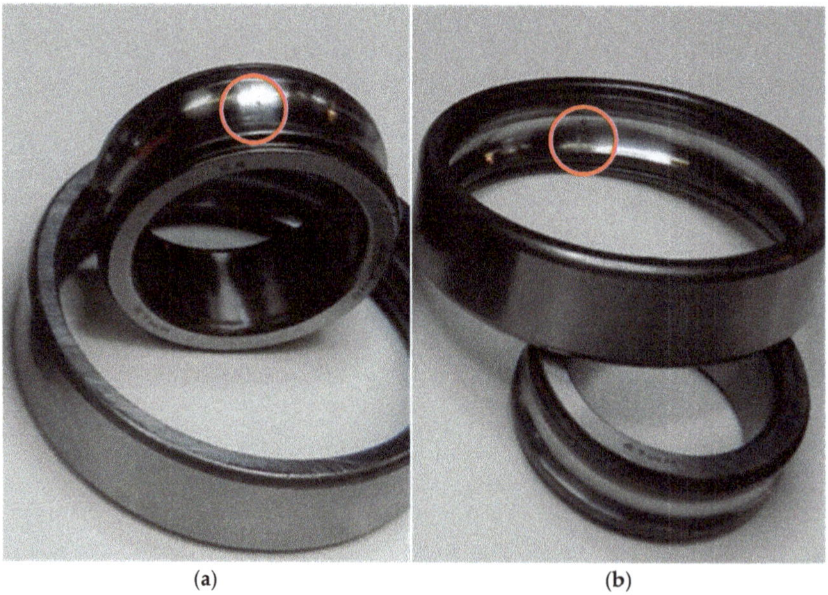

Figure 11. Setting scheme for bearing faults: (**a**) inner fault and (**b**) outer fault.

Figure 12 presents the ODFB selection and demodulation outcomes of three methods for the inner race fault signals. The employed method identified distinct ODFBs: FK selected 47,500–48,000 Hz, Autogram selected 6000–7500 Hz, and RIAutogram selected a lower frequency range, 1500–3000 Hz. Envelope spectra obtained using Autogram and FK failed to reveal fault frequencies due to noise interference. In contrast, RIAutogram successfully identified the fault characteristic frequencies as well as the second and third harmonics, which shows the advantageous capability of bearing fault diagnosis.

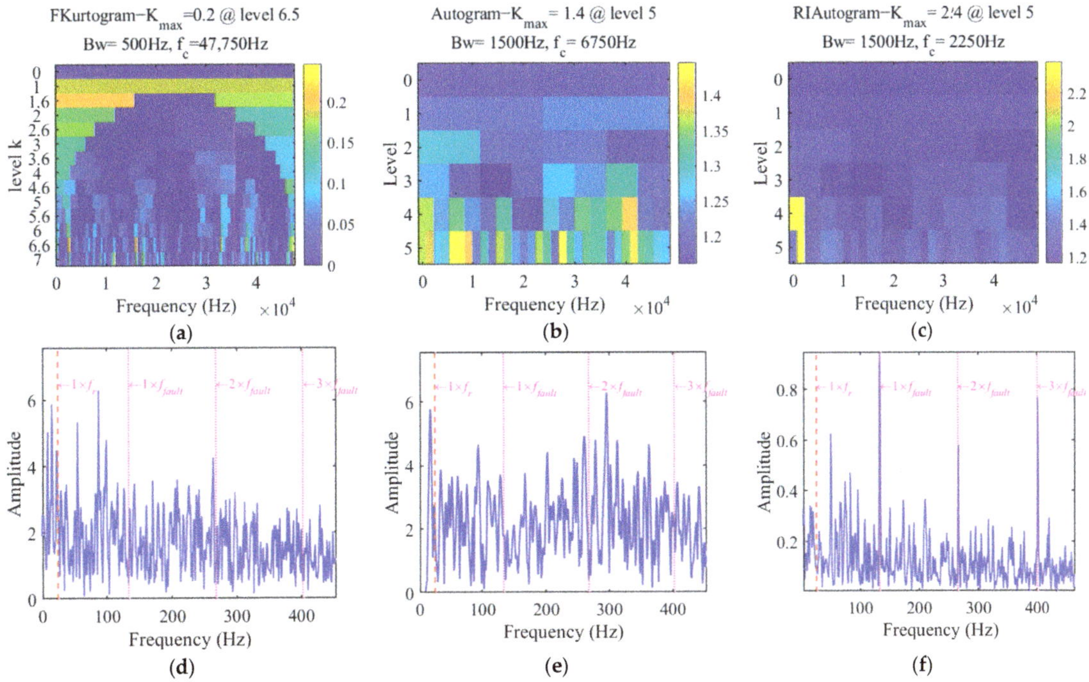

Figure 12. Processing results of inner ring fault vibration signal: (**a**,**d**) FK, (**b**,**e**) Autogram, and (**c**,**f**) RIAutogram.

For the outer race faults, FK obtains an ODFB of 3000–3750 Hz, which is not useful for detecting the outer race defect, as shown in Figure 13d. Autogram selects the frequency range from 37,500 Hz to 39,000 Hz for demodulation analysis, and the corresponding spectra results in Figure 13e are unable to be used to extract bearing fault information. As illustrated in Figure 13c,f, RIAutogram selects an ODFB of 3000–4000 Hz, and the demodulated spectrum reveals the BPFI and its high-order harmonics. In addition, the side-band of the rotation frequency of the fundamental BPFI is visible, showing the promising diagnostic performance.

Two experimental case studies show that RIAutogram can effectively determine the ODFB for effective and efficient bearing fault diagnosis. FK and Autogram are unable to diagnose the early bearing faults under extremely low SNRs. The frequency band identified by FK is very wide, which contains some fault information but simultaneously includes large quantities of noise, shielding the fault features. A relatively narrow frequency band can be picked in Autogram, but the center frequency is not appropriate due to the unrobust sparsity statistics. By combining the advantages of the physical resonance modulation mechanism, RIAutogram enables the accurate and robust selection of the optimal frequency band that contains the richest fault information under strong background noise for early bearing fault diagnosis.

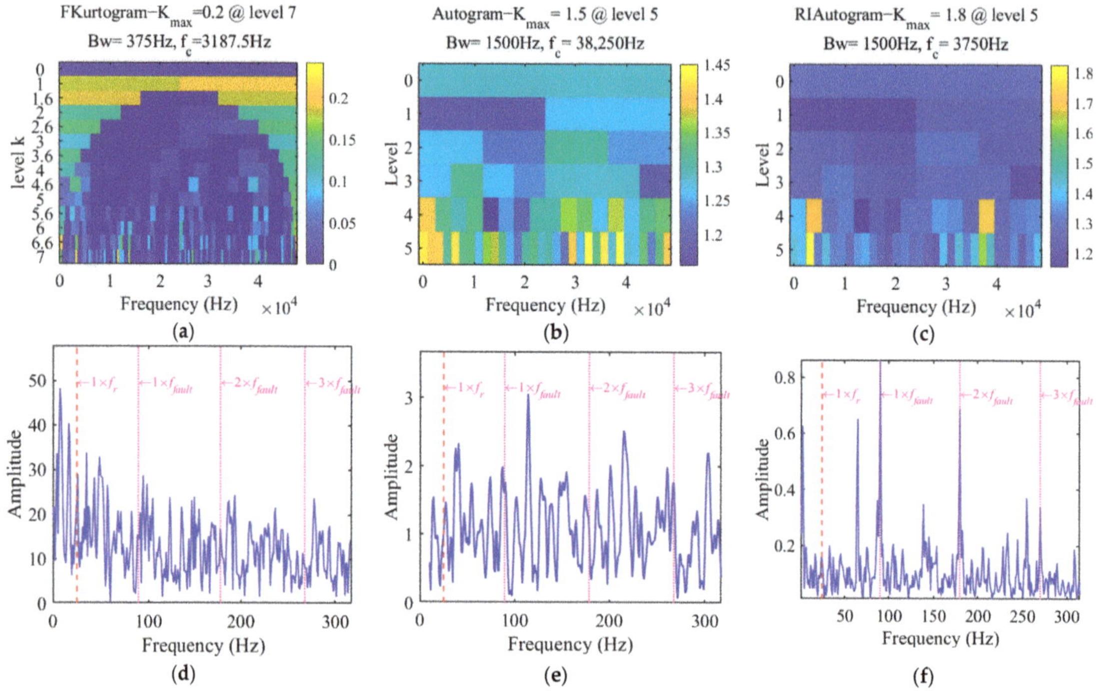

Figure 13. Processing results of outer ring fault vibration signal: (**a**,**d**) FK, (**b**,**e**) Autogram, and (**c**,**f**) RIAutogram.

4.3. Performance Metrics and Comparative Evaluation of the Algorithms

In this section, the diagnostic performance of the RIAutogram method will be effectively quantified by utilizing the properties of precision of fault detection, processing time, and memory usage. The data for the evaluation are generated from the experimental bench mentioned in Section 4.2, measuring the diagnostic performance of RIAutogram by processing signals of different lengths. The length of the signal is taken as 1 s, 15 s, and 30 s. For each signal length, 30 tests were conducted, comprising 15 tests using outer race fault signals and 15 tests using inner race fault signals. Processing time and memory usage are averaged over all test results. The diagnostic performance test results of the three methods are shown in Table 3, Table 4, and Table 5, respectively.

Table 3. Test results of RIAutogram diagnostic performance.

Signal Length	Precision (%)	Processing Time (s)	Memory Usage (MB)
1 s	83.3	3.72	192.9
15 s	90.0	31.5	2683.7
30 s	93.3	42.13	5574.8

Table 4. Test results of Fast Kurtogram diagnostic performance.

Signal Length	Precision (%)	Processing Time (s)	Memory Usage (MB)
1 s	63.3	0.22	0.8
15 s	73.3	0.96	18.4
30 s	90	1.85	36.7

Table 5. Test results of Autogram diagnostic performance.

Signal Length	Precision (%)	Processing Time (s)	Memory Usage (MB)
1 s	76.7	1.26	167.69
15 s	86.7	27.81	2599.2
30 s	93.3	39.1	5198.3

The results of the diagnostic performance comparison of the three methods show that RIAutogram is the method of choice for high-precision diagnostic scenarios due to its excellent accuracy and stable long-signal adaptation. In contrast, the FK method is more suitable for occasions with higher real-time requirements. In the diagnostic results, the running speed and memory occupation of the RIAutogram method are slightly insufficient compared with Autogram. This is attributed to the fact that RIAutogram incorporates resonant information, building upon Autogram, which inherently increases computational complexity. However, anti-noise and accuracy are guaranteed, which facilitates early fault diagnosis.

5. Conclusions

For rotating machinery systems, early fault detection is of great significance. The rotating machinery system is often accompanied by strong noise interference in practical engineering conditions, which causes failure of condition monitoring. Most of the similar methods focus on the impulse properties in the data but neglect the physical resonant modulation mechanism. A RIAutogram method is proposed in this paper to deliver physical information guided optimal frequency band selection. The complementary advantages of resonant frequency identification and impulsive characteristics make the proposed approach robust and effective for the accurate diagnosis of bearings with weak faults in low SNR vibration signals. The RIAutogram is validated by both simulation and experimental studies to be superior to FK and Autogram in the fault diagnosis of bearings. In addition, the RIAutogram method, which is based on the physical resonance mechanism guiding the band selection, is able to shield the interference of incoherent noise well. This is a significant advantage in low signal-to-noise environments such as railroad train bearings.

Author Contributions: Conceptualization, M.L. and Y.X.; methodology, Y.X.; software, M.L. and Y.X.; validation, M.L., Y.S. and Y.X.; formal analysis, M.L., Y.S. and Y.X.; investigation, M.L. and Y.X.; resources, Y.S. and Y.X.; data curation, M.L. and Y.X.; writing—original draft preparation, M.L.; writing—review and editing, Y.S. and Y.X.; visualization, M.L. and Y.X.; supervision, Y.S.; project administration, Y.S. and Y.X.; funding acquisition, Y.S. and Y.X. All authors have read and agreed to the published version of the manuscript.

Funding: This work was supported in part by the National Natural Science Foundation of China (grant No. 52205099), in part by the Natural Science Foundation of Hunan Province (grant 2024JJ6215), and in part by the Hunan Provincial Key Research and Development Program of China (grant 2023GK2026).

Data Availability Statement: The publicly available datasets in Section 4.1 are available at "https://mb.uni-paderborn.de/kat/forschung/kat-datacenter/bearing-datacenter (accessed on 7 July 2024)". The test data in Section 4.2 are unavailable due to privacy or ethical restrictions.

Conflicts of Interest: The authors declare no conflicts of interest.

Abbreviations

The following abbreviations are used in this manuscript:

AC	Auto-correlation.
BPFI	Ball pass frequency on inner race.
BPFO	Ball pass frequency on outer race.
DWPT	Discrete wavelet packet transform.
FDD	Frequency domain decomposition.
FK	Fast Kurtogram.
MF	Modal frequency.
MODWPT	Maximum overlapping discrete wavelet packet transform.
NF	Natural frequency.
NR	Nomalised rate.
ODFB	Optimal demodulation frequency band.
PUD	Paderborn University Dataset.
RIAutogram	Resonance identification guided Autogram.
SE	Squared envelope.
SK	Spectral Kurtosis.
SNR	Signal-to-noise ratio.
SSI	Stochastic subspace identification.
SSI-COV	Covariance driven stochastic subspace identification.
SVD	Singular value decomposition.

References

1. Chen, B.; Cheng, Y.; Cao, H.; Song, S.; Mei, G.; Gu, F.; Zhang, W.; Ball, A.D. Generalized Statistical Indicators-Guided Signal Blind Deconvolution for Fault Diagnosis of Railway Vehicle Axle-Box Bearings. *IEEE Trans. Veh. Technol.* **2024**, *74*, 2458–2469. [CrossRef]
2. Chen, B.; Cheng, Y.; Allen, P.; Wang, S.; Gu, F.; Zhang, W.; Ball, A.D. A Product Envelope Spectrum Generated from Spectral Correlation/Coherence for Railway Axle-Box Bearing Fault Diagnosis. *Mech. Syst. Signal Process.* **2025**, *225*, 112262. [CrossRef]
3. Chen, B.; Smith, W.A.; Cheng, Y.; Gu, F.; Chu, F.; Zhang, W.; Ball, A.D. Probability Distributions and Typical Sparsity Measures of Hilbert Transform-Based Generalized Envelopes and Their Application to Machine Condition Monitoring. *Mech. Syst. Signal Process.* **2025**, *224*, 112026. [CrossRef]
4. Zheng, J.; Wang, X.; Pan, H.; Tong, J.; Zhang, J.; Liu, Q. The Traverse Symplectic Correlation-Gram (TSCgram): A New and Effective Method of Optimal Demodulation Band Selection for Rolling Bearing. *IEEE Trans. Instrum. Meas.* **2021**, *70*, 1–15. [CrossRef]
5. Dwyer, R. Detection of Non-Gaussian Signals by Frequency Domain Kurtosis Estimation. In Proceedings of the ICASSP '83. IEEE International Conference on Acoustics, Speech, and Signal Processing, Boston, MA, USA, 14–16 April 1983; Institute of Electrical and Electronics Engineers: Boston, MA, USA, 1983; Volume 8, pp. 607–610.
6. Antoni, J.; Randall, R.B. The Spectral Kurtosis: Application to the Vibratory Surveillance and Diagnostics of Rotating Machines. *Mech. Syst. Signal Process.* **2006**, *20*, 308–331. [CrossRef]
7. Lei, Y.; Lin, J.; He, Z.; Zi, Y. Application of an Improved Kurtogram Method for Fault Diagnosis of Rolling Element Bearings. *Mech. Syst. Signal Process.* **2011**, *25*, 1738–1749. [CrossRef]
8. Wang, Y.; He, Z.; Zi, Y. Enhancement of Signal Denoising and Multiple Fault Signatures Detecting in Rotating Machinery Using Dual-Tree Complex Wavelet Transform. *Mech. Syst. Signal Process.* **2010**, *24*, 119–137. [CrossRef]
9. Antoni, J. Fast Computation of the Kurtogram for the Detection of Transient Faults. *Mech. Syst. Signal Process.* **2007**, *21*, 108–124. [CrossRef]
10. Xu, Y.; Tian, W.; Zhang, K.; Ma, C. Application of an Enhanced Fast Kurtogram Based on Empirical Wavelet Transform for Bearing Fault Diagnosis. *Meas. Sci. Technol.* **2019**, *30*, 035001. [CrossRef]
11. Wang, L.; Liu, Z.; Miao, Q.; Zhang, X. Time–Frequency Analysis Based on Ensemble Local Mean Decomposition and Fast Kurtogram for Rotating Machinery Fault Diagnosis. *Mech. Syst. Signal Process.* **2018**, *103*, 60–75. [CrossRef]
12. Liang, K.; Zhao, M.; Lin, J.; Ding, C.; Jiao, J.; Zhang, Z. A Novel Indicator to Improve Fast Kurtogram for the Health Monitoring of Rolling Bearing. *IEEE Sens. J.* **2020**, *20*, 12252–12261. [CrossRef]
13. Miao, Y.; Zhao, M.; Lin, J. Improvement of Kurtosis-Guided-Grams via Gini Index for Bearing Fault Feature Identification. *Meas. Sci. Technol.* **2017**, *28*, 125001. [CrossRef]

14. Tian, T.; Tang, G.; Wang, X.; Sun, J. An Improved Envelope Spectrum via Hoyer Index-Gram for Bearing Fault Extraction. *Meas. Sci. Technol.* **2024**, *35*, 076124. [CrossRef]
15. Barszcz, T.; JabŁoński, A. A Novel Method for the Optimal Band Selection for Vibration Signal Demodulation and Comparison with the Kurtogram. *Mech. Syst. Signal Process.* **2011**, *25*, 431–451. [CrossRef]
16. Antoni, J. The Infogram: Entropic Evidence of the Signature of Repetitive Transients. *Mech. Syst. Signal Process.* **2016**, *74*, 73–94. [CrossRef]
17. Smith, W.A.; Borghesani, P.; Ni, Q.; Wang, K.; Peng, Z. Optimal Demodulation-Band Selection for Envelope-Based Diagnostics: A Comparative Study of Traditional and Novel Tools. *Mech. Syst. Signal Process.* **2019**, *134*, 106303. [CrossRef]
18. Moshrefzadeh, A.; Fasana, A. The Autogram: An Effective Approach for Selecting the Optimal Demodulation Band in Rolling Element Bearings Diagnosis. *Mech. Syst. Signal Process.* **2018**, *105*, 294–318. [CrossRef]
19. Wang, X.; Zheng, J.; Pan, H.; Liu, Q.; Wang, C. Maximum Envelope-Based Autogram and Symplectic Geometry Mode Decomposition Based Gear Fault Diagnosis Method. *Measurement* **2021**, *174*, 108575. [CrossRef]
20. Magalhães, F. Explaining Operational Modal Analysis with Data from an Arch Bridge. *Mech. Syst. Signal Process.* **2011**, *25*, 1431–1450. [CrossRef]
21. Liu, F.; Zhang, H.; He, X.; Zhao, Y.; Gu, F.; Ball, A.D. Correlation Signal Subset-Based Stochastic Subspace Identification for an Online Identification of Railway Vehicle Suspension Systems. *Veh. Syst. Dyn.* **2020**, *58*, 569–589. [CrossRef]
22. Magalhães, F.; Cunha, Á.; Caetano, E. Online Automatic Identification of the Modal Parameters of a Long Span Arch Bridge. *Mech. Syst. Signal Process.* **2009**, *23*, 316–329. [CrossRef]
23. Cheynet, E.; Jakobsen, J.B.; Snæbjörnsson, J. Damping Estimation of Large Wind-Sensitive Structures. *Procedia Eng.* **2017**, *199*, 2047–2053. [CrossRef]
24. Xu, Y.; Zhen, D.; Gu, J.X.; Rabeyee, K.; Chu, F.; Gu, F.; Ball, A.D. Autocorrelated Envelopes for Early Fault Detection of Rolling Bearings. *Mech. Syst. Signal Process.* **2021**, *146*, 106990. [CrossRef]
25. Lessmeier, C.; Kimotho, J.K.; Zimmer, D.; Sextro, W. Bearing DataCenter—Konstruktions- Und Antriebstechnik (KAt) | Universität Paderborn. Available online: https://mb.uni-paderborn.de/kat/forschung/kat-datacenter/bearing-datacenter (accessed on 31 July 2024).

machines

MDPI

Article

A Novel Bearing Fault Diagnosis Method Based on Improved Convolutional Neural Network and Multi-Sensor Fusion

Zhongyao Wang [1,2], Xiao Xu [3,*], Dongli Song [3], Zejun Zheng [3] and Weidong Li [1]

1 School of Mechanical Engineering, Dalian Jiaotong University, Dalian 116028, China; wzy7889388@126.com (Z.W.); li@djtu.edu.cn (W.L.)
2 CRRC Changchun Railway Vehicles Co., Ltd., Changchun 130062, China
3 State Key Laboratory of Rail Transit Vehicle System, Southwest Jiaotong University, Chengdu 610031, China; sdlcds@swjtu.edu.cn (D.S.); zhengjuner@my.swjtu.edu.cn (Z.Z.)
* Correspondence: xuxiao@my.swjtu.edu.cn

Abstract: Bearings are key components of modern mechanical equipment. To address the issue that the limited information contained in the single-source signal of the bearing leads to the limited accuracy of the single-source fault diagnosis method, a multi-sensor fusion fault diagnosis method is proposed to improve the reliability of bearing fault diagnosis. Firstly, the feature extraction process of the convolutional neural network (CNN) is improved based on the theory of variational Bayesian inference, which forms the variational Bayesian inference convolutional neural network (VBICNN). VBICNN is used to obtain preliminary diagnosis results of single-channel signals. Secondly, considering the redundancy of information contained in multi-channel signals, a voting strategy is used to fuse the preliminary diagnosis results of the single-channel model to obtain the final results. Finally, the proposed method is evaluated by an experimental dataset of the axlebox bearing of a high-speed train. The results show that the average diagnosis accuracy of the proposed method can reach more than 99% and has favorable stability.

Keywords: rolling bearing; fault diagnosis; convolutional neural network; variational Bayesian inference; decision-level fusion

1. Introduction

Bearings are essential components of modern equipment, which directly affect the safety and reliability of equipment [1,2]. In large-scale equipment with harsh working conditions, such as high-speed trains and wind turbines, the harsh operating environment and high-load conditions cause bearings to become one of the key components that are easy to be damaged [3,4]. Therefore, it is of great significance to research effective approaches to diagnose the failure of bearings.

Bearing fault diagnosis can usually be divided into three main steps: data acquisition, feature extraction, and pattern recognition [5]. Due to the convenience of measuring mechanical vibration, bearing fault diagnosis methods based on vibration signals have received a lot of attention [6,7]. Traditional methods of bearing fault diagnosis mainly focus on signal analysis and processing [8–10]. This kind of research often adopts Fourier transform, envelope analysis, spectral kurtosis, wavelet transform, and empirical modal decomposition to process the vibration signal. Then, different features are extracted from the processed vibration signals for detecting the health status of bearings, and the common features can be categorized into time-domain features, frequency-domain features, and time-frequency-domain features. Existing studies usually input the features into

Machines **2025**, *13*, 216 https://doi.org/10.3390/machines13030216

intelligent classification algorithms to realize the automatic identification of the health status of bearings, and the typical classification algorithms are BP neural networks [11], support vector machines [12,13], and Bayesian classifiers [14]. Although the traditional fault diagnosis methods based on signal processing have been widely used, they rely heavily on manual extraction of fault features, leading to the problems of weak adaptive ability.

Recently, deep learning has developed rapidly which has attracted the attention and application of a large number of scholars [15]. Deep learning models have flexible structures that can extract deep features from input data, eliminating the step of manually extracting and selecting features [16]. The convolutional neural network (CNN) is a typical deep learning model that has been applied in the field of bearing fault diagnosis [17–22]. Janssens et al. [17] established a feature learning model based on CNNs that can automatically learn useful features for bearing condition monitoring and fault diagnosis from input data and compared it with the traditional method based on feature extraction by experimental data collected from a bearing test bed. In the comparison experiments, both methods used vibration data collected simultaneously and of the same length, and multiple samples were obtained by adding windows to the vibration data to meet the needs of the classification algorithms and CNNs in terms of the number of training samples. The window used in the traditional method contains 60 s of vibration data and overlaps by 50 percent with its neighboring window, while the window used in the proposed method contains 1 s of vibration data and does not overlap. The experiments on the 4-classification recognition task of bearing health state show that the performance of the feature learning method based on CNN is significantly better than that of the traditional method based on feature extraction, improving the recognition accuracy of bearing faults by about 6%. Zhang et al. [18] revealed a certain similarity between CNN and human discriminative laws by analyzing the Western Reserve University bearing fault dataset, which provides some proof for the application of CNN models within the field of bearing fault diagnosis. Ruan et al. [21] developed a physics-guided CNN (PGCNN) for bearing fault diagnosis. The size of the CNN's input and the size of the CNN's kernel depend on the physical characteristics of the bearing acceleration signal, and the validation results using experimental data show that the PGCNN is better than the standard CNN. These research studies have achieved satisfactory results with experimental data having fewer disturbances. However, bearings of complex equipment operate in complex environments, where single-source signals often contain a lot of noise and there is a risk of sensor faults.

In contrast, multi-source signals contain redundant and complementary information, promising to improve the accuracy and reliability of diagnosis. Currently, multi-source information fusion methods have been applied in fault diagnosis of rotating machinery. Based on the level of fusion, it can be grouped into data-level fusion [23–25], feature-level fusion [26–28] and decision-level fusion [29–31]. Data-level fusion directly fuses raw data from multiple sensors and then performs fault diagnosis based on the fused data. Yang et al. [23] proposed a multi-source data-level fusion method based on autoencoder (AE). The method directly fuses vibration signals and motor current signals at the data layer for wind turbine gearbox fault diagnosis. Feature-level fusion fuses the features extracted from the raw data and then performs pattern recognition based on the fused features. Wang et al. [27] calculated 18 statistical features based on the vibration signals collected from each sensor and proposed a hybrid Gaussian variational self-encoder that fuses the same features from different vibration signals at the feature level to realize bearing fault diagnosis and gear fault diagnosis. Decision-level fusion fuses the preliminary diagnosis results obtained based on each single sensor data. The voting method is a classical strategy of decision fusion. Li et al. [30] proposed a bearing fault diagnosis method based on voting strategy and AE. The proposed method used AEs to construct multiple base models and designed a weighted voting strategy based on the classification performance of each base model. Eventually, it realizes bearing fault diagnosis by fusing the pre-diagnosis results of the base models at the decision level. The above research has provided more reliable results and verified the validity of multi-source information fusion.

As mentioned earlier, CNN has the advantages of adaptive feature extraction and pattern recognition, allowing for end-to-end diagnosis. However, it can be affected by incomplete diagnostic information from a single signal. Multi-source information fusion can overcome this issue, decision-level fusion being particularly effective for integrating the pre-diagnosis results obtained based on CNN to guarantee the accuracy and reliability of the diagnosis results. Based on these insights, a bearing fault diagnosis method based on improved CNN and multi-source information fusion is proposed in this paper. The method adopts a framework of diagnosis with decision-level fusion. Firstly, a variational Bayesian inference convolutional neural network (VBICNN) is constructed based on variational Bayesian inference with modifications to the feature extraction method of the CNN, assuming that the latent features extracted by the model obey some prior distribution. The VBICNN is used to train a series of basic diagnosis models based on single-channel signals to provide multiple preliminary diagnosis results. Then, a weighted voting method is used to assign the weights of different basic diagnosis models and calculate the scores of different fault types. Ultimately, the fault type with the highest score is selected as the final prediction.

The rest of this paper is organized as follows. Section 2 briefly describes the theoretical foundations of the methods involved in this paper. Section 3 presents the proposed methodology, explaining the framework, methodology, and implementation process. Section 4 provides an experimental study of axlebox bearings for high-speed trains. Sections 5 and 6 include the discussion and conclusion, respectively.

2. Theoretical Background

2.1. Convolutional Neural Network

CNN is a common model in the field of deep learning, which was proposed by LeCun et al. [32]. Different from some traditional machine learning algorithms such as support vector machines and BP neural networks, CNNs do not need to extract features manually and can realize "end-to-end" fault diagnosis [33]. The structure of CNN mainly consists of the input layer, convolutional layer, pooling layer, fully connected layer, and output layer [34], as shown in Figure 1.

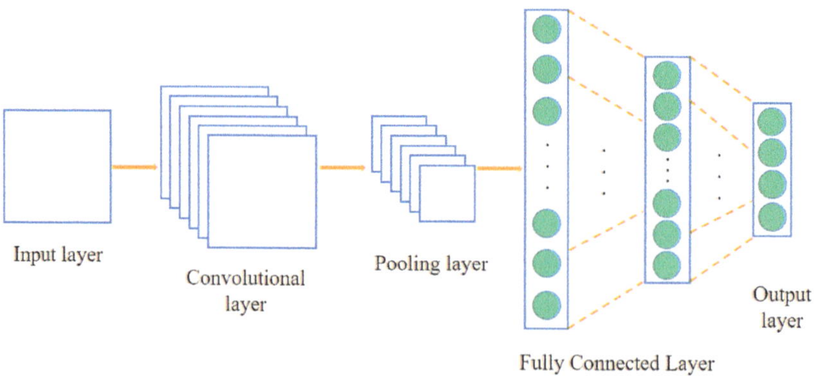

Figure 1. Structure of CNN.

The convolutional layer serves to extract features mainly from a localized region on the feature map, where various convolution kernels correspond to different feature extractors [35]. The output of the convolutional layer is [34]:

$$Z^p = W^p \otimes X + b^p = \sum_{d=1}^{D} W^{p,d} \otimes X^d + b^p \tag{1}$$

where $\otimes$ denotes the cross-correlation operation, and $W^P \in R^{m \times n \times D}$ denotes the convolution kernel. The output of the convolutional layer needs to be activated by an activation function and the output Y^P can be expressed as [34]:

$$Y^P = f(Z^P) \tag{2}$$

where $f(\cdot)$ denotes the activation function. In this paper, the linear rectification function ReLU is used as the activation function.

The pooling layer is usually designed after the convolutional layer and its main role is feature dimensionality reduction. The pooling layer can reduce the number of parameters while extracting the main features, reducing the amount of computation. Average pooling and max pooling are two common forms of calculating the pooling layer [35]. In this paper, we use max pooling, which can be described as follows [35]:

$$Y^d_{m,n} = \max_{i \in R^d_{m,n}} x_i \tag{3}$$

where x_i is the activation value of each neuron in the region $R^d_{m,n}$ on the feature map.

The fully connected layer is usually set following the convolutional and pooling layers, and its main role is to classify the features based on the features extracted through the convolutional layers and pooling layers. In this paper, the cross-entropy loss function is used and the parameters of each layer of the model are continuously optimized and adjusted with the goal of minimizing the loss function. The expression of the cross-entropy loss function is as follows [21]:

$$L_e(y, \hat{y}) = -\sum_{i=1}^{C} y_i \log(\hat{y}_i) \tag{4}$$

where y_i is the true label value of the training sample, and $\hat{y}$ is the predicted label of the model output.

2.2. Variational Bayesian Inference

Variational Bayesian inference is a method that combines Bayesian inference and variational inference [36]. Its purpose is to estimate the posterior distribution of latent variables. According to Bayes' theorem, the posterior distribution $p(z \mid x)$ of the latent variable z for sample x is as follows [37]:

$$p(z|x) = \frac{p(x|z)p(z)}{p(x)} = \frac{p(x|z)p(z)}{\int p(x,z)dz} \tag{5}$$

where $p(x \mid z)$ represents the likelihood function of sample x given the latent variable z, $p(z)$ represents the prior distribution of the latent variable z, $p(z)$ represents the marginal likelihood function of sample x, and $p(x,z)$ represents the joint probability distribution of sample x and latent variable z. Obviously, the true posterior distribution $p(z \mid x)$ of the latent variables is difficult to compute. According to the basic thought of variational inference, it is possible to approximate the complex distribution to be inferred by applying a known simple distribution $q(z)$ [37]. The similarity of the two distributions is measured using the Kullback–Leibler (KL) divergence as follows [37]:

$$D_{KL}(q(z)||p(z|x)) = \int q(z) \log \frac{q(z)}{p(z|x)} dz \tag{6}$$

To avoid direct computation of the true posterior distribution, Equation (6) can be derived as follows [37]:

$$D_{KL}(q(z)||p(z|x)) = \int p(z) \ln \frac{q(z)}{p(z,x)} dz + \ln p(x) \tag{7}$$

where the first term on the right-hand side of the equation is the variational free energy. When the two distributions are equal, the KL divergence takes the value of 0 [38]. Therefore, the objective of the optimization problem is transformed into [39]:

$$\underset{q}{\text{argmax}} \int q(z) \ln \frac{p(z,x)}{p(z)} dz \tag{8}$$

3. Proposed Method

As mentioned earlier, in complex and changing operating environments, a single monitoring signal for a bearing has the potential problem of incomplete diagnostic information. Existing studies have shown that multi-source signals mounted at different locations or in different directions contain more redundant information that is more beneficial for fault diagnosis [30,31]. Therefore, this paper proposes a new diagnosis method based on multi-source information fusion for bearing fault diagnosis. The proposed method consists of two main parts: the basic diagnostic model construction based on improved CNN, and the decision fusion based on a weighted voting strategy, as shown in Figure 2.

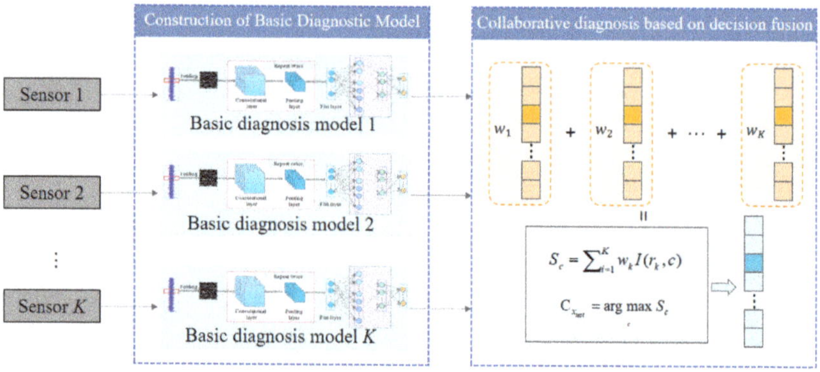

Figure 2. Framework of the proposed bearing fault diagnosis model. Where the red boxes represent the segments intercepted from the raw signals as inputs to the model.

3.1. VBICNN

The structure of the CNN employed in this paper is shown in Figure 3a. The input to the model is the folded vibration signal, and the folding method is referred to in the literature [31]. The model contains two convolution layers and a pooling layer. And the output of the last pooling layer is flattened to obtain a one-dimensional vector. This vector is denoted as the flattened layer and is connected to the fully connected layer. In a standard CNN model, there can be one fully connected layer or multiple fully connected layers. In this paper, a fully connected layer is set up for the last feature extraction before the classifier.

Denote the output of the flattened layer neuron as x and the output of the fully connected layer neuron as z. According to the formula for fully connected layer neurons in a standard CNN, the output of the fully connected layer is [34]:

$$z = f_{FC}(W_{FC}x + b_{FC}) \tag{9}$$

where W_{FC} and b_{FC} are the weight matrix and bias vector from the flat layer to the fully connected layer, respectively, and $f_{FC}(\cdot)$ is the activation function of the fully connected layer. To avoid confusion, note the final extracted feature as F. In a standard CNN, the output of the fully connected layer is the final extracted feature, $F = z$. Finally, the feature F is connected to a Softmax classifier for bearing fault type identification.

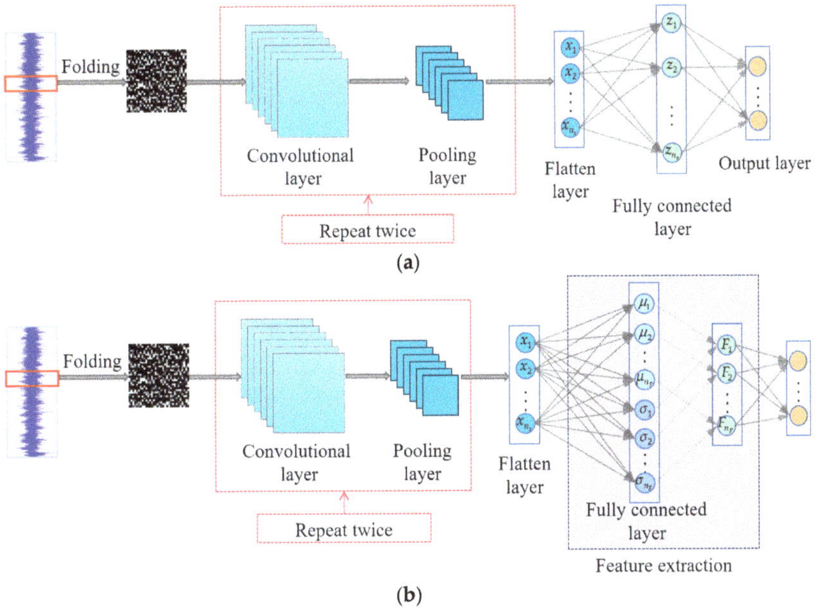

Figure 3. Schematic diagram of the models: (**a**) CNN; (**b**) VBICNN. Where n_x is the number of neurons in the flattened layer, n_z is the number of neurons in the fully connected layer in the CNN, n_F is the number of features extracted by the VBICNN, n_z is equal to n_F, and the red boxes represent the segments intercepted from the raw signals as inputs to the model.

Further, in order to enhance the nonlinear mapping ability of the model and the deep feature extraction ability of the fully connected layer, variational Bayesian inference is introduced into the feature extraction of the fully connected layer of the CNN model, forming the VBICNN model, as shown in Figure 3b. In the model training process, the fully connected layer is used to estimate the posterior distribution of the feature F. This means that the neurons of the fully connected layer represent the parameter values of the posterior distribution of the feature F and no longer the value of the feature, $F \neq z$. Following the ideology of variational Bayesian inference, a simple known distribution such as a Gaussian distribution can be used to approximate the true posterior distribution. So, assume that the feature $F = \{F_i\}$ follows Gaussian distribution, $z_i \sim N(\mu_i, \sigma_i^2)$. Then, the output z of the fully connected layer is:

$$z = [z_1, \cdots, z_{n_F}, z_{n_F+1}, \cdots, z_{2n_F}]^{\mathrm{T}} = [\mu_1, \cdots, \mu_{n_F}, \sigma_1, \cdots, \sigma_{n_F}]^{\mathrm{T}} \tag{10}$$

where n_F is the number of features. Combined with Equation (9), the formula for the parameter values is obtained:

$$[\mu_1, \ldots, \mu_{n_F}, \sigma_1, \ldots, \sigma_{n_F}]^T = f(\mathbf{W}\mathbf{z}^{F-1} + \boldsymbol{b}) \tag{11}$$

And then random sampling is performed to generate the feature ***F***. Finally, the feature F is input into the Softmax classifier for bearing fault type identification.

It can be found that on the basis of the optimization objective of the standard CNN model, the VBICNN model adds the constraints of simple distribution and true posterior distribution in variational Bayesian inference. Therefore, the optimization objective of variational Bayesian inference is added to the cross-entropy damage function shown in Equation (4), to form the loss function of the VBICNN model as follows:

$$L_{\mathrm{VBCNN}}(\boldsymbol{x}, \boldsymbol{y}) = L_e(\hat{\boldsymbol{y}}, \boldsymbol{y}) + \beta \cdot D_{KL}(q(z)||p(z|\boldsymbol{x})) \tag{12}$$

where β is the weight of the error in the distribution of learning features.

3.2. Decision-Level Fusion Method Based on Weighted Fusion Strategy

Assume that the signals come from K channels and that there are C fault types to be recognized. The signals of each channel are divided into training, validation, and test data. After the base VBICNN models are obtained using the training data, validation samples are used to determine the weights of each base VBICNN. In this paper, we assign weights based on the overall validation accuracy of the base model.

Noting that VBICNN$_k$ is the kth base model trained with the kth channel signal, its validation accuracy Acc_k on the validation set is calculated as follows [31]:

$$Acc^k = \frac{T_k}{N_k} \tag{13}$$

where N_k is the number of validation samples for the kth channel, and T_k is the number of validation samples in which the diagnosis is correct. Based on this, the weight matrix of all VBICNNs is obtained [40]:

$$\boldsymbol{w} = [w_1, w_2, \cdots, w_K]^T = \left[\frac{Acc^1}{\sum_{i=1}^K Acc^i}, \frac{Acc^2}{\sum_{i=1}^K Acc^i}, \ldots, \frac{Acc^K}{\sum_{i=1}^K Acc^i}\right]^T \tag{14}$$

Combining the diagnostic results of each channel further, the score for each fault type is calculated as follows [31]:

$$S_c = \sum_{i=1}^K w_k I(r_k, c) \quad c = 1,\ 2, \cdots, C \tag{15}$$

$$I(r_k, c) = \begin{cases} 1 & r_k = c \\ 0 & r_k \neq c \end{cases} \tag{16}$$

where is the preliminary diagnosis of the kth VBICNN.

Ultimately, the type with the highest score is selected as the final result of the diagnosis, as follows [31]:

$$C_{x_{test}} = \arg\max_c S_c \tag{17}$$

3.3. Flow of the Proposed Method

The process of the proposed method is summarized in Figure 4. The specific steps are as follows.

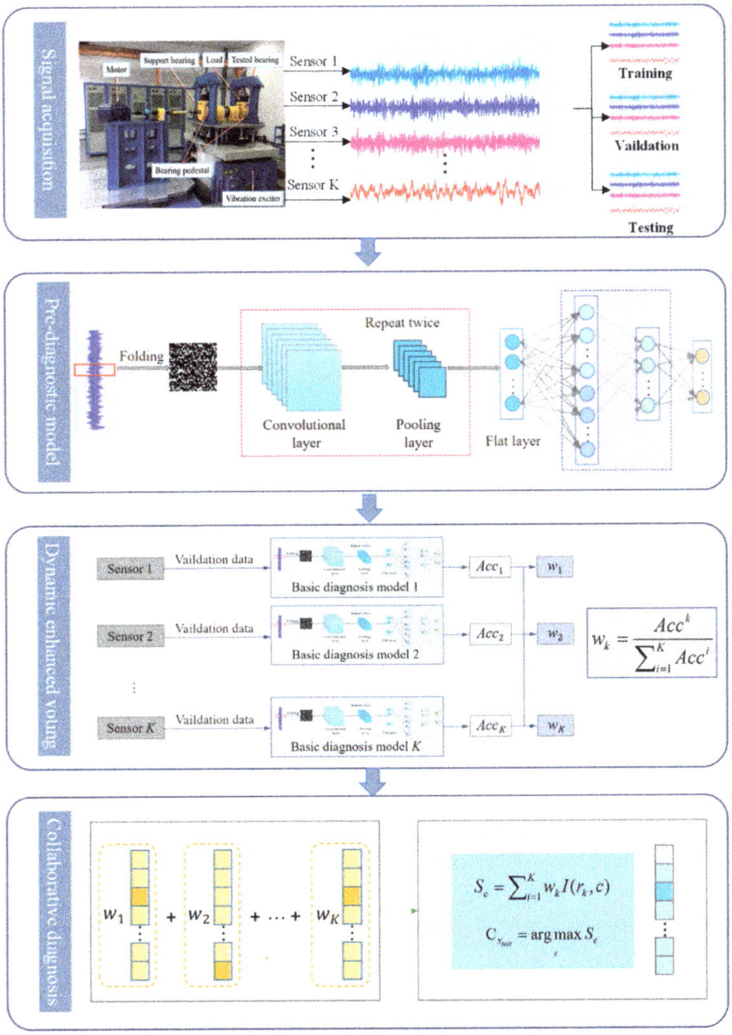

Figure 4. Framework and main steps of the proposed method.

Step 1: Data acquisition. Collect multi-channel vibration signals of mechanical equipment in different operating conditions. The data from each channel is divided into training samples, validation samples and testing samples.

Step 2: Construction of a series of basic VBICNNs. Set the structure and hyperparameters of the VBICNN, and input training samples into the VBICNN to update the parameters.

Step 3: Assigning weights to the basic VBICNNs. Validation samples are fed into the trained basic VBICNNs to calculate the validation accuracy of each VBICNN. Based on the validation accuracy, calculate the weight matrix $\boldsymbol{w}$ of basic VBICNNs.

Step 4: Validation of the proposed method by testing samples. Input the testing samples into a series of trained basic VBICNNs to obtain a series of pre-diagnostic results. Calculate the score of each fault type according to the weight matrix $\boldsymbol{w}$ to obtain the final diagnosis result.

4. Experimental Study

In order to verify the feasibility and effectiveness of the proposed method, multi-channel vibration signals from the high-speed train bearing test bench of Southwest Jiaotong University were used for the experiments.

4.1. Experimental Setup

The test bed for the axlebox bearing of the high-speed train of Southwest Jiaotong University is shown in Figure 5. The test device is composed of a motor, support bearing, tested bearing, bearing pedestal, and exciter. The exciter is used to simulate the wheel-rail excitation in the real train operating environment. In the data collection process, the real operating conditions can be simulated by applying the rotating speed, static load force, and excitation frequency to the bearing.

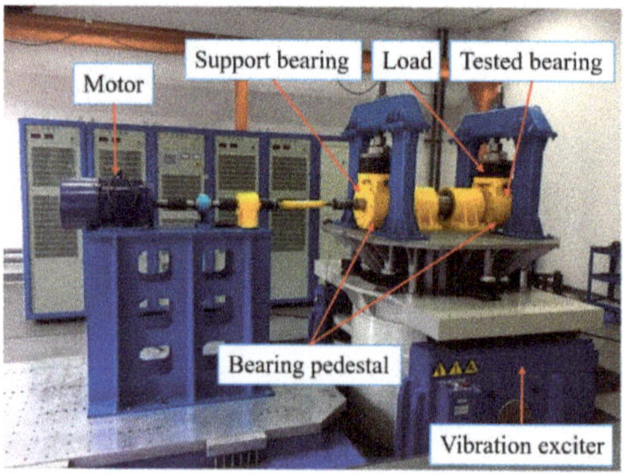

Figure 5. Test-bed for axlebox bearing of high-speed train.

The parameters of the axlebox bearings used are shown in Table 1. In this case, a tri-axis vibration accelerometer was used to acquire the signals in the (x, y, z) direction at the surface of the tested bearing. Noted in order as channels 1–3. In the process of data acquisition, the speed is 1100 rpm, the load is 1200 kg, the excitation frequency is 10 Hz, and the sampling frequency is 25.6 kHz. In addition, A total of nine bearing fault types were considered, including normal, inner race fault, outer race fault, cage fault, roller fault, poor grease uniformity, compound faults (inner race and outer race), compound faults (outer race and roller) and compound faults (outer race and grease), as shown in Figure 6. The labels for these 9 fault types are noted in order as 1–9. There are 400 samples for each fault type: 280 training samples, 40 validation samples, and 80 test samples.

Table 1. Parameters of the axlebox bearing.

Inner race Diameter	Pitch Circle Diameter	Roller Diameter	Roller Number	Contact Angle (°)
130 mm	184 mm	27 mm	17	0

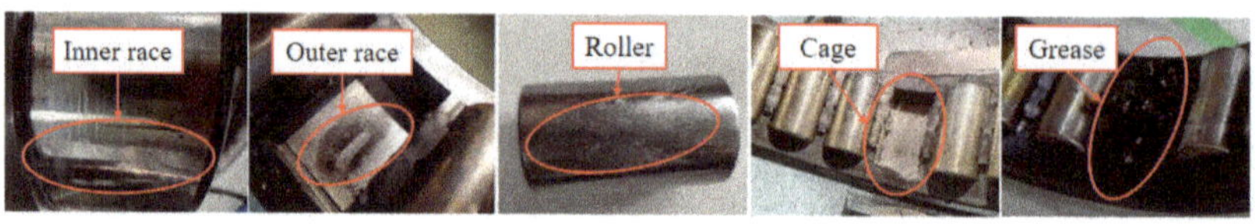

Figure 6. Five kinds of fault bearings.

4.2. Structural Design of the Model

In this case, the VBICNN of each channel uses the same structure, as shown in Table 2. The structure of VBICNN was obtained by adjusting the structure and parameters of the CNN models that have been repeatedly validated in the industry, consisting of two convolutional layers, two pooling layers, and a fully connected layer. The number of extracted features is 32. The model is trained using the Adam optimizer with a minimum batch size of 128 and an initial learning rate of 0.01. The β in Equation (12) is set to 0.001.

Table 2. Parameters of the VBCNN.

Description	Number of Nodes	Config
Input layer	$32 \times 32 \times 1$	-
The first convolutional layer	$32 \times 32 \times 8$	5@8/Stride(1,1)/Relu
The first pooling layer	$8 \times 8 \times 8$	Stride(4,4)
The second convolutional layer	$8 \times 8 \times 16$	5@16/Stride(1,1)/Relu
The second pooling layer	$2 \times 2 \times 16$	Stride(2,2)
Flatten layer	64	-
feature extraction layer	32	Tahn
Output layer	9	-

4.3. Ablation Experiment and Results

To validate the effectiveness and necessity of the method proposed in this paper, ablation experiments were carried out.

First, single-channel signals are used as inputs to the models to train the CNN model and the VBICNN model, which obtained three trained CNNs and three trained VBICNNs. By comparing the diagnostic accuracies of CNNs and VBICNNs, the necessity of an improved CNN model using Bayesian variational inference is verified. Then, a decision-level fusion strategy is used to collaborate the diagnostic results of single-channel signals to validate the necessity of collaborative diagnosis of multi-channel signals. Considering the possibility of chance in the experimental results, 10 repeated experiments were carried out and the results are shown in Figure 7, with the average accuracy and standard deviation detailed in Table 3.

Table 3. Fault diagnosis results using different models and different signals.

Model	Fusion Strategy	Testing Accuracy	
		Average Accuracy	Standard Deviations
CNN	Channel 1	94.01%	1.0925
	Channel 2	96.78%	0.9729
	Channel 3	93.61%	2.0020
VBCNN	Channel 1	97.38%	0.8283
	Channel 2	97.74%	0.4630
	Channel 3	97.04%	0.3823
Proposed method	Channel All	99.37%	0.3829

Comparing (a) with (b), (c) with (d), and (e) with (f) in Figure 7, it can be found that the diagnosis accuracy of the VBICNN is always higher than that of the CNN when a single-channel signal is used as the model input. Observing the diagnosis results of CNN and VBICNN in Table 3, the average accuracy provided by VBICNN in channels 1 to 3 is 97.38%, 97.74%, and 97.04% in that order, all of which are higher than the results using CNN, and the variance is much smaller, which indicates that the model is more stable. Meanwhile, comparing the proposed method with the VBICNN model with a single channel. As shown in Table 3, the accuracy of the proposed method is 99.28%, which is significantly higher than the VBICNN with a single channel.

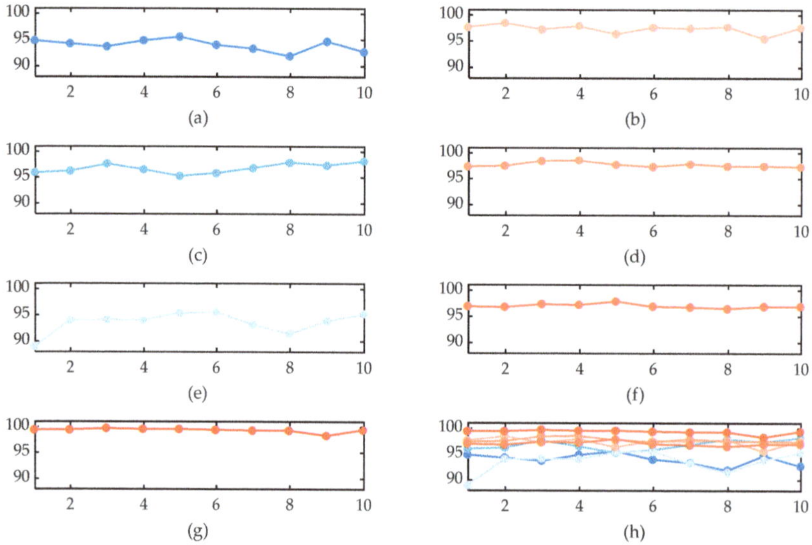

Figure 7. Results of repeated diagnostics using different channel signals for different fault diagnostic models (horizontal axis is the number of runs, vertical axis is the test accuracy): (**a**) CNN by channel 1; (**b**) VBCNN by channel 1; (**c**) CNN by channel 2; (**d**) VBCNN by channel 2; (**e**) CNN by channel 3; (**f**) VBCNN by channel 3; (**g**) proposed method using all channels; (**h**) comparison of results, where different colors represent different models using different channel signals, and the color meanings are consistent with (**a**–**g**).

4.4. Comparative Experiments and Results

In order to verify the effectiveness and superiority of the proposed approach, it was compared with other fault diagnosis approaches. There are 9 approaches were compared, specifically:

- Approach 1: A fault diagnosis approach based on the data-level fusion driven by principal component analysis (PCA) and CNN. Fuse the signals of all channels by PCA. And input the fusion result into CNN for classification.
- Approach 2: A fault diagnosis approach based on the data-level fusion driven by PCA and VBICNN. Input the fused signals of Approach 1 into VBICNN for classification.
- Approach 3: A fault diagnosis approach based on the data-level fusion driven by linear weighting and CNN. Fuse the signals of all channels by a similarity-based linear weighting method. And input the fusion result into CNN for classification.
- Approach 4: A fault diagnosis approach based on the data-level fusion driven by linear weighting and VBICNN. Input the fused signals of Approach 3 into VBICNN for classification.
- Approach 5: A diagnosis approach based on statistical feature fusion. Extract the statistical features of the signals of each channel. The features of all channels were combined with the Softmax classifier for classification. The statistical features used are consistent with the reference [41].
- Approach 6: A fault diagnosis approach based on the feature-level fusion driven by sparse auto-encoder (SAE). Extract the features of each channel signal by using SAE. The features of all channels were combined with the Softmax classifier for classification.

- Approach 7: A fault diagnosis approach based on the feature-level fusion driven by variational auto-encoder (VAE). Extract the features of each channel signal by using VAE. The features of all channels were combined with the Softmax classifier for classification.
- Approach 8: A decision-level fusion method based on the average voting. Based on the pre-diagnosis results of each channel obtained by VBICNN, use the average voting to assign the weights of each VBICNN for fusion.
- Approach 9: A decision-level fusion method based on weighted voting. Adopt the second fusion strategy proposed in the reference [40] to assign the weights of each VBICNN for fusion.

Ten repeated runs were performed to calculate the average testing accuracy and standard deviation. The results are shown in Figure 8. Approaches 1–9 provided average accuracies of 89.26%, 91.14%, 91.33%, 94.54%, 97.81%, 94.20%, 94.89%, 99.25%, 99.22%, with standard deviations of 2.0972, 0.8601, 1.4347, 0.5243, 1.0139, 0.4118, in order, 0.4914, 0.3993, and 0.3884. The average accuracy of the proposed method was 99.36% with a standard deviation of 0.3829, which shows that the proposed method is more accurate and reliable.

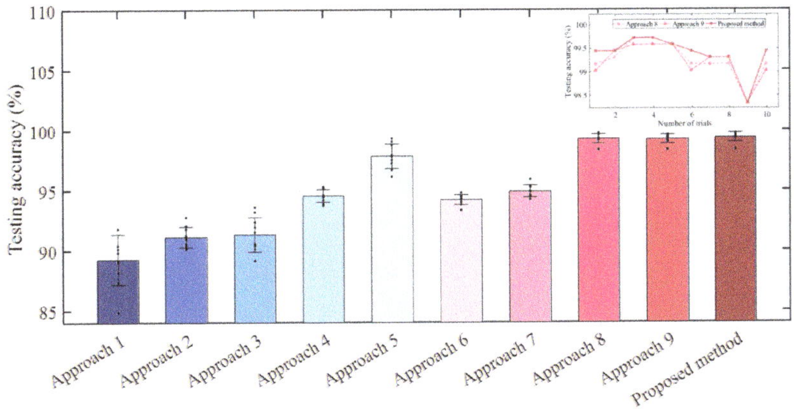

Figure 8. Diagnosis results for different approaches.

5. Discussion

5.1. The Necessity of Multi-Channel Information Fusion Diagnosis

A randomly selected set of results from the ablation experiments was used to calculate the confusion matrix for VBICNN. The results are shown in Figure 9, where the horizontal and vertical axes are the predicted and true labels, respectively. As can be seen from the figure, the signals from different channels have different diagnosis accuracies for each fault type. For instance, channel 1 is excellent at recognizing fault types 2, 5, 6, and 8, with 100% diagnostic accuracy, and there is underreporting of other fault types. Channel 2 recognizes fault type 4 with 100% accuracy but is unable to recognize other fault types with 100% accuracy. Channel 3 can diagnose all verification samples for fault types 5 and 9, but there are omissions for other fault types. In parallel, Figure 9 also reflects that the information of each channel signal is complementary. For instance, Channel 1 and Channel 2 appear to be unable to identify fault type 9, whereas Channel 3 makes a significant contribution. Channel 3 has a low accuracy for type 4, but this can be corrected by sensor 2. This phenomenon occurs because the signals from different channels carry different information about the bearing. Reviewing the multi-channel signals introduced in Section 4.1, channels 1–3 collected vibration data in three orthogonal directions on the bearing surface. The existing analysis results of the bearing dynamic response show that there are differences in the vibration response in different directions [42,43]. Thus, the multi-channel vibration signals carry different information, resulting in differences in diagnosis accuracy. Meanwhile, research on the behavioral analysis of high-speed bearing-rotor systems has shown that two orthogonal vibration signals, within the same cross-section, exhibit steady-state eddy characteristics when the system is in smooth operation [44]. So, the mutually orthogonal signals used in the experimental case are closely correlated, which makes the multi-channel diagnosis results complementary. Based on the above, the fusion of information from channels 1–3 is beneficial in guaranteeing the accuracy and reliability of the diagnosis results. According to the results of the ablation experiments, the optimal diagnostic accuracy and average accuracy of VBICNN with a single channel are 98.61% and 97.38%, respectively, and those of the proposed method are 99.72% and 99.36%, which shows that the fusion of multi-channels improves the accuracy of diagnosis.

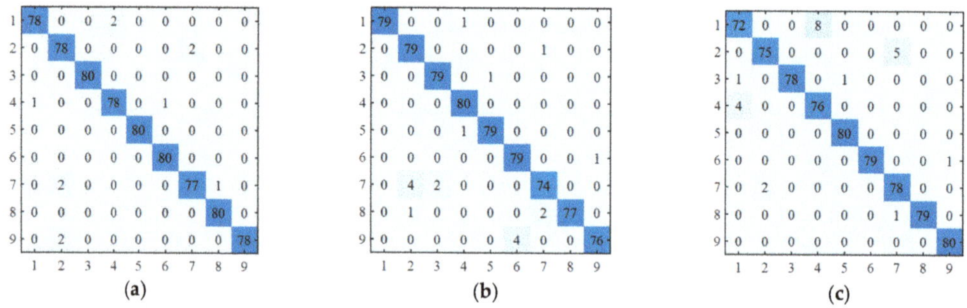

Figure 9. Confusion matrices of the basic diagnosis mode with different channel signals: (**a**) Channel 1; (**b**) Channel 2; (**c**) Channel 3.

5.2. The Effectiveness of Variational Bayesian Inference

Comparing the model containing variational Bayesian inference and the model without variational Bayesian inference. There are five sets of comparison results, namely, CNN and VBICNN with channels 1–3 in the ablation experiment, and CNN and VBICNN with PCA data fusion signals and weighted data fusion signals in the comparison experiment. The comparison results are shown in Figure 10. As can be seen from the figure, the model improved with variational Bayesian inference has a significant advantage in terms of diagnosis accuracy, both in the case of single-channel signals as inputs and multi-channel fusion signals as inputs. At the same time, the results of 10 repeated experiments have less fluctuation.

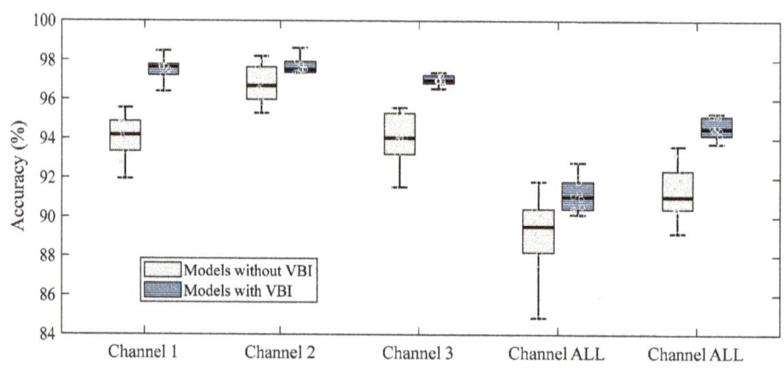

Figure 10. Comparison of diagnostic results.

6. Conclusions

To address the issue of incomplete diagnosis information from a single signal, a new multi-sensor fusion fault diagnosis approach is proposed in this paper. Firstly, a basic diagnosis model based on VBICNN is constructed. Secondly, vibration signals from multiple channels were used to train a series of basic VBICNNs to provide multiple pre-diagnosis results. Then, a voting strategy based on verification accuracy is used to assign weights to each model, obtaining more reliable collaborative diagnostic results. In order to validate the proposed method, ablation experiments, and comparative experiments have been carried out. The following conclusions can be drawn:

(1) The comparison with the diagnosis method based on a single-channel signal confirmed that multi-channel signals contain complementary information, which enables the acquisition of more accurate and stable results.
(2) The comparison with the models without variational Bayesian inference confirms that variational Bayesian inference has the advantage of dealing with high-dimensional uncertain data, which could improve the classification performance of CNN.
(3) Comparison with other multichannel fusion approaches confirms that the proposed approach has better performance in multi-channel vibration signal fusion diagnosis. The comparison approaches include several data-level fusion methods, feature-level fusion methods, and decision-level fusion methods.

Although the proposed method demonstrates excellent performance on experimental data with complex interference, the weighted voting strategy used for information fusion is relatively simple and employs a fixed validation accuracy to determine the weights. To enhance collaboration among multi-channel signals, it is necessary to investigate dynamic fusion strategies to learn the optimal weights during the training process. Therefore, we will explore the dynamic weight assignment strategy of the proposed method in the future. Meanwhile, it should be noted that the proposed method is applicable not only to bearings but also to rotating components such as gears. For other anomaly detection problems, such as the Tennessee Eastman process, the framework of the proposed methodology can also be adopted, but a more appropriate basic diagnosis model needs to be reconstructed according to the specific task.

Author Contributions: Conceptualization, Z.W. and X.X.; methodology, Z.W. and X.X.; software, Z.W. and X.X.; validation, X.X. and Z.Z.; formal analysis, Z.W. and Z.Z.; investigation, Z.W. and W.L.; resources, D.S.; data curation, D.S.; writing—original draft preparation, Z.W.; writing—review and editing, X.X.; visualization, Z.W. and X.X.; supervision, D.S.; project administration, D.S.; funding acquisition, D.S. and Z.W. All authors have read and agreed to the published version of the manuscript.

Funding: This research was funded by the National Natural Science Foundation of China under Grant 52472368; the Major Science and technology projects in Sichuan China, grant number 2023ZDZX0009; and the research and development project of China National Railway Group Co., Ltd., grant number K2022J004.

Data Availability Statement: The data presented in this study are available on request from the corresponding author.

Acknowledgments: The authors want to express their gratitude to Ping He at Southwest Jiaotong University, Jianxiong Dong at Chongqing University, Zhiheng Chen and Cong Deng at CRRC Zhuzhou Locomotive Co., Ltd. for their support on the rail vehicle bogie axlebox bearing test rig.

Conflicts of Interest: Author Zhongyao Wang was employed by the company CRRC Changchun Railway Vehicles Co., Ltd. The remaining authors declare that the research was conducted in the absence of any commercial or financial relationships that could be construed as a potential conflict of interest.

References

1. Alshorman, O.; Irfan, M.; Abdelrahman, R.B.; Abdelrahman, B.; Masadeh, M.; Alshorman, A.; Sheikh, M.A.; Saad, N.; Rahman, S. Advancements in condition monitoring and fault diagnosis of rotating machinery: A comprehensive review of image-based intelligent techniques for induction motors. *Eng. Appl. Artif. Intell.* **2024**, *130*, 107724. [CrossRef]
2. Liu, L.; Cheng, Y.; Song, D.; Zhang, W.; Tang, G.; Luo, Y. A lightweight network with adaptive input and adaptive channel pruning strategy for bearing fault diagnosis. *IEEE Trans. Instrum. Meas.* **2024**, *73*, 3510911. [CrossRef]
3. Peng, D.; Wang, H.; Liu, Z.; Zhang, W.; Zuo, M.; Chen, J. Multibranch and multiscale CNN for fault diagnosis of wheelset bearings under strong noise and variable load condition. *IEEE Trans. Ind. Inform.* **2017**, *16*, 4949–4960. [CrossRef]
4. Strombergsson, D.; Marklund, P.; Berglund, K.; Larsson, P.E. Bearing monitoring in the wind turbine drivetrain: A comparative study of the FFT and wavelet transforms. *Wind Energy* **2020**, *23*, 1380–1393. [CrossRef]
5. Sahraoui, M.A.; Rahmoune, C.; Meddour, I.; Bettahar, T.; Zair, M. New criteria for wrapper feature selection to enhance bearing fault classification. *Adv. Mech. Eng.* **2023**, *15*, 16878132231183862. [CrossRef]
6. Kumar, D.; Mehran, S.; Shaikh, M.Z.; Hussain, M.; Chowdhry, B.S.; Hussain, T. Triaxial bearing vibration dataset of induction motor under varying load conditions. *Data Brief* **2022**, *42*, 108315. [CrossRef]
7. Chen, B.; Cheng, Y.; Allen, P.; Wang, S.; Gu, F.; Zhang, W.; Ball, A.D. A product envelope spectrum generated from spectral correlation/coherence for railway axle-box bearing fault diagnosis. *Mech. Syst. Signal Proc.* **2025**, *225*, 112262. [CrossRef]
8. Igba, J.; Alemzadeh, K.; Durugbo, C.; Eiriksson, E.T. Analysing RMS and peak values of vibration signals for condition monitoring of wind turbine gearboxes. *Renew. Energy* **2016**, *91*, 90–106. [CrossRef]
9. Li, Y.; Zhang, W.; Xiong, Q.; Lu, T.; Mei, G. A novel fault diagnosis model for bearing of railway vehicles using vibration signals based on symmetric alpha-stable distribution feature extraction. *Shock Vib.* **2016**, *2016*, 5714195. [CrossRef]
10. Chen, B.; Song, D.; Zhang, W.; Cheng, Y. A novel spectral coherence-based envelope spectrum for railway axle-box bearing damage identification. *Struct. Health Monit.* **2023**, *22*, 879–896. [CrossRef]
11. Hou, H.; Zhang, X.; Wang, X.; Wan, Y.; Shi, H.; Liu, W.; Wang, P. Location of underground multilayer media based on BP neural network and near-field electromagnetic signal. *IEEE Sens. J.* **2025**, *25*, 725–736. [CrossRef]
12. Wang, X.; Jiang, H. Gearbox fault diagnosis based on refined time-shift multiscale reverse dispersion entropy and optimised support vector machine. *Machines* **2023**, *11*, 646. [CrossRef]
13. Ziani, R.; Felkaoui, A.; Zegadi, R. Bearing fault diagnosis using multiclass support vector machines with binary particle swarm optimization and regularized Fisher's criterion. *J. Intell. Manuf.* **2017**, *28*, 405–417. [CrossRef]
14. Yao, C.; Li, N.; Feng, Z.; Chen, D. Fault diagnosis based on rough set attribute reduction and bayesian classifier. *China Mech. Eng.* **2015**, *26*, 1969–1977.
15. Dong, S.; Wang, P.; Abbas, K. A survey on deep learning and its applications. *Comput. Sci. Rev.* **2021**, *40*, 100379. [CrossRef]
16. LeCun, Y.; Bengio, Y.; Hinton, G. Deep learning. *Nature* **2015**, *521*, 436–444. [CrossRef]
17. Janssens, O.; Slavkovikj, V.; Vervisch, B.; Stockman, K.; Loccufier, M.; Verstockt, S.; Van de Wallk, R.; Van Hoecke, S. Convolutional neural network based fault detection for rotating machinery. *J. Sound Vibr.* **2016**, *377*, 331–345. [CrossRef]
18. Zhang, J.; Yang, Z.; Chen, X.; Zhai, Z.; Liu, Y. Interpretability discussion on convolutional neural network in bearing fault diagnosis. *Bearing* **2020**, *7*, 54–60.
19. Mohiuddin, M.; Islam, M.S.; Islam, S.; Miah, M.S.; Niu, M. Intelligent Fault Diagnosis of Rolling Element Bearings Based on Modified AlexNet. *Sensors* **2023**, *23*, 7764. [CrossRef]
20. Ruan, D.; Wang, J.; Yan, J.; Gühmann, C. CNN parameter design based on fault signal analysis and its application in bearing fault diagnosis. *Adv. Eng. Inform.* **2023**, *55*, 101877. [CrossRef]
21. Salunkhe, V.G.; Khot, S.M.; Jadhav, P.S.; Yelve, N.P.; Kumbhar, M.B. Experimental Investigation Using Robust Deep VMD-ICA and 1D-CNN for Condition Monitoring of Roller Element Bearing. *J. Comput. Inf. Sci. Eng.* **2024**, *24*, 124501. [CrossRef]
22. Xu, Y.; Feng, K.; Yan, X.; Sheng, X.; Sun, B.; Liu, Z.; Yan, R. Cross-modal fusion convolutional neural networks with online soft-label training strategy for mechanical fault diagnosis. *IEEE Trans. Ind. Inform.* **2024**, *20*, 73–84. [CrossRef]
23. Yang, S.; Wang, Y.; Li, C. Wind turbine gearbox fault diagnosis based on an improved supervised autoencoder using vibration and motor current signals. *Meas. Sci. Technol.* **2021**, *32*, 114003. [CrossRef]
24. Xie, T.; Huang, X.; Choi, S. Intelligent mechanical fault diagnosis using multisensor fusion and convolution neural network. *IEEE Trans. Ind. Inform.* **2022**, *18*, 3213–3223. [CrossRef]
25. Chen, X.; Shao, H.; Xiao, Y.; Yan, S.; Cai, B.; Liu, B. Collaborative fault diagnosis of rotating machinery via dual adversarial guided unsupervised multi-domain adaptation network. *Mech. Syst. Signal Proc.* **2023**, *198*, 119427. [CrossRef]
26. Stief, A.; Ottewill, J.; Baranowski, J.; Orkisz, M. A PCA and two-stage bayesian sensor fusion approach for diagnosing electrical and mechanical faults in induction motors. *IEEE Trans. Ind. Electron.* **2019**, *66*, 9510–9520. [CrossRef]
27. Wang, C.; Xin, C.; Xu, Z.; Qin, M.; He, M. Mix-VAEs: A novel multisensor information fusion model for intelligent fault diagnosis. *Neurocomputing* **2022**, *492*, 234–244. [CrossRef]

28. Mejbel, B.G.; Sarow, S.A.; Al-Sharify, M.T.; Al-Haddad, L.A.; Ogaili, A.A.F.; Al-Sharify, Z.T. A Data Fusion Analysis and Random Forest Learning for Enhanced Control and Failure Diagnosis in Rotating Machinery. *J. Fail. Anal. Prev.* **2024**, *24*, 2979–2989. [CrossRef]
29. Huo, Z.; Martinez-Garcia, M.; Zhang, Y.; Shu, L. A Multisensor Information Fusion Method for High-Reliability Fault Diagnosis of Rotating Machinery. *IEEE Trans. Instrum. Meas.* **2022**, *71*, 3500421. [CrossRef]
30. Li, X.; Jiang, H.; Niu, M.; Wang, R. An enhanced selective ensemble deep learning method for rolling bearing fault diagnosis with beetle antennae search algorithm. *Mech. Syst. Signal Proc.* **2020**, *142*, 106752. [CrossRef]
31. Xu, X.; Song, D.; Wang, Z.; Zheng, Z. A novel collaborative bearing fault diagnosis method based on multisignal decision-level dynamically enhanced fusion. *IEEE Sens. J.* **2024**, *24*, 34766. [CrossRef]
32. Lecun, Y.; Bottou, L.; Bengio, Y.; Haffner, P. Gradient-based learning applied to document recognition. *Proc. IEEE* **1998**, *86*, 2278–2324. [CrossRef]
33. Zhao, X.; Wang, L.; Zhang, Y.; Han, X.; Deveci, M.; Parmar, M. A review of convolutional neural networks in computer vision. *Artif. Intell. Rev.* **2024**, *57*, 99. [CrossRef]
34. Yasin, M.; Sarigül, M.; Avci, M. Logarithmic Learning Differential Convolutional Neural Network. *Neural Netw.* **2024**, *172*, 106114. [CrossRef]
35. Momeny, M.; Latif, A.M.; Sarram, M.A.; Sheikhpour, R.; Zhang, Y. A noise robust convolutional neural network for image classification. *Results Eng.* **2021**, *10*, 100225. [CrossRef]
36. Fox, C.; Roberts, S. A tutorial on variational Bayesian inference. *Artif. Intell. Rev.* **2012**, *38*, 85–95. [CrossRef]
37. Zhang, C.; Bütepage, J.; Kjellström, H.; Mandt, S. Advances in Variational Inference. *IEEE Trans. Pattern Anal. Mach. Intell.* **2019**, *41*, 2008–2026. [CrossRef]
38. Cao, Y.; Jan, N.M.; Huang, B.; Wang, Y.; Pan, Z.; Gui, W. No-delay multimodal process monitoring using Kullback-Leibler divergence-based statistics in probabilistic mixture models. *IEEE Trans. Autom. Sci. Eng.* **2023**, *20*, 167–178. [CrossRef]
39. Lin, H.; Hu, C. Variational inference based distributed noise adaptive Bayesian filter. *Signal Process.* **2021**, *178*, 107775. [CrossRef]
40. Chao, Q.; Gao, H.; Tao, J.; Wang, Y.; Zhou, J.; Liu, C. Adaptive decision-level fusion strategy for the fault diagnosis of axial piston pumps using multiple channels of vibration signals. *Sci. China-Technol. Sci.* **2022**, *65*, 470–480. [CrossRef]
41. Chen, Z.; Li, W. Multisensor feature fusion for bearing fault diagnosis using sparse autoencoder and deep belief network. *IEEE Trans. Instrum. Meas.* **2017**, *66*, 1693–1702. [CrossRef]
42. Ohta, H.; Sugimoto, N. Vibration characteristics of tapered roller bearings. *J. Sound Vibr.* **1996**, *190*, 137–147. [CrossRef]
43. Li, N.; Zhang, J.; Meng, X.; Han, Q.; Zhai, J. Dynamic response and failure analysis of bearing under the impact of vibration excitation. *Eng. Fail. Anal.* **2023**, *154*, 107640. [CrossRef]
44. Zhang, J.; Han, D.; Xie, Z.; Huang, C.; Rao, Z.; Song, M.; Su, Z. Nonlinear behaviors analysis of high-speed rotor system supported by aerostatic bearings. *Tribol. Int.* **2022**, *170*, 107111. [CrossRef]

machines

MDPI

Article

MTAGCN: Multi-Task Graph-Guided Convolutional Network with Attention Mechanism for Intelligent Fault Diagnosis of Rotating Machinery

Bo Wang † **and Shuai Zhao** *,†

School of Information and Artificial Intelligence, Nanchang Institute of Science & Technology, Nanchang 330108, China; wb_wh0118@163.com
* Correspondence: zs1025198686@163.com
† These authors contributed equally to this work.

Abstract: Deep learning (DL)-based methods have shown great success in multi-category fault diagnosis due to their hierarchical networks and automatic feature extraction. However, their superior performance is mostly based on single-task learning, which makes them unsuitable for increasingly sophisticated engineering environments. In this paper, a novel multi-task graph-guided convolutional network with an attention mechanism for intelligent fault diagnosis, named MTAGCN, is proposed. Most existing fault diagnosis models are commonly bounded by a single diagnosis objective, especially when handling multiple tasks jointly. To address this limitation, a new multi-task fault diagnosis framework is designed, incorporating an attention mechanism between the task-specific module and task-shared modules. This framework enables multiple related tasks to be learned jointly while improving diagnostic and identification performance. Moreover, it is observed that most existing DL-based methods share incomplete fault representations, leading to unsatisfactory fault diagnosis. To overcome this issue, a graph convolutional network (GCN)-based fault diagnosis framework is introduced, which not only captures structural characteristics but also enhances diagnostic effectiveness. Comprehensive experiments based on three case studies demonstrate that the proposed MTAGCN outperforms state-of-the-art (SOTA) methods, striking a good balance between accuracy and multi-task learning.

Keywords: fault diagnosis; multi-task learning; graph convolutional networks; attention mechanism

1. Introduction

As fundamental components of rotating machinery, bearings, gears, couplings, and other related parts are essential in modern industry [1–3]. However, harsh factory conditions make them highly prone to wear and performance degradation under prolonged working time. In the era of big data, traditional signal processing methods are increasingly limited due to insufficient accuracy, inefficiency, and dependence on expert knowledge and denoising [4]. At the same time, the rapid development of deep learning (DL)-based techniques with sensor technology and the Internet of Things (IoT) has brought new opportunities for fault diagnosis. Most importantly, intelligent fault diagnosis (IFD)-based methods, which are known for their efficiency and accuracy, have received great attention from academia and industry in recent years and have been significantly applied in rotating machinery fault diagnosis [5].

 https://doi.org/10.3390/machines13050347

As an efficient fault diagnosis tool, machine learning (ML)-based methods have received much attention from academia and have shown significant potential in rotating machinery fault diagnosis. Generally, the implementation process consists of three key steps: initial data acquisition, feature engineering with expert knowledge, and final health condition classification [6,7]. Chen et al. proposed a hierarchical ML-based framework with only two layers to enable fault diagnosis in rotating machinery [8]. Zhang et al. proposed a novel ML-based framework by integrating reverse feature elimination (RFE) and extreme learning machine (ELM) to achieve fault diagnosis of shield machines [9]. However, manual feature extraction in ML-based methods heavily relies on expert knowledge, which can lead to significant time and labor costs. Meanwhile, they are commonly constrained by limited data, making them unsuitable for generalization and self-learning capabilities under complex engineering environments [10]. In recent years, there has been growing interest in developing advanced methods for feature extraction and health condition identification in rotating machinery using large-scale monitoring data.

In recent years, deep learning (DL)-based diagnostic models have been developed to automatically learn fault-based features from monitoring data through deep hierarchical architectures. Compared to ML-based methods, DL-based models can directly establish the relationship between the input samples and health conditions of rotating machinery, which utilizes the learned features to assess the fault types or severities of rotating machinery. Several approaches, including deep belief networks (DBNs) [11], deep autoencoders (DAEs) [12], convolutional neural networks (CNNs) [13], and graph convolutional networks (GCNs) [14], have been explored for fault diagnosis. For example, Karimi et al. proposed a multi-source domain adaptation method for fault diagnosis, leveraging the attention mechanism, domain attribute loss, and knowledge fusion to enhance cross-domain diagnostic performance [15]. Shi et al. designed the improved multi-sensor DBNs (MSIDBNs) with redefined pretraining and finetuning stages to effectively extract features in rolling mill fault diagnosis [16]. Yu et al. introduced a novel DL-based approach by improving one-dimensional and two-dimensional CNNs (I1DCNNs and I2DCNNs) with group normalization (GN) and global average pooling (GAP) for enhanced intelligent fault diagnosis [17]. Han et al. built a novel multi-source heterogeneous information fusion framework by utilizing improved DL-based methods to enhance the effectiveness and robustness of intelligent fault diagnosis [18].

Though DL-based methods have made significant advancements in intelligent fault diagnosis of rotating machinery, some drawbacks of these methods still remain. On the one hand, the majority of current DL-based methods assume that they can extract features only for a single task to achieve fault diagnosis while ignoring task-invariant features. On the other hand, most existing DL-based feature extraction methods rely mainly on class labels while overlooking the underlying structure characteristics of samples. This may lead to incomplete feature representations in deep networks.

Multi-task learning (MTL), as a practical and powerful technique for mitigating the first limitation, has been used to learn the task-shared features from various diagnostic tasks. Recently, MTL-based methods have received growing interest in handling multiple tasks jointly. MTL-based methods can enhance model generalization by integrating useful information from multiple tasks, sharing features across tasks, and training within a unified framework. MTL-based techniques have been widely used in various fields such as medical image classification, fault diagnosis, question answering, object detection, and so on. Zhao et al. proposed a novel MTL-based self-supervised transfer learning paradigm by integrating the multi-perspective feature transfer to achieve fault diagnosis [19]. Zhang et al. proposed an auxiliary prior knowledge-informed framework based on MTL for few-shot fault diagnosis [20]. Zheng et al. proposed an MTL-based framework with deep inter-task interactions to achieve breast tumor segmentation and classification [21]. Zhang et al. proposed a novel uncertainty bidirectional guidance MTL-based framework (UBGM) to enhance segmentation and classification in medical image analysis [22]. However, these methods cannot adequately consider the interactions of different diagnostic tasks at different levels. Therefore, the design of the MTL-based methods still has much room for improvement.

To address the limitations of DL-based methods, a novel multi-task graph-guided convolutional network with an attention mechanism (MTAGCN) has been proposed for intelligent fault diagnosis of rotating machinery. In MTAGCN, multi-task diagnosis is achieved by effectively modeling the data structure, feature extraction, and attention mechanism (AM) in a unified deep network. The fault severity labels and fault type labels are modeled by two task-specific modules, respectively. In order to adjust the data structure from the graph data, task-invariant features are mined from the raw vibration signals using convolutional blocks (Conv). Then, the block is utilized to build instance graphs of the feature structure characteristics. After that, the task-shared module consisting of Conv and graph blocks is used to mine different levels of features. AM should be used to enable connections between task-shared modules and task-specific modules that capture the significance of task-shared characteristics and select valuable and representative characteristics for a particular diagnostic task. The outline of our proposed MTAGCN is shown in Figure 1. The task-shared module extracts valuable features from input vibration signals, while two task-specific modules further process these features to output diagnostic information, including fault type and severity.

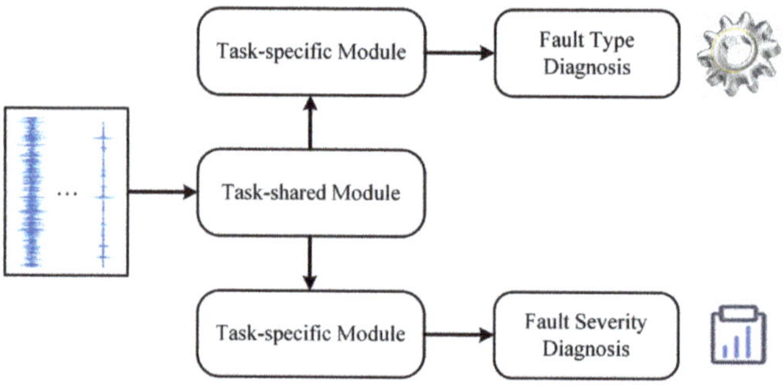

Figure 1. Overview of the proposed MTAGCN.

The main contributions of this article are listed as follows.

(1) The multi-task GCN is proposed to achieve end-to-end fault diagnosis by modeling the fault type diagnosis task and fault severity diagnosis task simultaneously in a unified DL-based framework.
(2) The instance graphs are constructed to mine structural characteristics in the task-shared module. By integrating the AM modules at multiple levels, the proposed MTAGCN can more effectively capture critical task-specific information in specific task modules.
(3) Extensive experiments are conducted to show the feasibility and effectiveness of our proposed method.

The remaining part of the paper proceeds as follows: In Section 2, related works about CNNs, GCNs, AM, and MTL are illustrated. In Section 3, the proposed MTAGCN is described. The feasibility and effectiveness of the proposed MTAGCN on MTL fault diagnosis are validated by three datasets in Section 4. In Section 5, a set of comparison experiments of MTAGCN is carried out. The summary and directions for future research are presented in Section 6 of this paper.

2. Related Works

2.1. Convolutional Neural Networks

CNNs, as one of the classic DL-based architectures, are widely used for processing data characterized by a grid-like structure, such as images and signals [23,24]. Their core components include convolutional layers, pooling layers, and fully connected layers. CNNs have been extensively applied in tasks such as image classification, object detection, and semantic segmentation due to their strong capabilities in feature extraction and representation. The detailed architecture of CNNs is illustrated in Figure 2.

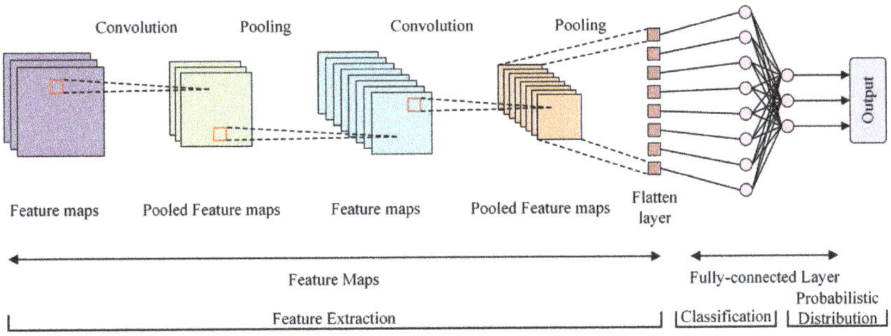

Figure 2. Detailed architecture of CNNs.

Convolutional layers are the most fundamental component of CNNs and are used to extract coarser features by sliding the input data through a learnable filter or kernel. CNNs can realize equivariant representations through the learnable filters in the input data, while greatly reducing the input dimensionality. To learn about finer features, we usually stack convolutional layers at different depths of the network. Given that the input feature maps $x_i^{l-1} \in \mathbb{R}^{H \times W}$, where H and W stand for the height and width, respectively, they are input to the convolutional layer, defined as:

$$x_c^l = \sigma\left(\sum_{i=1}^{C^{l-1}} k_{i,c}^l * x_i^{l-1} + b_c^l\right) \tag{1}$$

where x_c^n represents the output feature of the l-th layer. $\sigma(\cdot)$ is the activation function (i.e., ReLU). $k_{i,c}^n$ denotes the kernel function. W and C refer to the size of the kernel and the number of channels, separately. b_c^n means the bias of the l-th layer.

The pooling layers are then followed by the convolutional layers to further reduce the dimensionality of the feature map and prevent overfitting. Max-pooling, as one of the common pooling operations, is used in this paper and is defined as follows:

$$P_n = \max_{x_c^n \in S}\left(x_c^l\right) \tag{2}$$

where P_n means the pooled feature maps of the l-th layer. $\max(\cdot)$ denotes the max-pooling operation.

After these feature maps undergo several convolution and pooling operations, they are flattened into a one-dimensional vector and input into the fully connected layer for final fault diagnosis, as expressed below:

$$h^m = \sigma^m\left((W^m)^T * x^{m-1} + b^m\right) \tag{3}$$

where W^m and b^m are the weight matrix and bias of the m-th layer in the fully connected layer, respectively.

Finally, the probability distributions from the fully connected layer are converted through a softmax function, which is expressed as:

$$\hat{y}_i = \begin{bmatrix} P(y_i = 1) \\ P(y_i = 2) \\ \vdots \\ P(y_i = k) \end{bmatrix} = \frac{1}{\sum_{i=1}^{k} e^{h_{oi}}} \begin{bmatrix} e^{h_{oi}(1)} \\ e^{h_{oi}(2)} \\ \vdots \\ e^{h_{oi}(k)} \end{bmatrix} \tag{4}$$

where $h_{oi} \in \mathbb{R}^{k\times 1}$ means the output feature maps of input samples x_i.

2.2. Graph Convolutional Networks

GCNs, as a novel architecture for DL-based methods, are specialized for processing graph-based data and non-Euclidean data [25]. Compared to the CNNs mentioned above, GCNs not only extract node features but also capture the structural relationship of nodes. By aggregating information from neighboring nodes, GCNs can effectively learn fault-based representations. The detailed architecture of GCNs is illustrated in Figure 3.

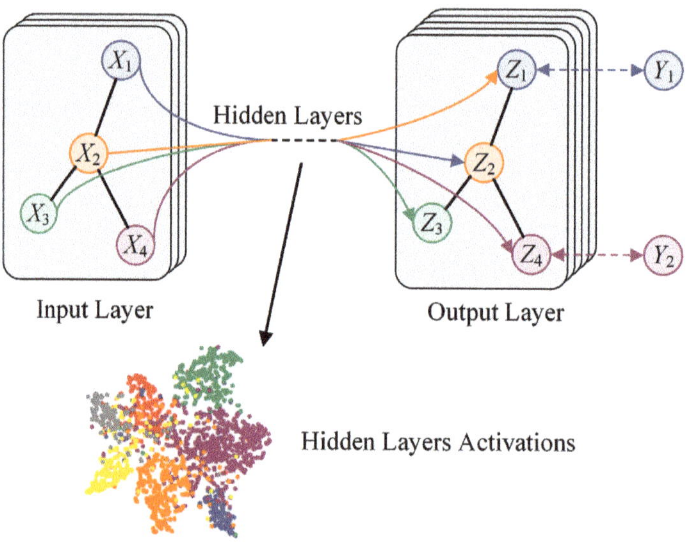

Figure 3. Detailed architecture of GCNs.

A graph $G(V, E, F)$ consists of a set of nodes V, a set of edges E, and a set of features F. GCNs update node representations by aggregating information from neighboring nodes, enabling effective processing of graph-structured data. The mathematical equations for the graph convolution operation can be written as follows:

$$H^{(l+1)} = \sigma\left(\tilde{D}^{-\frac{1}{2}}\tilde{A}\tilde{D}^{-\frac{1}{2}}H^{(l)}W^{(l)}\right) \tag{5}$$

with $\tilde{D}_{ii} = \sum_j \tilde{A}_{ij}$ and $\tilde{A} = A + I$, where $\tilde{D}$ is the degree matrix after adding self-loops and $\tilde{D}$ means the adjacency matrix after adding self-loops. $H^{(l)} \in \mathbb{R}^{N \times d^{(l)}}$ is the node feature matrix at the l-th layer. N is the number of nodes. $d^{(l)}$ is the feature dimension at the l-th layer. $W^{(l)} \in \mathbb{R}^{d^{(l)} \times d^{(l+1)}}$ is the learnable weight matrix at the l-th layer. In this paper, Chebyshev graph convolutional networks (ChebNet) are introduced to capture more complex structural features of graphs.

For an undirected graph G, the normalized Laplacian matrix is written as follows:

$$L = I_n - D^{-1/2} A D^{-1/2} \tag{6}$$

where A denotes the adjacency matrix. D means the degree matrix. I_n is the identity matrix. To apply Chebyshev polynomials, it is necessary to rescale the eigenvalues of the Laplacian matrix L to the interval $[-1, 1]$, resulting in:

$$\tilde{L} = \frac{2L}{\lambda_{max}} - I_n \tag{7}$$

where λ_{max} denotes the largest eigenvalue of the Laplacian matrix L.

Hence, the convolution operation in ChebNet can be expressed as:

$$g_\theta * x = \sum_{k=0}^{K-1} \theta_k T_k(\tilde{L}) x \tag{8}$$

where $x \in \mathbb{R}^N$ denotes the input signal defined on the graph. $\theta \in \mathbb{R}^K$ represents the filter parameters. T_k is the Chebyshev polynomial of order k. $T_k(\tilde{L})$ denotes the k-th order Chebyshev polynomial evaluated on the rescaled graph Laplacian matrix $\tilde{L}$.

For multi-channel input features, the layer-wise propagation rule of ChebNet can be formulated as:

$$H^{(l+1)} = \sigma\left(\sum_{k=0}^{K-1} T_k(\tilde{L}) H^{(l)} W_k^{(l)}\right) \tag{9}$$

where $H^{(l)} \in \mathbb{R}^{N \times F_l}$ denotes the node feature matrix at layer l. $W_k^{(l)} \in \mathbb{R}^{F_l \times F_{l+1}}$ is the learnable weight matrix associated with the k-th order Chebyshev polynomial. σ is a non-linear activation function.

2.3. Attention Mechanism

The visual attention mechanism (AM), widely used in deep learning (DL)-based models, is inspired by the capacity of the human visual system to focus on salient regions while filtering out irrelevant information [26,27]. AM dynamically assigns greater weights to the most informative regions of feature maps, thereby enhancing model performance. By weighting different regions or feature maps according to their importance, AM enables the model to prioritize critical information, improving both interpretability and accuracy. Most importantly, AM can be classified into three types: spatial attention, channel attention, and spatial-channel attention.

Spatial attention mechanisms focus on identifying and assigning varying weights to different spatial locations within the input, depending on task-specific requirements. Jaderberg et al. introduced the spatial Transformer module, which enables convolutional neural networks (CNNs) to efficiently learn spatial invariance by performing adaptive transformations on feature maps [28]. Channel attention mechanisms, on the other hand, prioritize important regions within feature maps by emphasizing the relative significance of different channels in the feature representations. Wang et al. proposed the efficient channel attention (ECA) module, which enhances CNN performance by leveraging a local cross-channel interaction strategy, avoiding dimensionality reduction [29]. Spatial-channel attention mechanisms combine the strengths of both spatial and channel attention, allowing deep learning models to extract more informative and discriminative feature representations. He et al. introduced the multi-scale attention (EMA) module, which efficiently preserves per-channel information while reducing computational costs by grouping channels and enhancing spatial semantics within each group [30]. These attention mechanisms are often integrated into deep learning architectures (e.g., CNNs or Transformers) to enhance the model's capacity to capture long-range dependencies and fine-grained information.

3. Proposed Method

The proposed MTAGCN consists of a task-shared module and two task-specific modules: fault type diagnosis (FTD) and fault severity diagnosis (FSD). The task-shared module is responsible for extracting task-invariant, structural, and discriminative features from the input data. Specifically, shared features in the task-shared module are propagated through convolutional (Conv) blocks, allowing the model to extract deep features at multiple levels and capture fine-grained information. The task-shared module and the two task-specific modules are interconnected through the proposed attention mechanism (AM), which acts as a feature selector, learning and extracting task-specific features. The detailed structure of the proposed MTAGCN is shown in Figure 4.

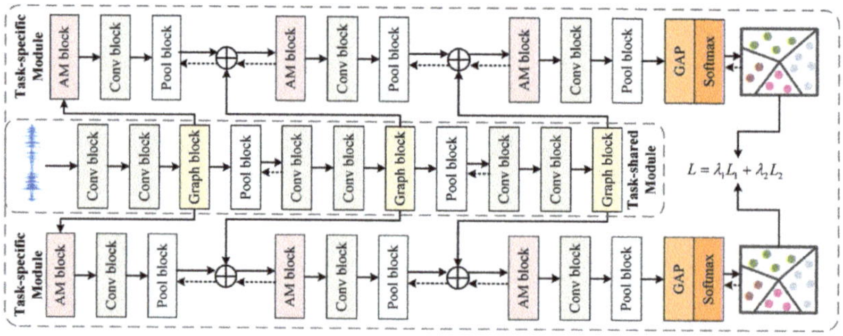

Figure 4. Detailed architecture of the proposed method.

Figure 4 illustrates the MTL-based architecture, which comprises three parallel branches. The top and bottom branches are task-specific pathways, each consisting of attention modules (AM blocks), convolutional blocks, and pooling layers with skip connections to enhance information flow. The middle branch functions as a shared feature extraction backbone, processing input waveform signals through alternating convolutional blocks, graph blocks, and pooling layers. The overall architecture integrates task-specific losses using the formula $L = \lambda_1 L_1 + \lambda_2 L_2$ and outputs to their respective classification matrices via global average pooling (GAP) and Softmax layers. This design demonstrates an effective multi-task learning paradigm that preserves task specificity while enabling efficient feature sharing.

3.1. Task-Shared Module

The raw vibration signals, without any preprocessing, are directly used as input to the MTAGCN, thereby enabling an end-to-end fault diagnosis framework. The dimensionality of each input signal is 1024 with no overlap, represented as $x \in \mathbb{R}^{1\times 1024}$. To analyze the structural relationships within the input features, the extracted features are transformed into instance graphs. Initially, a convolutional layer (Conv) is applied to extract meaningful features from the raw vibration data. The detailed structure of the Conv block is illustrated in Figure 5a. This Conv block consists of a one-dimensional convolutional layer (f_{1DConv}), followed by batch normalization (f_{bn}), and a rectified linear unit activation function (f_{ReLU}).

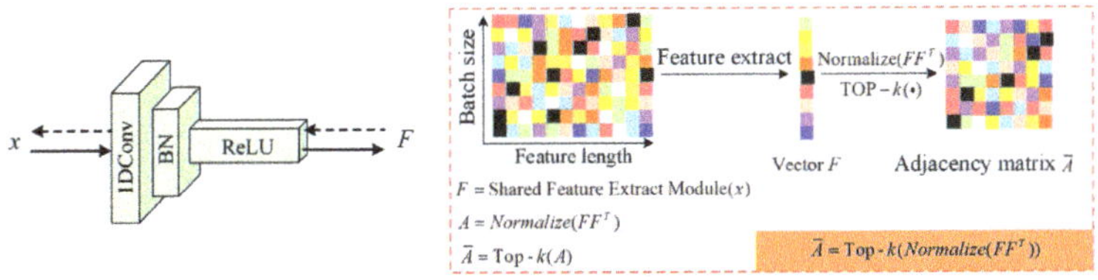

(a) Conv block (b) Instance Graph Generation

Figure 5. Detailed process of the Conv block and instance graph generation. (**a**) Conv block. (**b**) Instance graph generation.

The raw vibration signals are fed into the Conv block, which can be mathematically expressed as follows:

$$\begin{cases} f_{1DConv} = 1DConv(x) \\ f_{bn}(f_{1DConv}, B) = \gamma \frac{f_{1DConv} - \mu(B)}{\sqrt{\sigma(B)^2}} + \beta \\ F = \max(0, f_{bn}) \end{cases} \tag{10}$$

where B denotes the mini-batch. The parameters γ and β are learnable scaling and shifting factors in the batch normalization process. The mean $\mu(B)$ and variance $\sigma^2(B)$ of the mini-batch are computed as follows:

$$\mu(B) = \frac{1}{m}\sum_{i=1}^{m} x_i \tag{11}$$

$$\sigma(B) = \sqrt{\frac{1}{m}\sum_{i=1}^{m};(x_i - \mu(B))^2} \tag{12}$$

where m represents the size of mini-batch B.

Next, the graph block is employed after the Conv block to construct the adjacency matrix A based on the extracted features F. To facilitate the generation of instance graphs, a top-k sorting mechanism is introduced to identify the k nearest neighbors, as illustrated in Figure 5b. The detailed formulation of the graph block is given as follows:

$$\begin{cases} \tilde{F} = MLP(F) \\ A = \text{normlize}(\tilde{F}\tilde{F}^T) \\ \bar{A} = \text{Top} - k(\text{A}) \end{cases} \tag{13}$$

where A denotes the constructed adjacency matrix, and $\tilde{F}$ represents the output of the multilayer perceptron (MLP). The function normalize$(\cdot)$ is used to normalize the product of $\tilde{F}$ and its transpose. To enhance computational efficiency, a sparse adjacency matrix $\bar{A}$ is derived by applying the Top-$k(\cdot)$ operation, which retains only the top k largest values in each row of A. This approach effectively reduces the computational overhead while preserving the most significant relational information.

It is noteworthy that the task-shared module is composed of six Conv blocks and three Graph blocks arranged in an alternating sequence. This architecture design enables the network to simultaneously capture both local temporal features (through Conv blocks) and global structural relationships (through Graph blocks) from the input vibration signals. By integrating these complementary feature extraction mechanisms, the MTAGCN framework can effectively learn discriminative representations that enhance fault diagnosis accuracy across different operating conditions.

3.2. Attention Mechanism Module

The detailed architecture of SE-Net is illustrated in Figure 6.

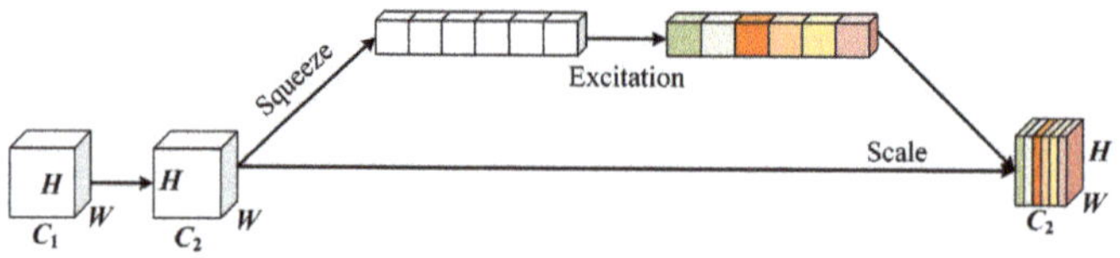

Figure 6. The detailed process of the SE-Net.

Assuming the input data for the AM module is defined as $x_1 \in \mathbb{R}^{C_1 \times W \times H}$, where C_1, W, and H denote the number of channels, width, and height, respectively. A convolutional operation is then applied to extract deeper features, which can be formulated as follows:

$$x_2 = F_{tr}(x_1) \tag{14}$$

where $x_2 \in \mathbb{R}^{C_2 \times W \times H}$ represents the deeper features with channel dimension C_2, width W, and height H. $F_{tr}(\cdot)$ denotes the convolution operation. Subsequently, global information is aggregated by a squeeze operation, resulting in a scalar vector $x_3 \in \mathbb{R}^{1 \times 1 \times C_2}$ that captures the global receptive field. The squeezing process is formulated as follows:

$$x_2 = F_{sq}(x_1) \tag{15}$$

where $F_{sq}(\cdot)$ denotes the global average pooling.

Then, two fully connected (FC) layers are employed to perform the excitation operation, enabling the model to capture inter-channel dependencies within the feature maps.

$$\begin{cases} s = \sigma(W_2 \delta(W_1 x_2)) \\ \sigma = \frac{1}{1+e^{-x}} \end{cases} \tag{16}$$

where W_1 and W_2 refer to the weight parameters of the two FC layers, separately. $\sigma(\cdot)$ denotes the non-linear activation function, which is ReLU. s represents the weights of the feature maps.

Finally, the scaling operation is applied to re-calibrate the features x_1, as expressed below:

$$\tilde{x} = s x_1 \tag{17}$$

Most importantly, the AM module not only learns task-specific features from different layers of the task-shared module but also identifies important features to enhance diagnostic performance. It enables the proposed MTAGCN to dynamically learn task-specific features within the task-shared module for use in the two task-specific modules. We integrated AM modules at multiple layers within the proposed MTAGCN to capture multi-level features. Specifically, incorporating AM modules at different levels in the task-shared module allows the two task-specific modules to capture rich, representative features.

3.3. Task-Specific Module

As illustrated in Figure 4, the task-specific modules include the fault type diagnosis (FTD) module and the fault severity diagnosis (FSD) module, both of which share the same architecture. Each task-specific module consists of three AM modules, three 1×1 Conv blocks, three max-pooling layers, and a classifier.

The task-specific modules facilitate the extraction and optimization of features tailored to specific fault diagnosis tasks (i.e., fault types or severities) by leveraging task-shared features through the AM module. The key advantage of these task-specific modules lies in their ability to mine the most discriminative and representative features for fault diagnosis, thereby enhancing the efficiency of multi-task learning (MTL) through the task-shared module.

3.4. Loss Function and Optimization

In general, the proposed MTAGCN framework is trained using a loss function that incorporates the two task-specific modules, L_1 and L_2, simultaneously. The objective function can be expressed as:

$$L = \lambda_1 L_1 + \lambda_2 L_2 \tag{18}$$

where λ_1 and λ_2 refer to the impact factors for balancing the two task-specific modules, respectively.

Herein, the L_1 and L_2 can be expressed as

$$L_1(X, Y_1) = -\frac{1}{pq}\sum\nolimits_{p,q} Y_1(p,q)\log \hat{Y}_1(p,q) \tag{19}$$

$$L_2(X, Y_2) = -\frac{1}{pq}\sum\nolimits_{p,q} Y_2(p,q)\log \hat{Y}_2(p,q) \tag{20}$$

where Y and $\hat{Y}$ refer to the true and predicted labels of rotating machinery, respectively.

To further improve the fault diagnostic and identification performance of the proposed MTAGCN, the parameters can be fine-tuned and iteratively updated using the Adaptive Moment Estimation (Adam) optimization algorithm proposed by Kingma and Ba [31]. This enables optimal parameter tuning and, thus, improves the effectiveness and feasibility of the proposed MTAGCN in MTL-based diagnostic tasks. The following equations represent the update process:

$$\begin{cases} r^{(\text{epoch}+1)} = r^{(0)}\sqrt{(1-\beta_2^{\text{epoch}})/1-\beta_1^{\text{epoch}}} \\ m_1^{(\text{epoch}+1)} = \beta_1 m_1^{(\text{epoch})} + (1-\beta_1)\Delta\theta_{ft}^{(\text{epoch})} \\ m_2^{(\text{epoch}+1)} = \beta_2 m_2^{(\text{epoch})} + (1-\beta_2)(\Delta\theta_{ft}^{(\text{epoch})})^2 \\ \theta_{ft}^{(\text{epoch}+1)} = \theta_{ft}^{(\text{epoch})} - r^{(\text{epoch}+1)} m_1^{(\text{epoch}+1)} / \sqrt{m_2^{(\text{epoch}+1)} + \varepsilon} \end{cases} \tag{21}$$

where the hyperparameters β_1, β_2, and ε are used to balance the optimization process, with typical values of 0.9, 0.999, and 10^{-8}, respectively. epoch indicates the total number of training iterations. The variables m_1 and m_2 are initialized to 0, corresponding to the first and second moment estimates, respectively. Meanwhile, $r^{(\text{epoch}+1)}$ denotes the updated learning rate after $(\text{epoch} + 1)$ iteration, and $r^{(0)}$ denotes the initial learning rate, which can be customized by the user.

4. Experiments

4.1. Case 1: Cylindrical Roller Bearing (CRB) Dataset

The CRB dataset is widely employed in bearing fault detection research and serves as a standard for evaluating the performance and practicality of the proposed approach in comparison to other state-of-the-art (SOTA) models [32]. As depicted in Figure 7, the experimental setup consists of a motor, an accelerometer, two bearings, and a load cell. Vibration data from the test bearing were recorded with an accelerometer at a sampling rate of 70,000 Hz. The CRB dataset includes four distinct bearing health statuses: normal (N), inner race fault (IRF), outer race fault (ORF), and roller fault (RF). The specifications of the cylindrical rolling bearing are shown in Table 1. For each faulty bearing, five different fault diameters were introduced using electrical discharge machining (EDM), as described in Table 2. The experiments were conducted with a shaft rotation speed of 2050 rpm and a load of 200 N. The different fault types and their severity levels define two distinct task requirements. In total, twelve rolling bearing health conditions were identified, alongside the normal condition. For simplicity, the thirteen health conditions are denoted as N, IRF-I, IRF-II, IRF-III, IRF-IV, IRF-V, ORF-I, ORF-II, ORF-III, ORF-IV, ORF-V, RF-I, RF-II, RF-III, RF-IV, and RF-V, where IRF-I represents the smallest fault diameter in the inner ring fault category. Detailed diagrams of the different faults are illustrated in Figure 8.

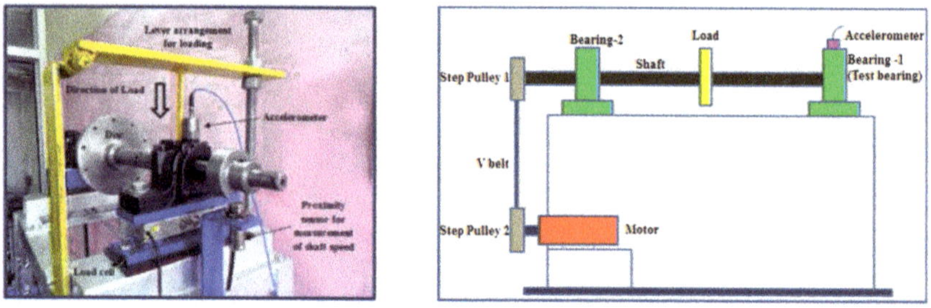

(a) (b)

Figure 7. Experimental benchmark of the CRB dataset. (**a**) Typical photograph. (**b**) Schematic diagram.

Table 1. Specifications of the cylindrical rolling bearing.

Parameters	Value
Diameter (Inner race)	25 mm
Diameter (Outer race)	52 mm
Pitch Diameter (D)	38.9 mm
Ball Diameter (d)	7.5 mm
Number of rolling elements (N)	13
Contact angle (φ)	0°

Table 2. Defect width of the cylindrical roller bearing.

Bearing Component	Label	Defect Width (mm)
Inner race	IRF-I	0.43 mm
	IRF-II	1.01 mm
	IRF-III	1.56 mm
	IRF-IV	2.03 mm
Outer race	ORF-I	0.42 mm
	ORF-II	0.86 mm
	ORF-III	1.55 mm
	ORF-IV	1.97 mm
Roller	RF-I	0.49 mm
	RF-II	1.16 mm
	RF-III	1.73 mm
	RF-IV	2.12 mm

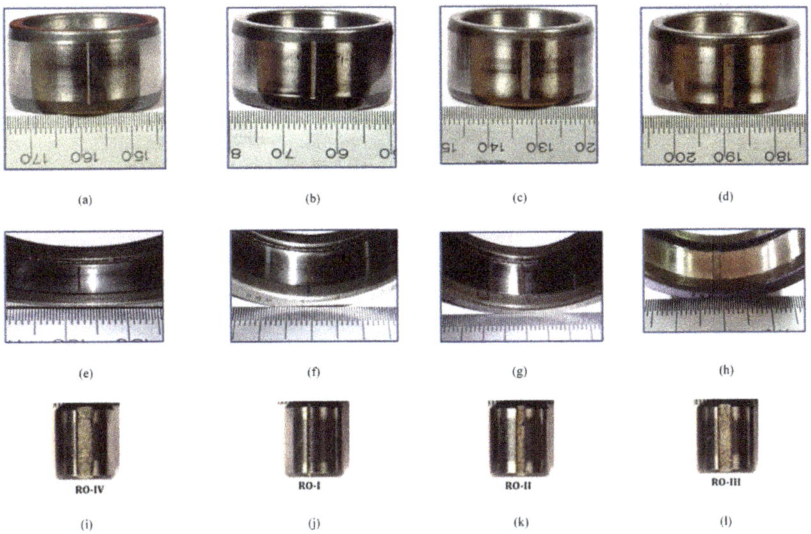

Figure 8. The bearing faults in the CRB dataset. (**a–d**) Inner race fault. (**e–h**) Outer race fault. (**i–l**) Roller fault.

4.2. Case 2: HUSTbearing Dataset

To further demonstrate the advancement and advantages of the proposed MTAGCN under different rotational speed conditions, we conducted experiments using the HUSTbearing dataset. This dataset includes bearing data collected from a simulation platform under varying rotational speeds, fault types, and severity levels [33]. The parameters of the tested bearings are described in Table 3.

Table 3. Parameters of the tested bearings.

Parameters	Value
Shaft Diameter	38.52 mm
Ball Diameter	7.94 mm
Pitch Diameter (D)	9
Number of rolling balls (N)	9

As illustrated in Figure 9, the experimental platform consists of a speed controller, motor, shaft, acceleration sensor, bearing module, and data acquisition board. Unprocessed vibration signals were collected from the test bearings using accelerometers mounted on the bearing casing under four distinct operating conditions (i.e., 65 Hz, 70 Hz, 75 Hz, and 80 Hz), with a sampling rate of 25.6 kHz and a constant load maintained across all speeds. To simulate various fault types and severities, single-point defects were introduced into the inner ring, outer ring, and rolling elements of the test bearings, as shown in Figure 10. The bearing conditions are categorized into four groups: normal (N), inner ring fault (IRF), outer ring fault (ORF), and roller ball fault (RBF). Each of the IRF, ORF, and RBF conditions includes two severity levels, denoted as I and II. In total, the HUSTbearing dataset comprises seven distinct bearing health conditions.

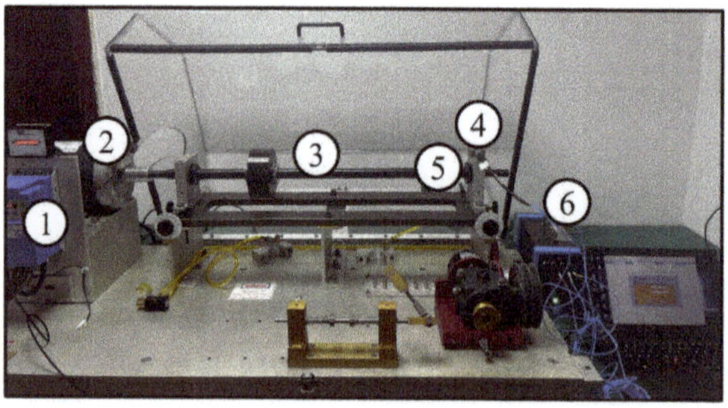

Figure 9. Experimental platform of the HUSTbearing dataset.

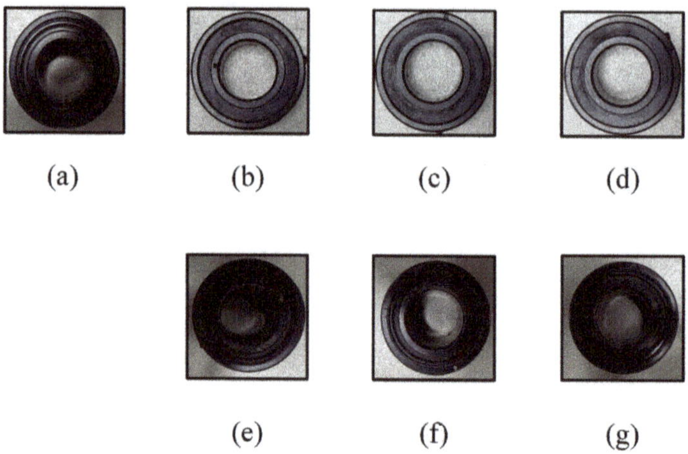

Figure 10. The bearing faults in the HUSTbearing dataset. (**a**) Normal. (**b**) Inner ring fault. (**c**) Outer ring fault. (**d**) Roller ball fault. (**e**) Inner ring fault. (**f**) Outer ring fault. (**g**) Roller ball fault.

Based on the HUSTbearing dataset, four diagnostic tasks were constructed under varying rotational speeds, as summarized in Table 4.

Table 4. Fault diagnosis based on the HUSTbearing dataset.

Rotating Speed	Load	Samples	Tasks	
			T_1	T_2
65 Hz	Equal	200 × 7	N	N
70 Hz		200 × 7	IRF	I severity
75 Hz		200 × 7	ORF	II severity
80 Hz		200 × 7	RBF	

4.3. Case 3: MCC5-THU Gearbox Dataset

In previous studies, vibration data were typically generated using simulated bearing platforms [34]. However, these case studies often fall short of capturing the complexity of real-world rotating machinery, thus limiting the generalization capability of diagnostic models. To further evaluate the robustness and generalizability of the proposed method, we employed a real gearbox test rig to collect vibration signals. The experimental setup of the MCC5-THU gearbox test rig is depicted in Figure 11. It consists of two three-axis vibration acceleration sensors, a frequency inverter, a torque sensor, a magnetic powder brake, a parallel gearbox, an epicyclic gearbox, a speed sensor, a motor, and a data acquisition system. In this experimental study, four types of gear failures were induced: tooth crack, gear wear, tooth break, and gear pitting, each with three levels of fault severity (i.e., light, medium, and severe), as summarized in Table 5. Consequently, the dataset includes twelve distinct gear health conditions. Vibration data were acquired using a three-axis accelerometer mounted on the gearbox, sampled at 12.8 kHz over a 60 s duration.

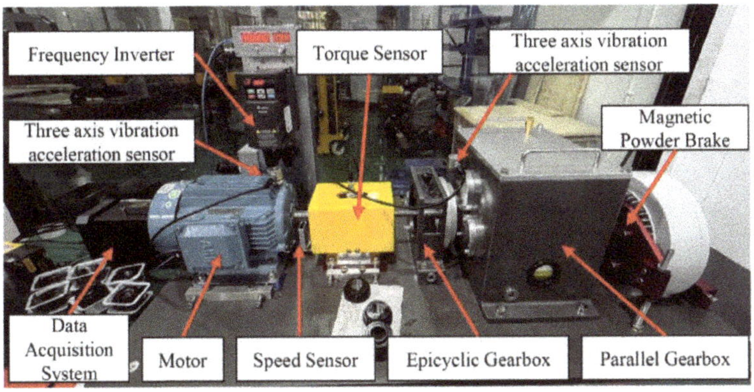

Figure 11. Experimental platform of the MCC5-THU Gearbox dataset.

Table 5. The details of gear fault parameters.

Fault Parameter of Gear	Fault Degree		
	Light	Medium	High
Teeth Crack Depth	1/4 of the teeth height	1/2 of the teeth height	3/4 of the teeth height
Gear Wear	1/3 of the teeth surface area	1/2 of the teeth surface area	Full teeth surface area
Teeth Break	1/4 of the teeth width	1/2 of the teeth width	3/4 of the teeth width
Gear Pitting	Fault diameter 0.5 mm	Fault diameter 1.0 mm	Fault diameter 1.5 mm

4.4. Baseline Methods

To demonstrate the practicality and effectiveness of the proposed MTAGCN, we conducted a series of comparative experiments across three rotating machinery case studies.

In these experiments, MTAGCN was benchmarked against four state-of-the-art (SOTA) methods, including two single-task learning (STL)-based approaches—MRFGCN and IDBN—and two multi-task learning (MTL)-based approaches—MTCNN and JLCNN.

4.5. Implementation Details

All experiments were implemented using the PyTorch 1.9.0 framework with Python 3.8.3. The computational workload was carried out on a workstation equipped with an Intel i5-10400F CPU, 32 GB of RAM, and an NVIDIA RTX 3070 GPU. In all three case studies, the acquired vibration signals were segmented into subsamples using a sliding window of 1024 points without overlap between consecutive segments. All models were trained for 300 epochs. For the proposed MTAGCN framework, the batch size was set to 32 during training and 16 during testing.

4.6. Diagnosis Results Analysis

The experimental results on the CRB dataset are summarized in Table 6 and illustrated in Figure 12.

Table 6. Diagnostic performance of two fault diagnosis tasks on the CRB dataset (T_1 means the fault type diagnosis (FTD) task; T_2 means the fault severity diagnosis (FSD) task).

Ours		MTCNN		JLCNN		MRFGCN		IDBN	
T_1	T_2	T_1	T_2	T_1	T_2	T_1	T_2	T_1	T_2
100	100	99.14 ± 0.21	95.64 ± 0.73	98.09 ± 0.45	93.78 ± 0.89	97.49 ± 0.44	89.14 ± 1.24	96.84 ± 0.51	87.26 ± 1.33

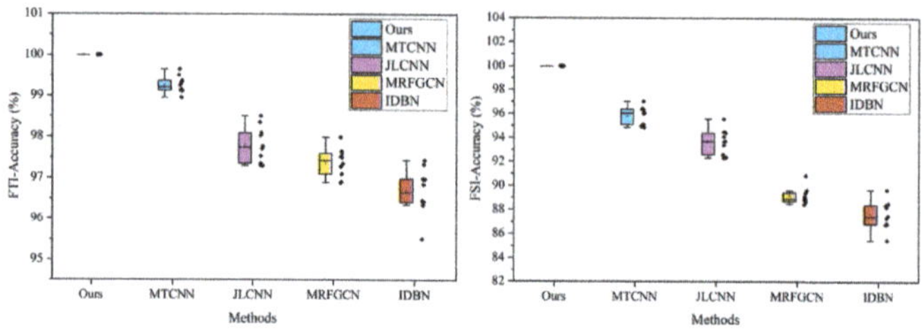

(a) (b)

Figure 12. Ten experimental diagnostic results of all methods under the CRB dataset. (**a**) FTI task. (**b**) FSI task.

As shown in Table 6 and Figure 12, the proposed MTAGCN achieves 100% diagnostic accuracy on both the FTI and FSI tasks, demonstrating its superior performance and robustness. While MTCNN and JLCNN attain over 95% accuracy on the FTI task, their performance on the FSI task drops to approximately 90%, which is notably lower than that of MTAGCN. Moreover, other benchmark methods, such as MRFGCN and IDBN, exhibit even lower accuracy on the FSI task, with results not exceeding 90%. Additionally, the proposed MTAGCN displays significantly smaller error bars compared to other state-of-the-art methods, further validating its enhanced stability and robustness on the CRB dataset.

As demonstrated in Table 7 and Figure 13, the proposed MTAGCN consistently surpasses other state-of-the-art methods in both the FTI and FSI tasks on the HUSTbearing dataset, underscoring its superior adaptability and diagnostic accuracy across varying rotational speed conditions.

Table 7. Diagnostic accuracy (%) of two fault diagnosis tasks under four rotating speeds (T_1 means the fault type diagnosis (FTD) task; T_2 means the fault severity diagnosis (FSD) task).

Speed	Ours		MTCNN		JLCNN		MRFGCN		IDBN	
	T_1	T_2	T_1	T_2	T_1	T_2	T_1	T_2	T_1	T_2
65 Hz	100	100	98.50	96.89	97.74	95.98	98.94	85.24	92.38	78.25
70 Hz	100	100	94.20	99.16	93.51	90.55	97.67	82.97	90.25	71.56
75 Hz	100	100	98.31	96.41	93.72	91.08	95.19	81.12	88.64	66.74
80 Hz	100	100	97.58	95.94	92.57	90.41	95.84	80.56	89.05	68.57
Average	100	100	97.15	97.10	94.39	92.01	96.91	82.47	90.08	71.28

As shown in Table 7 and Figure 13, the proposed MTAGCN significantly outperforms existing state-of-the-art (SOTA) methods in both the FTI and FSI tasks across all four rotating speed scenarios. While MTCNN and JLCNN exhibit competitive performance in the FTI task, MTAGCN consistently surpasses JLCNN in the FSI task throughout all experiments. This demonstrates the effectiveness of MTAGCN in handling multiple tasks simultaneously and improving diagnostic performance holistically. In contrast, JLCNN and other single-task learning (STL)-based methods exhibit relatively lower recognition accuracy in both tasks. Notably, MTAGCN achieves an average diagnostic accuracy of 100% across both tasks, outperforming all comparison methods. Although MRFGCN and IDBN perform well under the T_1 scenario, their accuracy drops considerably in T_2, primarily because STL-based methods treat tasks independently and fail to leverage shared features across tasks. Moreover, the diagnostic accuracy of most SOTA methods tends to decline as the rotation speed increases, highlighting the challenge of maintaining robustness under varying operational conditions.

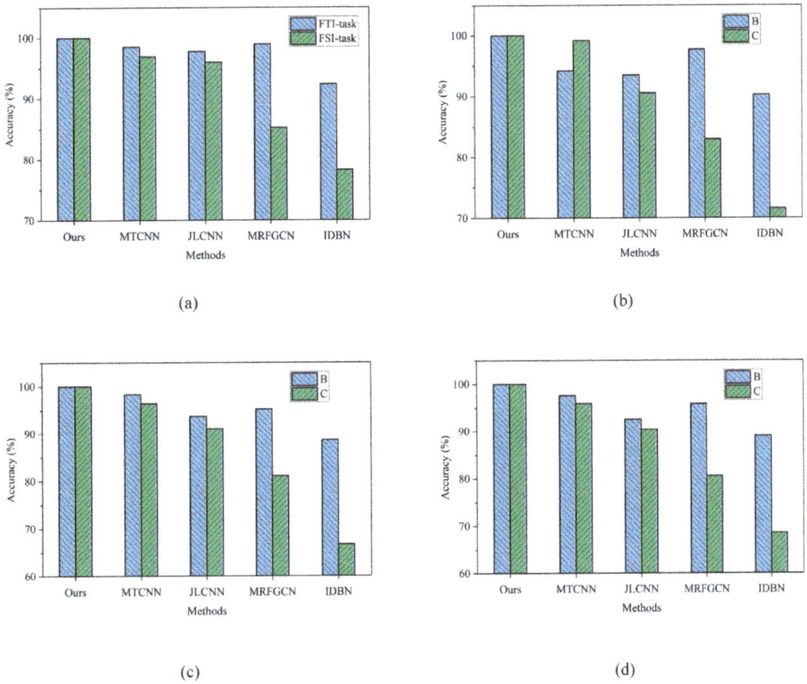

Figure 13. Diagrams for diagnostic results of all methods under four rotating speeds: (**a**) 65 Hz, (**b**) 70 Hz, (**c**) 75 Hz, and (**d**) 80 Hz.

The diagnostic results from the MCC5-THU are described in Table 8, and its corresponding histogram is illustrated in Figure 14.

Table 8. Diagnostic accuracy of two fault diagnosis tasks under the MCC5-THU dataset.

Ours		**MTCNN**		**JLCNN**		**MRFGCN**		**IDBN**	
T_1	T_2	T_1	T_2	T_1	T_2	T_1	T_2	T_1	T_2
99.51 ± 0.11	99.47 ± 0.14	98.64 ± 0.39	93.71 ± 0.87	97.64 ± 0.61	89.47 ± 0.92	95.27 ± 0.65	85.64 ± 1.41	95.29 ± 0.76	81.67 ± 1.54

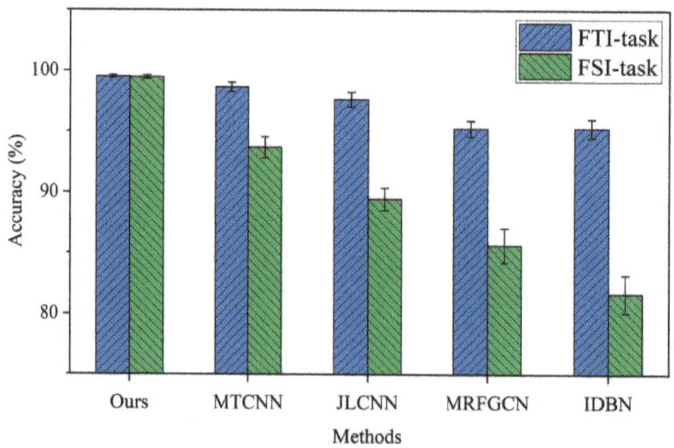

Figure 14. Diagnostic accuracy of two fault diagnosis tasks under MCC5-THU dataset.

4.7. Feature Visualization

In this section, t-distributed stochastic neighbor embedding (t-SNE) is employed to visualize high-dimensional features in a low-dimensional space, thereby providing deeper insight into the strengths and limitations of each method. The HUSTbearing dataset is selected as a representative case for feature visualization. The detailed t-SNE visualization results for all compared methods under the T_1 scenario at 65 Hz are illustrated in Figure 15.

As illustrated in Figure 15, the extracted features corresponding to the four domains (i.e., four fault categories) are well aligned and clearly separable. Furthermore, the multi-task learning (MTL)-based methods demonstrate superior capability in learning more discriminative and informative features compared to single-task learning (STL)-based methods, which is consistent with the diagnostic performance results. These observations confirm that the proposed method effectively captures task-invariant and class-discriminative features, which are essential for enhancing the performance of MTL-based fault diagnosis models.

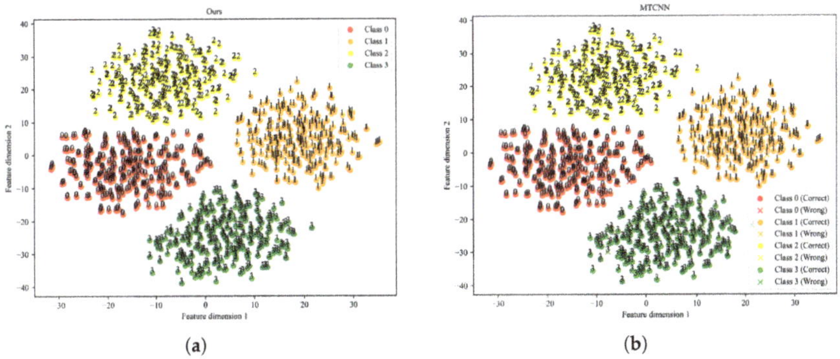

Figure 15. *Cont.*

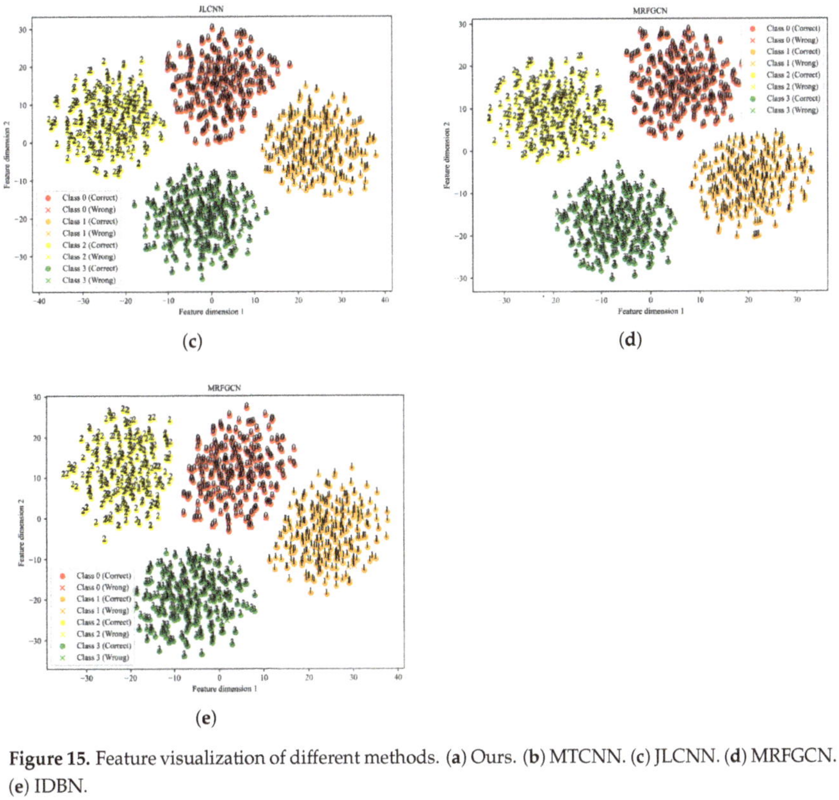

Figure 15. Feature visualization of different methods. (**a**) Ours. (**b**) MTCNN. (**c**) JLCNN. (**d**) MRFGCN. (**e**) IDBN.

4.8. Classification Performance

To evaluate the diagnostic effectiveness of the proposed method, confusion matrices are utilized to present the relationship between true and predicted labels in a structured format. The confusion matrix results for all methods on the CRB dataset are shown in Figure 16.

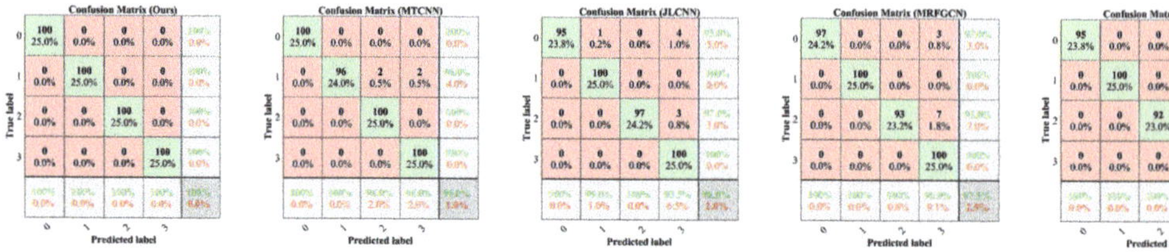

(a) (b) (c) (d) (e)

Figure 16. Confusion matrix results of all methods for the FTI task on the CRB dataset. (**a**) Ours. (**b**) MTCNN. (**c**) JLCNN. (**d**) MRFGCN. (**e**) IDBN.

It is evident that the proposed MTAGCN accurately classifies all samples in the FTI task. In comparison to other MTL-based methods (i.e., MTAGCN, MTCNN, and JLCNN), the STL-based methods (i.e., MRFGCN and IDBN) demonstrate lower performance in identifying health conditions.

As shown in Figure 17, our proposed MTAGCN achieves the highest classification performance. In contrast, the worst-performing method (i.e., IDBN) frequently misclassifies Label 2 into other labels. Compared to the FTI task, the diagnostic accuracies of all methods—except for MTAGCN—decrease in the FSI task.

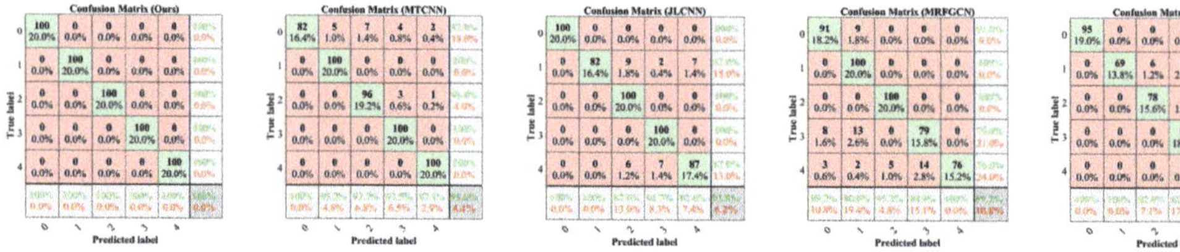

(a) (b) (c) (d) (e)

Figure 17. Confusion matrix results of all methods for the FSI task on the CRB dataset. (**a**) Ours. (**b**) MTCNN. (**c**) JLCNN. (**d**) MRFGCN. (**e**) IDBN.

5. Model Discussions

5.1. Detailed Structure and Complexity of MTAGCN

To illustrate the detailed structure and explore the complexity of the proposed MTAGCN, the corresponding parameters are presented in Table 9.

Table 9. Detailed structure and complexity of MTAGCN.

Component	Layer	Filter	Input Size	Output Size	Parameters	FLOPs
Task-shared	Conv1	16 × 7 × 1	1 × 1024	16 × 1024	192	229,376
	Conv2	16 × 7 × 1	16 × 1024	16 × 1024	1872	3,670,016
	Graph	16 × 7 × 1	16 × 1024	16 × 1024	1872	3,670,016
	MaxPool	4 × 1 (stride = 4)	16 × 1024	16 × 256	0	16,384
	Conv3	32 × 5 × 1	16 × 256	32 × 256	192	229,376
	Conv4	32 × 5 × 1	16 × 256	32 × 256	1872	3,670,016
	Graph	32 × 5 × 1	16 × 256	32 × 256	1872	3,670,016
	MaxPool	4 × 1 (stride = 4)	16 × 256	32 × 64	0	16,384
	Conv5	64 × 3 × 1	32 × 64	64 × 64	192	229,376
	Conv6	64 × 3 × 1	32 × 64	64 × 64	1872	3,670,016
	Graph	64 × 3 × 1	32 × 64	64 × 64	1872	3,670,016
Task-specific	Attention	k = 3	16 × 1024	16 × 1024	6	192
	Att Conv1	32 × 1 × 1	16 × 1024	32 × 1024	5280	2,621,440
	Att MaxPool	4 × 1 (stride = 4)	32 × 1024	32 × 256	0	16,384
	Attention	k = 3	32 × 256	32 × 256	0	16,384
	Att Conv2	64 × 1 × 1	32 × 256	64 × 256	5280	2,621,440
	Att MaxPool	4 × 1 (stride = 4)	64 × 256	64 × 64	0	16,384
	Attention	k = 3	64 × 64	64 × 64	6	192
	Att Conv3	128 × 1 × 1	128 × 64	128 × 64	5280	2,621,440
	Att MaxPool	4 × 1 (stride = 4)	128 × 64	128 × 16	0	16,384
Task-specific	Attention	k = 3	16 × 1024	16 × 1024	6	192
	Att Conv1	32 × 1 × 1	16 × 1024	32 × 1024	5280	2,621,440
	Att MaxPool	4 × 1 (stride = 4)	32 × 1024	32 × 256	0	16,384
	Attention	k = 3	32 × 256	32 × 256	0	16,384
	Att Conv2	64 × 1 × 1	32 × 256	64 × 256	5280	2,621,440
	Att MaxPool	4 × 1 (stride = 4)	64 × 256	64 × 64	0	16,384
	Attention	k = 3	64 × 64	64 × 64	6	192
	Att Conv3	128 × 1 × 1	128 × 64	128 × 64	5280	2,621,440
	Att MaxPool	4 × 1 (stride = 4)	128 × 64	128 × 16	0	16,384
Task FC Layers	AdaptiveAvgPool	1 × 1	128 × 16	128 × 1	0	4096
	Task 1 FC	128 × 4	128 × 1	4 × 1	516	1024
	Task 2 FC	128 × 4	128 × 1	4 × 1	516	1024

As observed from the results, the total number of FLOPs for the proposed MTAGCN is 15,497,888.

5.2. Ablation Study

In this section, an ablation study is conducted to evaluate the performance of each key innovation in our proposed method. The results of different variants are presented as follows: (1) Method 1 (M1), MTAGCN without GCNs, and (2) Method 2 (M2), MTAGCN without the attention mechanism (AM).

As shown in the data from Table 10, our proposed MTAGCN consistently outperforms the four variants in terms of accuracy. Specifically, MTAGCN surpasses M1 and M2 by 0.96% and 1.46% at 65 Hz, respectively. These significant improvements highlight the importance of incorporating GCNs and the attention mechanism (AM) into MTL-based fault diagnosis.

Table 10. Experimental results of the ablation study.

Tasks	M1	M2	Ours
65 Hz	99.04%	98.54%	100%
70 Hz	98.76%	97.69%	100%
75 Hz	97.89%	97.08%	100%

5.3. Evaluation Metrics

To further demonstrate the effectiveness and practicality of the proposed MTAGCN, four additional evaluation metrics—AUC, precision, recall, and F1-score—are introduced in this section. A detailed comparison of these evaluation metrics on the CRB dataset is presented in Table 11.

Table 11. Experimental results of the CRB dataset using five evaluation metrics.

Methods	Accuracy (%)	AUC	Precision (%)	Recall (%)	F1-Score (%)
Ours	100	100	100	100	100
MTCNN	99.14	97.48	96.85	96.91	96.88
JLCNN	98.09	95.62	96.17	96.64	96.40
MRFGCN	97.49	93.28	94.53	94.07	94.29
IDBN	96.84	93.61	93.07	92.51	92.79

The proposed MTAGCN achieves a mean accuracy, AUC, precision, recall, and F1-score of 100% on the CRB dataset, significantly outperforming other state-of-the-art methods. This superior performance can be attributed to the ability of MTAGCN to effectively mine representative features from task-shared modules, a challenge for other methods, which leads to a decline in their fault diagnostic performance.

6. Conclusions

In this paper, we propose the MTAGCN framework to model two specific tasks within a unified deep network, enabling multi-task learning (MTL) for simultaneous fault diagnosis of fault types and severities. Extensive experimental results demonstrate the effectiveness of the proposed method and its superiority over various state-of-the-art (SOTA) approaches. The key conclusions of this work are as follows: (1) the graph block allows features extracted by the Conv block to be automatically transformed into instance graphs, complementing the fault-related information; (2) the task-shared module learns common features, while the two task-specific modules capture task-invariant features; (3) experimental results from three case studies of rotating machinery show that the proposed MTAGCN can effectively mine diagnostic features for both tasks, outperforming comparison methods in multi-task-based fault diagnosis.

Future work on the MTAGCN framework will focus on (1) achieving domain adaptation to handle variable working conditions; (2) exploring multi-sensor data fusion to enhance robustness and anti-noise capabilities; and (3) advancing research in real-time fault diagnosis with the goal of enabling online multi-task learning.

Author Contributions: B.W.: Conceptualization, methodology, software, validation, formal analysis, investigation, resources, data curation, writing—original draft preparation. S.Z.: writing—review and editing, visualization, supervision, project administration, funding acquisition. All authors have read and agreed to the published version of the manuscript.

Funding: This research was funded by the research project funded by the Initial Scientific Research Foundation for Talented Scholars of Nanchang Institute of Science and Technology (Grant numbers NGRCZX-24-10), the Science and Technology Research Project of Jiangxi Provincial Department of Education (Grant No. GJJ2402814), the Initial Scientific Research Foundation for Talented Scholars of Nanchang Institute of Science and Technology (Grant No. NGRCZX-23-10), and the Science and Technology Research Project of Nanchang Institute of Science and Technology (Grant No. NGKJ-22-03).

Informed Consent Statement: Informed consent was obtained from all subjects involved in the study.

Data Availability Statement: The authors do not have permission to share private data.

Conflicts of Interest: The authors declare no conflicts of interest.

References

1. Yu, Y.; Karimi, H.R.; Gelman, L.; Cetin, A.E. MSIFT: A novel end-to-end mechanical fault diagnosis framework under limited & imbalanced data using multi-source information fusion. *Expert Syst. Appl.* **2025**, *274*, 126947.
2. Zhao, S.; Bao, L.; Hou, C.; Bai, Y.; Yu, Y. Multi-source domain adversarial graph convolutional networks for rolling mill health states diagnosis under variable working conditions. *Struct. Health Monit.* **2024**, *23*, 3505–3524. [CrossRef]
3. Shukla, K.; Holderbaum, W.; Theodoridis, T.; Wei, G. Enhancing gearbox fault diagnosis through advanced feature engineering and data segmentation techniques. *Machines* **2024**, *12*, 261. [CrossRef]
4. Liu, X.; Wen, S.; Liu, H.; Yu, F.R. CPL-SLAM: Centralized Collaborative Multi-Robot Visual-Inertial SLAM Using Point-and-Line Features. *IEEE Internet Things J.* **2025**. [CrossRef]
5. Yu, Y.; He, Y.; Karimi, H.R.; Gelman, L.; Cetin, A.E. A two-stage importance-aware subgraph convolutional network based on multi-source sensors for cross-domain fault diagnosis. *Neural Netw.* **2024**, *179*, 106518. [CrossRef]
6. Chen, B.; Smith, W.A.; Cheng, Y.; Gu, F.; Chu, F.; Zhang, W.; Ball, A.D. Probability distributions and typical sparsity measures of Hilbert transform-based generalized envelopes and their application to machine condition monitoring. *Mech. Syst. Signal Process.* **2025**, *224*, 112026. [CrossRef]
7. Chen, B.; Cheng, Y.; Allen, P.; Wang, S.; Gu, F.; Zhang, W.; Ball, A.D. A product envelope spectrum generated from spectral correlation/coherence for railway axle-box bearing fault diagnosis. *Mech. Syst. Signal Process.* **2025**, *225*, 112262. [CrossRef]
8. Chen, Q.; Wei, H.; Rashid, M.; Cai, Z. Kernel extreme learning machine based hierarchical machine learning for multi-type and concurrent fault diagnosis. *Measurement* **2021**, *184*, 109923. [CrossRef]
9. Liu, X.; Qi, W.; Wang, Y.; Shao, C.; Cong, Q. Fault diagnosis of shield machine based on RFE and ELM. *Res. Sq.* **2020**, *preprint*. [CrossRef]
10. Wen, S.; Liu, X.; Wang, Z.; Zhang, H.; Zhang, Z.; Tian, W. An improved multi-object classification algorithm for visual SLAM under dynamic environment. *Intell. Serv. Robot.* **2022**, *15*, 39–55. [CrossRef]
11. Huang, C.; Shi, Q.; Ding, W.; Mei, P.; Karimi, H.R. A robust MPC approach for platooning control of automated vehicles with constraints on acceleration. *Control Eng. Pract.* **2023**, *139*, 105648. [CrossRef]
12. Mei, P.; Karimi, H.R.; Xie, J.; Chen, F.; Ou, L.; Yang, S.; Huang, C. Battery state estimation methods and management system under vehicle–cloud collaboration: A Survey. *Renew. Sustain. Energy Rev.* **2024**, *206*, 114857. [CrossRef]
13. Ullah, A.; Younas, M.; Saharudin, M.S. Digital twin framework using real-time asset tracking for smart flexible manufacturing system. *Machines* **2025**, *13*, 37. [CrossRef]
14. Huang, G.; Lei, W.; Dong, X.; Zou, D.; Chen, S.; Dong, X. Stage-Based Remaining Useful Life Prediction for Bearings Using GNN and Correlation-Driven Feature Extraction. *Machines* **2025**, *13*, 43. [CrossRef]
15. Yu, Y.; Karimi, H.R.; Shi, P.; Peng, R.; Zhao, S. A new multi-source information domain adaption network based on domain attributes and features transfer for cross-domain fault diagnosis. *Mech. Syst. Signal Process.* **2024**, *211*, 111194. [CrossRef]
16. Yu, Y.; Shi, P.; Tian, J.; Xu, X.; Hua, C. Rolling mill health states diagnosing method based on multi-sensor information fusion and improved DBNs under limited datasets. *ISA Trans.* **2023**, *134*, 529–547. [CrossRef]
17. Shi, P.; Yu, Y.; Gao, H.; Hua, C. A novel multi-source sensing data fusion driven method for detecting rolling mill health states under imbalanced and limited datasets. *Mech. Syst. Signal Process.* **2022**, *171*, 108903. [CrossRef]
18. Han, D.; Zhang, Y.; Yu, Y.; Tian, J.; Shi, P. Multi-source heterogeneous information fusion fault diagnosis method based on deep neural networks under limited datasets. *Appl. Soft Comput.* **2024**, *154*, 111371. [CrossRef]
19. Zhao, L.; He, Y.; Dai, D.; Wang, X.; Bai, H.; Huang, W. A Novel Multi-Task Self-Supervised Transfer Learning Framework for Cross–Machine Rolling Bearing Fault Diagnosis. *Electronics* **2024**, *13*, 4622. [CrossRef]
20. Zhang, T.; Chen, J.; Ye, Z.; Liu, W.; Tang, J. Prior knowledge-informed multi-task dynamic learning for few-shot machinery fault diagnosis. *Expert Syst. Appl.* **2025**, *271*, 126439. [CrossRef]
21. Zheng, S.; Li, J.; Qiao, L.; Gao, X. Multi-task interaction learning for accurate segmentation and classification of breast tumors in ultrasound images. *Phys. Med. Biol.* **2025**, *70*, 065006. [CrossRef]
22. Wu, X.; Gou, G. Uncertainty bidirectional guidance of multi-task mamba network for medical image classification and segmentation. *Signal Image Video Process.* **2025**, *19*, 29. [CrossRef]
23. Trejo-Chavez, O.; Cruz-Albarran, I.A.; Resendiz-Ochoa, E.; Salinas-Aguilar, A.; Morales-Hernandez, L.A.; Basurto-Hurtado, J.A.; Perez-Ramirez, C.A. A CNN-based methodology for identifying mechanical faults in induction motors using thermography. *Machines* **2023**, *11*, 752. [CrossRef]
24. Kumar, R.; Kumar, P.; Vashishtha, G.; Chauhan, S.; Zimroz, R.; Kumar, S.; Kumar, R.; Gupta, M.K.; Ross, N.S. Fault identification of direct-shift gearbox using variational mode decomposition and convolutional neural network. *Machines* **2024**, *12*, 428. [CrossRef]
25. Peng, R.; Gong, C.; Zhao, S. Multi-Sensor Information Fusion with Multi-Scale Adaptive Graph Convolutional Networks for Abnormal Vibration Diagnosis of Rolling Mill. *Machines* **2025**, *13*, 30. [CrossRef]

26. Kim, C.S.; Kim, H.B.; Lee, J.M. Self-Explanatory Fault Diagnosis Framework for Industrial Processes Using Graph Attention. *IEEE Trans. Ind. Inform.* **2025**, *21*, 3396–3405. [CrossRef]
27. Snyder, Q.; Jiang, Q.; Tripp, E. Integrating self-attention mechanisms in deep learning: A novel dual-head ensemble transformer with its application to bearing fault diagnosis. *Signal Process.* **2025**, *227*, 109683. [CrossRef]
28. Jaderberg, M.; Simonyan, K.; Zisserman, A. Spatial transformer networks. *Adv. Neural Inf. Process. Syst.* **2015**, *28* .
29. Wang, Q.; Wu, B.; Zhu, P.; Li, P.; Zuo, W.; Hu, Q. ECA-Net: Efficient channel attention for deep convolutional neural networks. In Proceedings of the IEEE/CVF Conference on Computer Vision and Pattern Recognition, Seattle, WA, USA, 13–19 June 2020; pp. 11534–11542.
30. Ouyang, D.; He, S.; Zhang, G.; Luo, M.; Guo, H.; Zhan, J. Efficient multi-scale attention module with cross-spatial learning. In Proceedings of the ICASSP 2023–2023 IEEE International Conference on Acoustics, Speech and Signal Processing (ICASSP), Rhodes Island, Greece, 4–10 June 2023; pp. 1–5.
31. Kingma, D.P.; Ba, J. Adam: A method for stochastic optimization. *arXiv* **2014**, arXiv:1412.6980.
32. Kumar, A.; Zhou, Y.; Gandhi, C.P.; Kumar, R.; Xiang, J. Bearing defect size assessment using wavelet transform based Deep Convolutional Neural Network (DCNN). *Alex. Eng. J.* **2020**, *59*, 999–1012. [CrossRef]
33. Zhao, C.; Zio, E.; Shen, W. Domain generalization for cross-domain fault diagnosis: An application-oriented perspective and a benchmark study. *Reliab. Eng. Syst. Saf.* **2024**, *245*, 109964. [CrossRef]
34. Chen, S.; Liu, Z.; He, X.; Zou, D.; Zhou, D. Multi-mode fault diagnosis datasets of gearbox under variable working conditions. *Data Brief* **2024**, *54*, 110453. [CrossRef] [PubMed]

machines

MDPI

Article

Rolling Bearing Life Prediction Based on Improved Transformer Encoding Layer and Multi-Scale Convolution

Zhuopeng Luo [1,2], Zhihai Wang [1,2,*], Xiaoqin Liu [1,2] and Yingming Yang [1,2]

[1] Faculty of Mechanical and Electrical Engineering, Kunming University of Science and Technology, Kunming 650500, China; 20222203081@stu.kust.edu.cn (Z.L.); liuxqsmile@gmail.com (X.L.); 20222103016@stu.kust.edu.cn (Y.Y.)
[2] Yunnan Provincial Key Lab of Advanced Equipment Manufacturing Technology, Kunming University of Science and Technology, Kunming 650500, China
* Correspondence: wzh@kust.edu.cn

Abstract: To accurately and reliably characterize the degradation trend of rolling bearings and predict their life cycle, this paper proposes a bearing life prediction model based on an improved transformer encoder layer and multi-scale convolution. First, time-domain, frequency-domain, and time-frequency domain features are extracted from the vibration data covering the entire lifespan of the rolling bearings and passed through the transformer encoder layer. A novel dual-layer self-attention mechanism network structure is proposed to capture global information on the lifecycle progression of rolling bearings. Next, to further extract local temporal features within the bearing's life cycle, a multi-scale convolution module is proposed to reinforce the local information across the entire lifespan. This method fully exploits both the long-term trends and short-term dynamic variations in the health status of rolling bearings, effectively enhancing the accuracy of life predictions. Experimental results show that, even under conditions with interference features, the TransCN model outperforms mainstream advantage model in terms of prediction accuracy and generalizability. This approach offers a new solution for managing the fault risk of rotating machinery and reducing maintenance costs.

Keywords: rolling bearing; life prediction; transformer; dual-layer self-attention; multi-scale convolution

1. Introduction

Rolling bearings are mainly responsible for supporting rotating shafts and reducing friction [1] and play a vital role in ensuring the smooth and efficient operation of rotating machinery. However, high loads, insufficient lubrication, and contamination can lead to gradual degradation of bearings [2,3]. Once bearing failure occurs, it is easy to cause damage to the equipment, huge economic losses, and even casualties. Prognostics and Health Management (PHM) [4] techniques for predicting the Remaining Useful Life (RUL) of rolling bearings can reveal the deterioration process of the equipment, which can help to make reasonable maintenance decisions and reduce the safety risks of mechanical equipment [5]. It is also a key challenge and an important topic in the field of engineering [6–8].

The prediction methods of the Remaining Useful Life of rolling bearings are mainly divided into two categories: one is based on model-driven methods and the other is based on data-driven methods [4]. The model-driven method establishes a mathematical or physical model based on the failure mechanism of the bearing [9,10], which is not very generalizable due to the complexity of the actual working conditions and its randomness, and requires the use of personnel with a wealth of expert experience [11]. The data-driven approach mainly collects vibration data sets that characterize the health state of bearings and then obtains the complex nonlinear mapping relationship between the data sets and RUL by deep learning. Compared with the traditional model-driven methods, the data-driven approach is more flexible, can adapt to the nonlinear and complex patterns in the data, and can automatically learn complex feature representations from a large amount of data without human intervention [12]. Therefore, data-driven based methods have become the mainstream methods for bearing RUL prediction.

Convolutional Neural Networks (CNN) and Recurrent Neural Networks (RNN) and their variants have been widely used in bearing life prediction tasks [13–16] to improve the accuracy and reliability of bearing life estimation. However, traditional CNN and RNN can affect the accuracy of bearing life prediction due to their limitations, such as the gradient vanishing or gradient exposing problem of RNN, and the insufficient ability of CNN to capture long-term dependencies and insensitivity to temporal location. Temporal Convolutional Network (TCN) [17] can better solve the above problems.

In 2017 Vaswani A [18] proposed the transformer architecture, which overcomes the limitations of traditional deep networks in dealing with temporal data and focuses on the relationship between temporal data through the mechanism of self-attention, which is more advantageous in long sequence prediction tasks. Tang x et al. [19] combined with the time series information captured by the LSTM network and the self-attention mechanism of the transformer network, effectively solved the problem of information loss in the LSTM network and improved the prediction accuracy. Zhou Z et al. [20] used trigonometric and cumulative transformations to correct the monotonicity and trend of the input features and then used the transformer model to predict the RUL of the bearings. Wei Y et al. [21] proposed a conditional variational transformer architecture combining local feature learning and global context modeling for improving the prediction accuracy of the RUL of bearings. Deng L et al. [22] proposed a rolling bearing RUL prediction method based on the probabilistic prediction model DeepAR-transformer (D-former), which combines the autoregressive property of DeepAR and the effective feature extraction capability of the transformer to improve the processing capability of long-term series data. Zhang M et al. [23] proposed the WTE trans framework, which combines the weighted time embedding module and the shift window transformer to explicitly enhance the time correlation modeling and significantly improve the prediction accuracy of RUL. Although the above research results have achieved certain results, the dot product algorithm with self-attention may lack sensitivity to the local context, which limits its ability to capture the dynamic changes in the health status of rolling bearings locally [24].

In summary, this paper proposes a rolling bearing life prediction model (TransCN) that incorporates an improved transformer encoding layer and multi-scale convolution module. The model combines the advantages of both, in which the self-attention mechanism module captures the subtle interaction features through the global perspective and combines them with the local dependency features provided by the multi-scale convolution module, which enables the model to take more influencing factors into account in the task of bearing life prediction, thus improving the accuracy of the prediction. The contributions of this study can be summarized as follows:

- The dual-layer self-attention mechanism is proposed to improve the transformer encoding layer, and the self-iteration of the self-attention mechanism is formed by stacking the self-attention layer. The first-layer self-attention mechanism captures the basic feature during the bearing degradation process and grasps the overall trend of the life process. The secondary layer self-attention mechanism reprocesses the output of the first layer, intensifies key features, and suppresses noise interference, thereby capturing the potential fluctuations and complex dynamic changes during the bearing degradation process more accurately.

- A multi-scale convolution module was designed. Through two layers of the same convolution layers with different convolution kernel configurations, the information on different scales in the bearing degradation feature was extracted and fused. This is conducive to the model capturing the local information of the bearing at different operating stages more comprehensively, thereby reflecting the degradation state of the bearing more accurately and effectively improving the model's ability to recognize complex degradation patterns.
- The high performance and robustness of the TransCN model were verified through experiments conducted in a data environment containing interference features. The experimental results show that even in the presence of many interference features, the TransCN model can still maintain high prediction accuracy and stable generalization ability. This provides a more reliable prediction tool for failure risk management and maintenance cost reduction in rotating machinery.

The main work of this paper is as follows: the first part summarizes and analyzes the problems existing in the field of bearing life prediction; the second part introduces the multi-domain feature extraction for rolling bearings; the third part describes the details of the composition of the TransCN network; the fourth part conducts experimental verification using the PHM2012 public data set and discusses the limitations of the model proposed in this paper; and the fifth part summarizes the proposed method and experimental results.

2. Rolling Bearing State Feature Extraction

2.1. Multi-Domain Feature Extraction

It is difficult to comprehensively characterize the complex dynamic changes in the degradation stage of rolling bearings with a single feature. In order to more accurately assess the health process of bearings, this paper extracts the degradation features from multiple domains such as the time domain, the frequency domain, and the time-frequency domain, respectively.

In this paper, 19 common time domain features are extracted, as shown in Table 1, where n is the total number of sampling points and x_i is the observation value of the ith sampling.

Table 1. Time domain feature.

Feature	Formula
Average	$T_1 = \frac{1}{n}\sum_{i=1}^{n} x_i$
Absolute average	$T_2 = \frac{1}{n}\sum_{i=1}^{n} \lvert x_i \rvert$
Maximum	$T_3 = \max(x_i)$
Minimum	$T_4 = \min(x_i)$
Peak	$T_5 = \max\lvert x_i \rvert$
Peak-to-Peak	$T_6 = T_3 - T_4$
Root mean square	$T_7 = \sqrt{\frac{1}{n}\sum_{i=1}^{n} x_i^2}$

Table 1. *Cont.*

Feature	Formula
Root amplitude	$T_8 = \left(\frac{1}{n}\sum_{i=1}^{n}\sqrt{\lvert x_i \rvert}\right)^2$
Variance	$T_9 = \frac{1}{n-1}\sum_{i=1}^{n}(x_i - T_1)^2$
Standard deviation	$T_{10} = \sqrt{T_9}$
Kurtosis	$T_{11} = \frac{1}{n}\sum_{i=1}^{n} x_i^4$
Skewness	$T_{12} = \frac{1}{n}\sum_{i=1}^{n} x_i^3$
Waveform index	$T_{13} = \frac{T_7}{T_2}$
Peak index	$T_{14} = \frac{T_5}{T_7}$
Impulse index	$T_{15} = \frac{T_5}{T_9}$
Margin index	$T_{16} = \frac{T_5}{T_8}$
Clearance factor	$T_{17} = \frac{T_5}{T_7^2}$
Kurtosis factor	$T_{18} = \frac{T_{11}}{T_7^4}$
Variation index	$T_{19} = \frac{T_{10}}{T_1}$

To further quantify the complexity of the running state of the bearing [25], eight entropy features are extracted based on the time domain, which are log energy entropy, Renyi entropy, Tsallis entropy, Shannon entropy, approximate entropy, fuzzy entropy, sample entropy, and permutation entropy, and are recorded as $E_1 \sim E_8$. The calculation formula is as follows:

For time series $Y = [y(1), y(2), \ldots, y(n)]$, the calculation formulas of log energy entropy, Renyi entropy, Tsallis entropy, and Shannon entropy are as follows:

$$E_1 = \sum_{i=1}^{n} \log(y_i^2) \tag{1}$$

$$E_2 = \frac{1}{1-\alpha} \log\left(\sum_{i=1}^{n} P_{y_i}^{\alpha}\right), \alpha > 0 \text{ and } \alpha \neq 1 \tag{2}$$

$$E_3 = \frac{1}{\alpha - 1}\left(1 - \sum_{i=1}^{n} P_{y_i}^{\alpha}\right) \tag{3}$$

When $\alpha = 1$, Renyi entropy is transformed into Shannon entropy. The simplified Shannon entropy calculation formula is as follows:

$$E_4 = -\sum_{i=1}^{n} P(y_i) \log_2 P(y_i) \tag{4}$$

where $P(y_i)$ is the probability that the random variable Y takes the ith value.

For a given time series Y, it can be constructed as $y_m(n)$ [26].

$$y_m(n) = \{y(n), y(n+1), \ldots, y(n+m-1)\} \tag{5}$$

d_{nj}^m is defined as the distance between $y_m(n)$ and $y_m(j)$.

$$d_{nj}^m = \max(\lvert y(n+k) - y(j-k) \rvert), k = 0, 1, \ldots, m \tag{6}$$

C_n^m is the probability that d_{nj}^m is less than r under a given similarity tolerance r, where θ is the Heaviside function and m is the embedding dimension.

$$C_n^m(r) = \frac{\sum\limits_{j=1}^{N-m+1} \theta(d_{nj}^m - r)}{N-m+1} \tag{7}$$

The approximate entropy, fuzzy entropy, and sample entropy are as follows:

$$E_5 = \frac{1}{N-m+1} \sum_{n=1}^{N-m+1} lnC_n^m(r) \tag{8}$$

$$E_6 = \frac{1}{N-m+1} \sum_{n=1}^{N-m+1} C_n^m(r) \tag{9}$$

$$E_7 = \frac{1}{N-m} \sum_{n=1}^{N-m} C_n^m(r) \tag{10}$$

The m-dimensional vector A can be constructed by reconstructing the phase space of a given time series $y(n)$.

$$A = \begin{bmatrix} y_{(1)} & y_{(1+\lambda)} & \cdots & y_{(1+(m-1)\lambda)} \\ y_{(2)} & y_{(2+\lambda)} & \cdots & y_{(2+(m-1)\lambda)} \\ \vdots & \vdots & \ddots & \vdots \\ y_{(\gamma)} & y_{(\gamma+\lambda)} & \cdots & y_{(\gamma+(m-1)\lambda)} \end{bmatrix} \tag{11}$$

where λ is the delay time and γ is the number of reconstructed components.

Sorting the y in each reconstructed component in ascending order will lead to $m!$ different permutations. If the probability of occurrence of the ith arrangement is P_i, the arrangement entropy formula is as follows:

$$E_8' = -\sum_{i=1}^{\gamma} P_i ln P_i \tag{12}$$

Finally, the permutation entropy is normalized.

$$0 \leq E_8 = \frac{E_8'}{\ln(m!)} \leq 1 \tag{13}$$

Seven frequency domain features are extracted in this paper, as shown in Table 2. Where f_i is the signal frequency and X_i is the spectrum amplitude.

Table 2. Frequency domain features.

Feature	Formula
Mean Frequency	$F_1 = \frac{1}{n}\sum\limits_{i=1}^{n} X_i$
Frequency standard	$F_2 = \sqrt{\frac{1}{n-1}\sum\limits_{i=1}^{n} (X_i - F_1)^2}$
Frequency center of gravity	$F_3 = \frac{1}{n}\sum\limits_{i=1}^{n} f_i$
RMS Frequency	$F_4 = \sqrt{\frac{\sum\limits_{i=1}^{n} f_i^2 X_i}{\sum\limits_{i=1}^{n} X_i}}$

Table 2. *Cont.*

Feature	Formula
F_5	$F_5 = \sqrt{\frac{\sum_{i=1}^{n}(X_i - F_1)^3}{n\sqrt{F_4^3}}}$
F_6	$F_6 = \sqrt{\frac{\sum_{i=1}^{n}(X_i - F_1)^4}{nF_4^2}}$
F_7	$F_7 = \sqrt{\frac{\sum_{i=1}^{n}(f_i - F_3)^2 X_i}{n}}$

Time frequency domain analysis combines the advantages of time domain and frequency domain analysis and can capture the changes in signal in time and frequency. In this paper, a three-layer wavelet packet decomposition is adopted to extract eight frequency band energy features of rolling bearings [27] which are recorded as $TF_1 \sim TF_8$. The original signal can be recursively decomposed by the wavelet packet algorithm as follows.

$$x_{j+1,2k}(t) = \sum_m h(m-2n)x_{j,k}(t) \quad (14)$$

$$x_{j+1,2k+1}(t) = \sum_m g(m-2n)x_{j,k}(t) \quad (15)$$

$$g(n) = (-1)^n h(1-n) \quad (16)$$

where $x_{j,k}(t)$ represents the wavelet coefficient of the kth subband in the jth layer.

Therefore, the original signal can be expressed as follows:

$$x(t) = \sum_{k=0}^{2^j-1} x_{j,k}(t) \quad (17)$$

where j and k are the levels and subbands of wavelet packet decomposition, respectively.

2.2. *Feature Normalization Processing*

To eliminate the influence of each dimension of the feature, improve the convergence speed of the deep learning model, facilitate data analysis and processing, and avoid the problem of gradient exploring in the prediction process, this paper uses the max–min normalization method to map the extracted multi-domain feature set to the [0,1] interval. The calculation formula is as follows:

$$x^* = \frac{x - x_{\min}}{x_{\max} - x_{\min}} \quad (18)$$

where $x_{\max}$ is the maximum value of the feature and $x_{\min}$ is the minimum value of the feature.

3. Rolling Bearing Life Prediction Based on Improved Transformer

3.1. *Transformer Architecture*

Transformer networks typically consist of an encoder and a decoder, where the encoder converts an input sequence into a continuous vector representation, and the decoder converts these representations into an output sequence. The self-attention layer introduces a multi-head attention mechanism to capture different subspace information, and each "head" independently learns different aspects of the data to obtain richer information expression and improve the training efficiency [28]. At the same time, location coding is used to make the model understand the sequence of each time point. In addition, the transformer uses residual connection and layer normalization after each sub-layer to help mitigate deep network gradient vanishing and stabilize the training process. The transformer performs well when handling complex sequential tasks, especially in scenarios such as rolling bearing life prediction that require long-distance dependencies and parallel processing [29]. To fully obtain the context dependence and characteristics of time series, make the model lighter and more efficient, and maintain good prediction accuracy, only the transformer encoder is used in this paper.

3.2. TransCN Model Structure

To take into account the global and local state features of rolling bearings, this paper proposes a TransCN model, which can not only capture the global trend in the process of bearing degradation but also sensitively identify the local information of the running state of rolling bearings, which can provide a more comprehensive and accurate rolling bearing life prediction solution. The model is mainly composed of the bottom-level network (self-attention mechanism module) and the top-level network (multi-scale convolutional module). The model structure is shown in Figure 1.

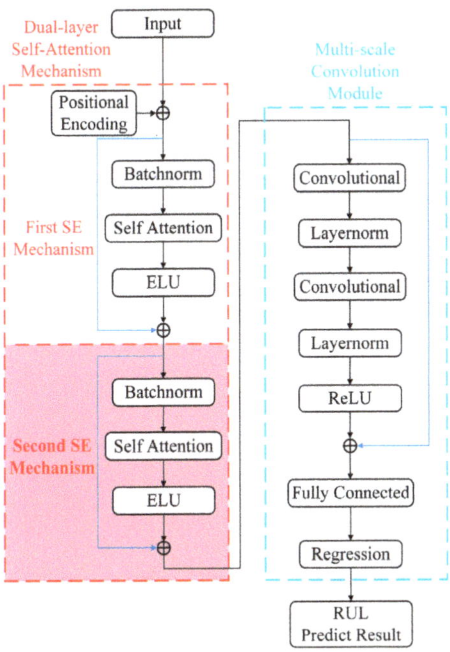

Figure 1. TransCN model structure.

3.2.1. Bottom-Level Network

The proposed bottom-level network consists of the input layer, positional encoding layer, and self-attention mechanism module. (Figure 1 red dashed box)

1. Input layer

This layer is responsible for accepting the multi-domain and multi-class feature sets of rolling bearing vibration, and the feature set comes from the bearing vibration signal information in the time domain, frequency domain, and time-frequency domain.

2. Positional encoding layer

The positional encoding can save the time sequence for each time step; the calculation formula is as follows:

$$PE_{(pos,2i)} = \sin\left(\frac{pos}{10000^{\frac{2i}{d_{model}}}}\right) \tag{19}$$

$$PE_{(pos,2i+1)} = \cos\left(\frac{pos}{10000^{\frac{2i}{d_{model}}}}\right) \tag{20}$$

where pos represents the position of the feature vector, i represents the dimension of the positional encoding, and d_{model} represents the dimension of the feature vector.

3. Residual connection

Residual connections preserve the original input information and allow information to flow directly from the input layer to the subsequent layers, adding the input of one layer directly to the output of that layer. The output of the network can then be represented as a nonlinear transformation of the input and a linear superposition, that is, $y = F(x) + x$. This alleviates the problem of difficult information transfer in deep networks and helps maintain a more direct flow of information. This not only provides a direct propagation path for the gradient, alleviates the problem of gradient vanishing or exploding in deep network training, but also prevents the model from overfitting, enables the model to learn the general feature law, and enhances the generalization ability of new data. Residual connection is used in dual-layer self-attention mechanism and multi-scale convolution module in TransCN (as shown by the blue line in Figure 1).

4. Attention mechanism module

The proposed attention mechanism module consists of batch normalization, self-attention mechanism, and ELU.

Self-attention connects two sequence positions to capture long-term dependencies in time series data. However, single-layer self-attention struggles to extract key degraded features from complex data. Thus, this study adds a second self-attention layer (pink shading in Figure 1) after the original in the transformer network. Batch normalization is applied to the input data to enhance model convergence and generalization, calculated as follows.

$$y_i = \gamma\frac{x_i - \mu}{\sqrt{\sigma^2 + \varepsilon}} + \beta \tag{21}$$

$$\sigma^2 = \frac{1}{B}\sum_{i=1}^{B}(x_i - u)^2 \tag{22}$$

$$\mu = \frac{1}{B}\sum_{i=1}^{B}x_i \tag{23}$$

where x_i is the feature vector of the input and ε is a small value that prevents division by zero. The parameters γ and β are trainable.

The self-attention layer captures the global degradation trend of rolling bearing life from a global perspective, aiding the model in understanding bearing state evolution and providing critical information for life prediction. The formula is as follows:

The input feature matrix is X, and Query, Key, and Value matrices are generated by three sets of linear transformations.

$$Q = XW^Q \tag{24}$$

$$K = XW^K \tag{25}$$

$$V = XW^V \tag{26}$$

where W^Q, W^K and W^V are trainable parameter matrices.

Attention weights are calculated by the dot product of Query and Key:

$$Attention(Q,K,V) = Softmax\left(\frac{QK^T}{\sqrt{d_k}}\right)V \tag{27}$$

where d_k is the dimension of Q and K matrices.

To capture attention for different features, a multi-head mechanism is used for parallel computation:

$$Multihead(Q,K,V) = Concat(head_1, head_2, \ldots, head_h)W^0 \tag{28}$$

$$head_i = Attention(Q_i, K_i, V_i) \tag{29}$$

where h is the number of attention heads, W^o is the output transformation matrix. It should be noted that the algorithms of each self-attention in the proposed dual-layer self-attention mechanism are the same.

To introduce nonlinearity and enhance convergence, the self-attention output is activated by ELU. The formula is as follows:

$$ELU(x) = \begin{cases} x, & x > 0 \\ \alpha(e^x - 1), & x \leq 0 \end{cases} \tag{30}$$

where α is a hyperparameter which usually takes the value 1.

The dual-layer self-attention mechanism offers significant advantages. First, the first layer captures basic state evolution features of rolling bearings (trends and periods), while the second layer extracts deeper features (subtle changes and irregular patterns). This structure enables the model to handle more complex inputs. Secondly, the second layer refines the first layer's output, enhancing focus on key fatigue features. For instance, in monitoring tasks, more accurately identifying fatigue features under different working conditions while suppressing less relevant noise. This targeted feature selecting enhances the model's effectiveness in dynamic behavior modeling, improves adaptability, meets diverse data analysis needs, boosts generalization, and ensures robust performance in complex industrial environments.

3.2.2. Top-Level Network

The proposed top-level network includes a convolutional layer, layer normalization, ReLU, fully connected layer and a regressive layer. (Figure 1 blue dashed box)

1. Convolutional layers

To take into account the local information of fatigue data of rolling bearings, this paper proposes that after the dual-layer self-attention mechanism, a one-dimensional same convolution layer with two layers of different convolution kerns is used to extract and merge the feature of bearing vibration signals at different scales to learn the complex nonlinear relationship between bearing vibration signals and their life course more effectively.

The one-dimensional same convolutional layer can flexibly control the receptive field by adjusting the size and step size of the convolution kernel while maintaining a low computational burden. Without increasing the difficulty of calculation, the design further introduces more nonlinearities by deepening the number of network layers to improve the expression ability of the model. In this paper, two same convolution layers can be used to further simplify the model structure and independently adjust the convolution kernel size and step size to adapt to different sequence lengths and feature extraction requirements, which helps to reduce the complexity of the model, reduce the risk of overfitting, and improve the generalization ability of the model, and extracting higher-level features layer by layer and fusion to improve the richness of feature expression. The model can capture and utilize the patterns in the data more effectively, which can further improve the accuracy of life prediction. The formula is as follows:

$$(f * g)(t) = \int_{-\infty}^{\infty} f(\tau)g(t-\tau)d\tau \tag{31}$$

where $f(t)$ is the input signal, $g(t)$ is the convolution kernel, $*$ represents the convolution operation, τ is the integral variable used for the displacement in the convolution operation, and $(f * g)(t)$ is the result of the convolution operation, that is, the feature extracted by the convolution operation.

2. Layer normalization

To reduce the influence of input distribution variation between layers, this paper performs layer normalization on input data to improve the training efficiency and stability of the model. Layer normalization utilizes the statistical information of a single sample and can maintain a stable gradient update effect when dealing with time series data. It is calculated as follows.

$$y_i = \gamma \frac{x_i - \mu}{\sqrt{\sigma^2 + \varepsilon}} + \beta \tag{32}$$

$$\sigma^2 = \frac{1}{B} \sum_{i=1}^{B} (x_i - u)^2 \tag{33}$$

$$\mu = \frac{1}{B} \sum_{i=1}^{B} x_i \tag{34}$$

where x_i is the feature vector of the input and ε is a small value that prevents division by zero. The parameters γ and β are trainable.

After normalization, the ReLU function is activated, and the life prediction value of the rolling bearing is obtained through the output of the regressive layer.

The primary purpose of the first convolution layer is to extract preliminary local features from the sequence. This layer is capable of capturing short-term patterns and trends in the sequence, and local features are crucial for understanding the real-time state of bearings, as they can reveal the dynamic changes in bearings at the microscopic level. The second convolution layer further refines the feature extraction process. This layer uses a different convolutional kernel configuration and is designed to capture more complex local features that are associated with a specific degradation pattern of the bearing. The introduction of the second convolutional layer enables the model to understand the bearing degradation process from a more fine-grained perspective, thereby enhancing the model's ability to identify subtle changes in the bearing degradation process. By convolving the two layers in series, the model can capture local features on multiple scales, thereby providing a more comprehensive understanding of the bearing degradation process. This multi-scale feature extraction method helps the model capture the dynamic changes in bearings on different time scales, thus improving the accuracy and robustness of the prediction. The concatenation of two layers of convolution not only enhances the expression ability of the feature but also improves the recognition ability of complex degenerated patterns through the fusion of features at different scales. To stabilize the training process and accelerate convergence, after each convolution operation, layer normalization is added. Layer normalization reduces internal covariate offsets by normalizing feature values for each sample, which helps prevent gradient vanishing or exposing problems, thus making deep network training more stable. Therefore, the model can adapt to the distribution change in data faster, thereby improving the training efficiency and model performance.

Finally, the ReLU activation function is used to introduce nonlinearity, which enables the model to identify and use the nonlinearity feature in the data to more accurately simulate the complex process of bearing degradation.

In summary, the life prediction model of rolling bearing proposed in this paper mainly has the following steps:

Step 1: the vibration signals of multiple bearings in the whole life cycle are collected, and the different bearings are divided into the training set and a test set.

Step 2: extract the multi-domain feature of the vibration signal, normalize the set of multi-domain features, and label it with a lifetime (Equations (18) and (35)).

Step 3: Build a dual-layer self-attention mechanism transformer encoding layer (Equations (24)–(29)). The first layer of attention captures the basic feature of the rolling bearing state, and the series structure of the second layer of attention forms the self-iteration of the attention mechanism to further extract the feature of the depth state.

Step 4: Construct the multi-scale convolution module (Equation (31)). The local feature of rolling bearing vibration signals in different time scales is extracted and fused.

Step 5: the output results of the model are smoothed (Equation (36)) to obtain the life prediction results.

As shown in Figure 2, the self-attention mechanism of the bottom-level network in the proposed model enables the model to identify and exploit long-term trends in the time series. Through the serial connection of self-attention mechanism, the feature is gradually extracted from the model at different abstraction levels. These features contain the global information of time series, providing a more comprehensive representation of time features for the proposed model, and laying a high-quality data foundation for further processing of subsequent top-level network. The top-level network uses the multi-scale convolution module to capture and fuse local features on different time scales, which enhances the sensitivity of the proposed model to regional changes in time series. Through convolution operation, the model can further refine and enhance the features transmitted from the bottom-level network, making the feature expression more abundant and accurate. By using this kind of division and cooperation mode of the top and bottom-level network series structure, the model can predict the life of the rolling bearing by using the global and local features at the same time and enhance the understanding ability of the model to the information of the bearing life process.

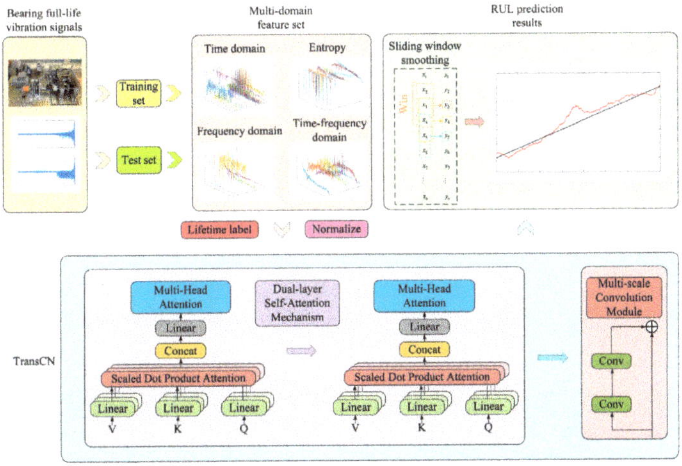

Figure 2. Proposed method flowchart.

4. Experimental Verification

4.1. Data Set Description

To verify the effectiveness of the proposed method, this paper uses the IEEE PHM 2012 public data set for testing [30]. The PRONOSTIA experimental platform is equipped with two DYTRAN 3035B (Dytran by HBK, Chatsworth, CA, USA) micro-accelerometers, which are mounted on the outer ring of the rolling bearing in a radial way at 90° to each other. The two accelerometers collect the bearing vibration signal at a sampling frequency of 25.6 kHz, and the sampling is carried out every 10 s and the data are collected for 0.1 s each time. Figure 3 shows the testbed as well as the sensor arrangement.

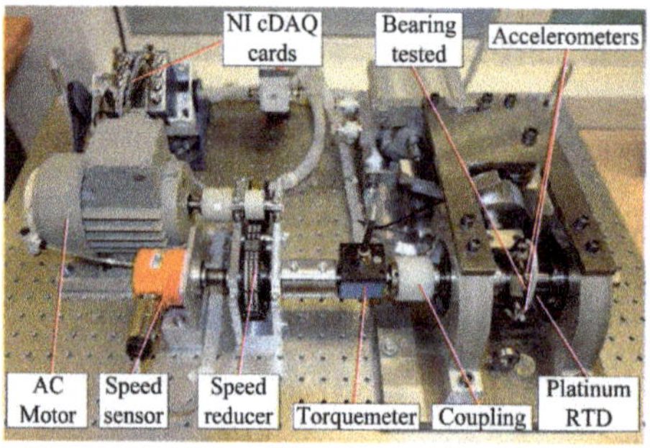

Figure 3. PRONOSTLA experimental platform.

This data set contains full-life vibration data of 17 bearings under three operating conditions. The specific operating conditions are shown in Table 3. In this paper, the data of working conditions 1 and 2 are selected to verify the model. In working condition 1, Bearing1_3 and Bearing1_7 are selected as the training set, and Bearing1_4 is selected as the test set. In condition 2, Bearing2_7 is selected as the training set and Bearing2_4 as the test set. In this paper, the horizontal vibration signal is selected for analysis.

Table 3. Bearing operating conditions.

Operating Conditions	Rotational Speed/rpm	Radial Force/kN	Bearing Number
1	1800	4.0	Bearing1_1~Bearing1_7
2	1650	4.2	Bearing2_1~Bearing2_7
3	1500	5.0	Bearing3_1~Bearing3_3

The label of each sample Is defined as: current running time/total running time, distributed between 0 and 1, where 0 indicates that the bearing is completely healthy and 1 indicates that the bearing is completely damaged.

$$y_i = \frac{t_i}{T_e} \tag{35}$$

where t_i is the current time, T_e is the failure time, and y_i is the sample label at the current time.

4.2. Experiment 1

The features of Bearing1_3 and Bearing1_7 horizontal vibration signals in the time domain, frequency domain, and time-frequency domain are extracted for network training. The horizontal vibration signals and features of Bearing1_3 and Bearing1_7 are shown in Figures 4–13. It can be seen that most features tend to be strong, such as T2–T10, E4,

E6, F1, F2, F7, and TF4 of Bearing1_3 and T2–T11, T13, E2, E3, E6, E8, F1, F2, and F7 of Bearing1_7. To test the effectiveness, robustness and generalization of the proposed algorithm, no denoising operation was carried out on the original signal in this paper, and some features with poor quality were also included as strong interference items in the test data (as shown by the red label in Figures 6–13). For example, Bearing1_3's T14–T16, E1–E3, F3–F6, TF2, and TF5–TF8 and Bearing1_7's T12, T14–T16, E1, TF1–TF3, TF5, and TF6 have very large burrs. T1 and T19 of Bearing1_3 and T1, T19, F3, F5, and F6 of Bearing1_7 do not change much during the whole life stage and do not reflect any information of bearing degradation.

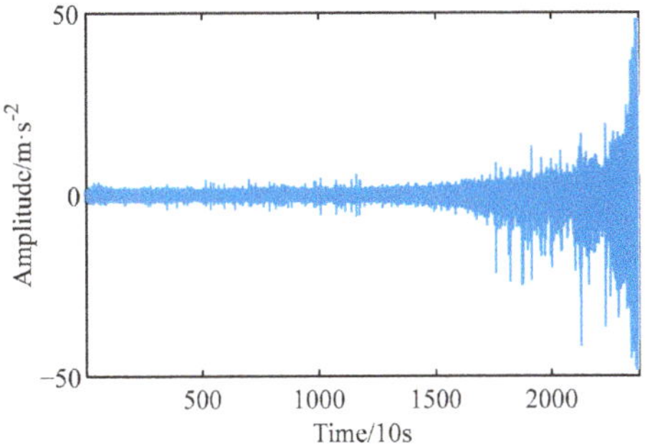

Figure 4. Bearing1_3 horizontal vibration signal.

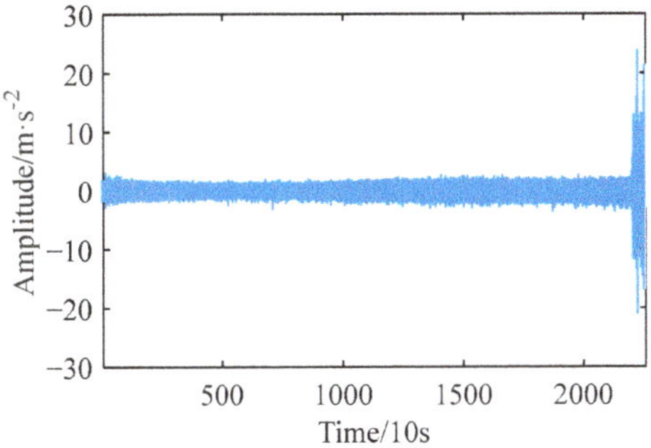

Figure 5. Bearing1_7 horizontal vibration signal.

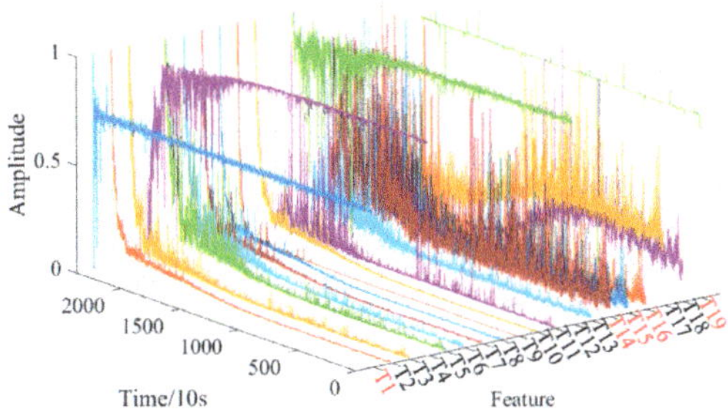

Figure 6. Bearing1_3 time domain features.

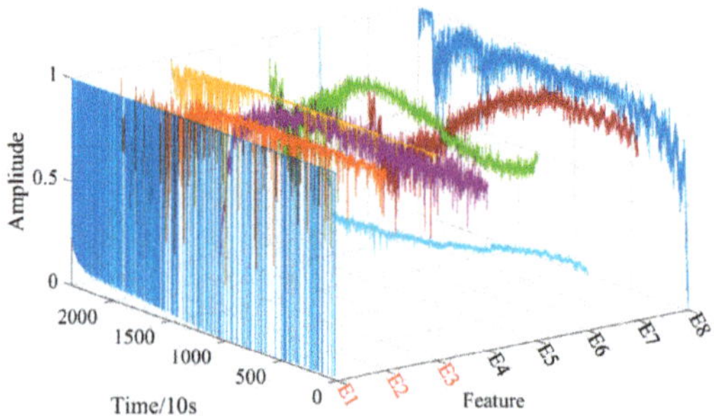

Figure 7. Bearing1_3 entropy feature.

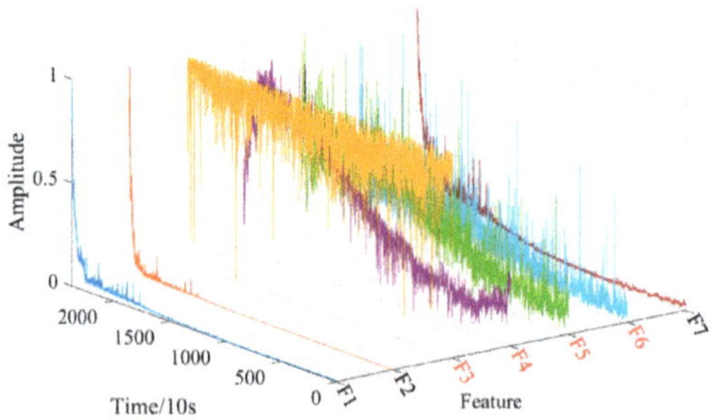

Figure 8. Bearing1_3 frequency domain features.

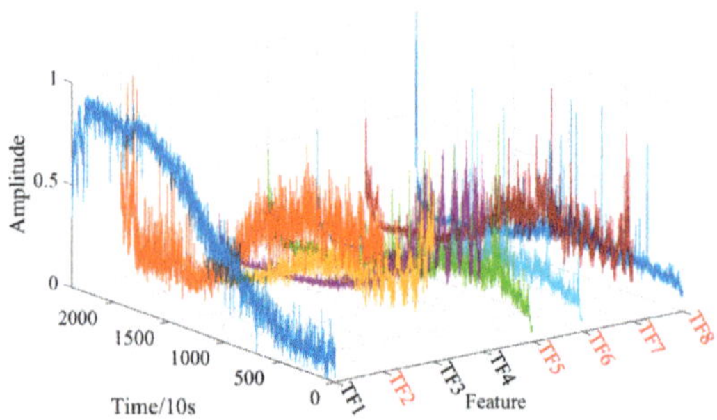

Figure 9. Bearing1_3 time-frequency domain features.

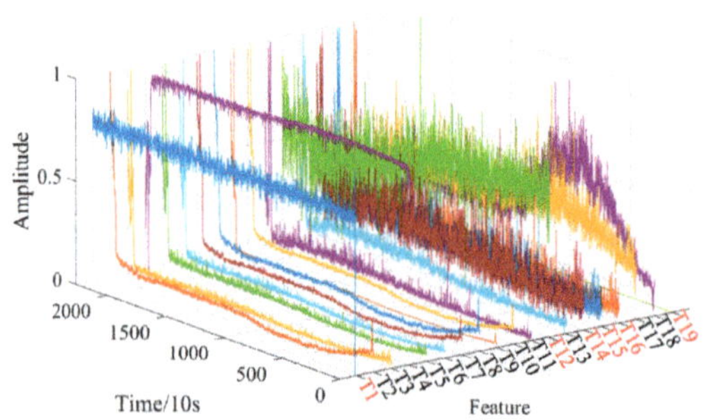

Figure 10. Bearing1_7 time domain features.

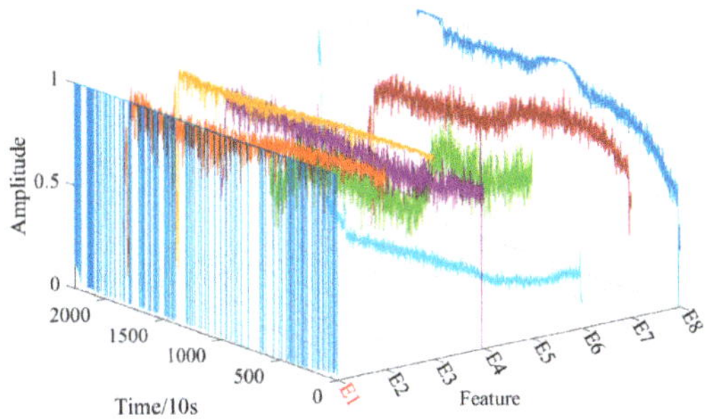

Figure 11. Bearing1_7 entropy feature.

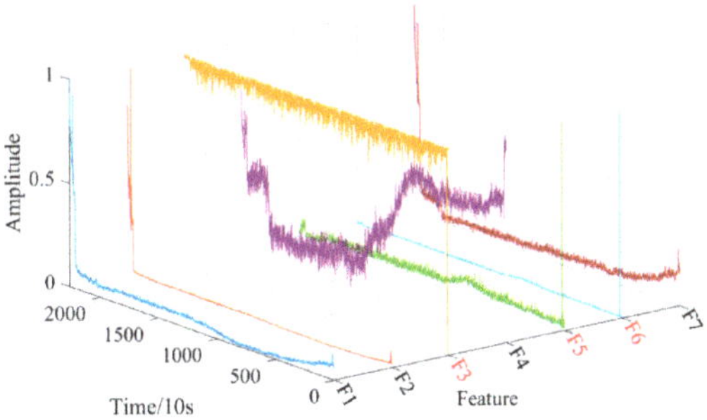

Figure 12. Bearing1_7 frequency domain features.

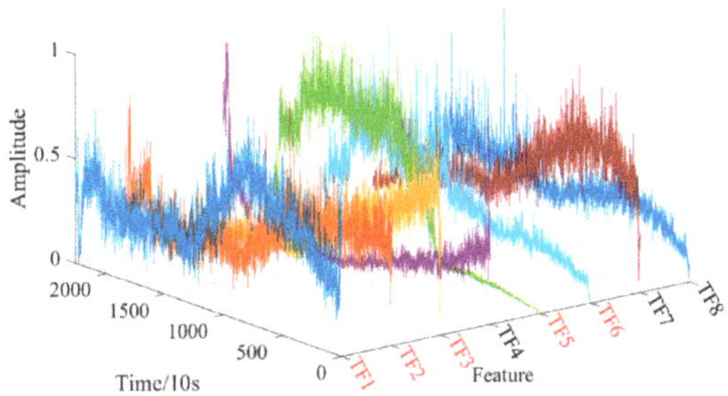

Figure 13. Bearing1_7 time-frequency domain features.

The multi-domain feature set of Bearing1_3 and Bearing1_7 is input into the network for training. After the network is trained, the multi-domain feature set of Bearing1_4 is input into the network to predict the bearing life. MAE and MSE are selected as the error indexes of the life prediction results. The network hyperparameter Settings are shown in Table 4.

Table 4. Experiment 1 hyperparameter setting.

Hyperparameters	Value
Epochs	50
BatchSize	100
Learning rate	0.00098
Multi-head attention head count	4
Size of the convolution kernel	4/5
The number of convolution kernels	42/42
Optimizer	Adam

To reduce the local oscillation of the predicted value, the moving average algorithm is used to smooth the life prediction results, so that the prediction results are more in line with the actual situation [14]. In this case, the window length is chosen to be 140, and the smoothing algorithm formula is as follows:

$$y_t = \begin{cases} \frac{1}{win} \sum_{i=1}^{t} x_i, & t < win \\ \frac{1}{win} \sum_{i=t-win+1}^{t} x_i, & win \leq t \leq n - win + 1 \\ \frac{1}{win} \sum_{i=1}^{t} x_i, & t > n - win + 1 \end{cases} \tag{36}$$

where x_i is the original series and y_t is the series after the moving average and the window size of win moving average. The results of the lifetime prediction are shown in Figure 14.

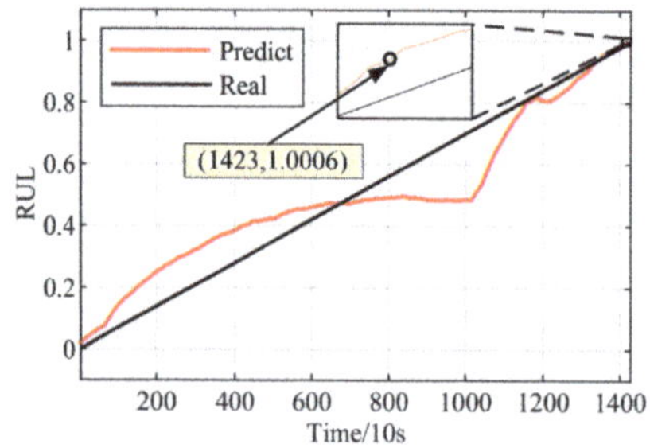

Figure 14. Bearing1_4 life prediction results.

According to Figure 14, fitting at the end of the full life of the rolling bearing shows that the proposed method predicts the bearing failure only 50 s in advance, accounting for 0.35% of the full life cycle of the bearing, which shows that the proposed model can accurately predict the bearing life. The error index of the life prediction result is MAE: 0.0767, MSE: 0.0090. On the whole, the life prediction result is basically consistent with the real life trend which reflects the degradation process of the bearing well.

To verify the effectiveness and generalization of the proposed model, this paper uses four current mainstream bearing life prediction models for comparison, as shown in Figure 15. The transformer encoder (blue curve) uses the self-attention mechanism to capture the global dependence and predicts the bearing degradation trend as a whole. However, the transformer encoder has relatively large errors in the early stage (before 1100 s) and poor compatibility with real life in the later stage (after 13,300 s), and predicts bearing failure 500 s in advance. It accounts for 3.50% of the whole life cycle of the bearing, and the error is large. Although TCN (light green curve) has advantages in capturing local features, it has shortcomings in extracting global features and predicting the overall trend. The prediction curve shows certain volatility, especially in the middle and late life (after 11,530 s), and there is a large deviation from the real life. Combining the advantages of the two methods, the proposed method is more comprehensive in feature extraction, and can better understand the complex dynamic changes in bearing degradation. The error metrics are shown in Table 5. Compared with CNN, GRU, TCN and transformer encoder, the MAE of the proposed method is reduced by 26.65%, 38.50%, 40.27% and 32.73%, and the MSE is reduced by 28.64%, 51.14%, 65.16% and 68.46%, respectively. It can be seen that the proposed method is superior to other methods.

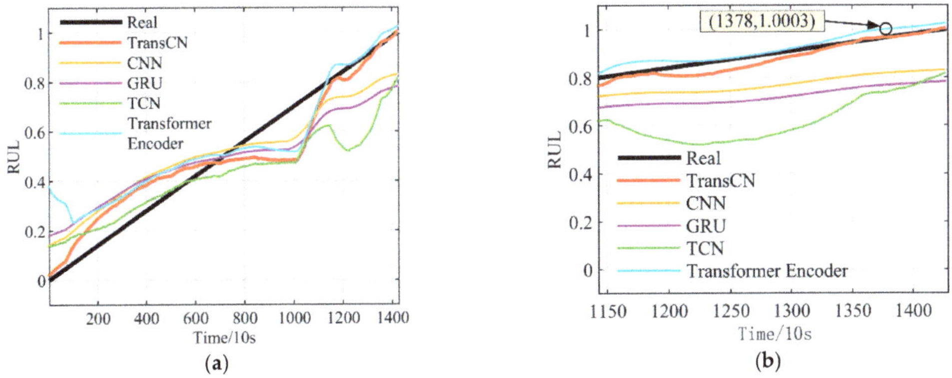

Figure 15. Experiment 1 comparison of multiple model prediction results: (**a**) comparison of the prediction results of the methods; (**b**) comparison results at the end of life.

Table 5. Experiment 1 error index of multi model prediction results.

Model	MAE	MSE
Method of this paper	**0.0767**	**0.0090**
CNN	0.1045	0.0126
GRU	0.1246	0.0184
TCN	0.1283	0.0258
Transformer Encoder	0.0964	0.0150

4.3. Experiment 2

To further verify the generalization of the proposed model, under the same conditions, a feature of Bearing2_7 horizontal vibration signal in the time domain, frequency domain and time-frequency domain are extracted for network training. Bearing2_7 horizontal vibration signal and feature are shown in Figures 16–20. It can be seen that most features show a certain overall growth or decline trend on the time axis, which is conducive to the life prediction of bearings, such as T2–T10, T17, T18, E4–E8, F1, F2, F4–F7, and TF1–TF4. At the same time, some features cannot reflect the degradation process of bearings (as shown in the red label in Figures 17–20), such as T13–T16, E1–E3 and F3, there are very large fluctuations, T1, T11, T12 and T19 change steadily throughout the life stage, which does not help predict the degradation state of bearings. The bad features are input into the model as interference items for training, to verify the robustness of the model.

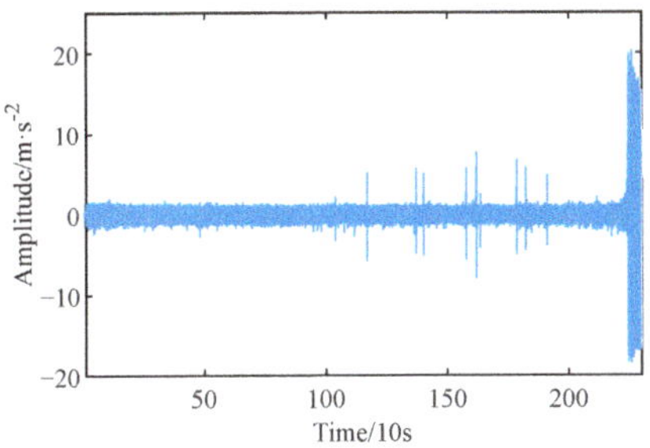

Figure 16. Bearing2_7 horizontal vibration signal.

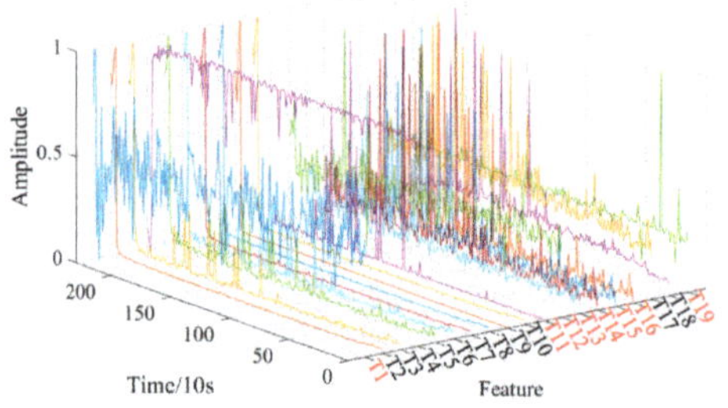

Figure 17. Bearing2_7 time domain features.

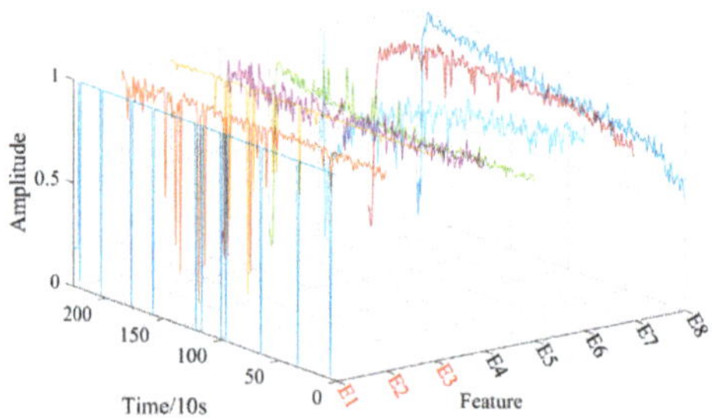

Figure 18. Bearing2_7 entropy feature.

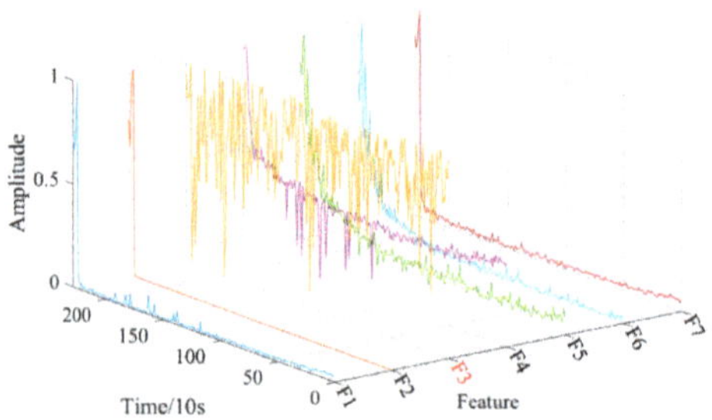

Figure 19. Bearing2_7 frequency domain features.

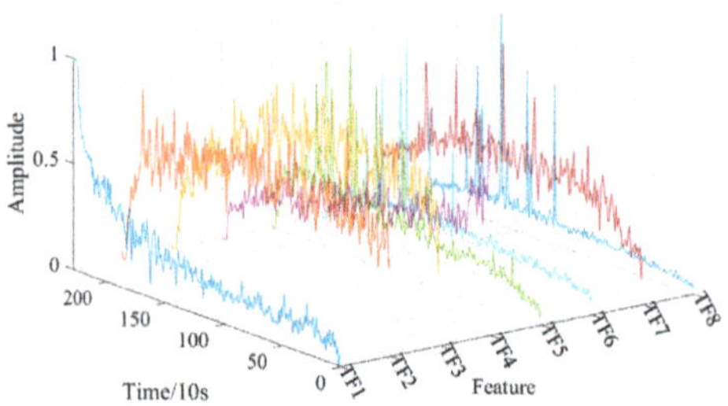

Figure 20. Bearing2_7 time-frequency domain features.

Bearing2_7's multi-domain feature set is input into the network for training. After the network training is complete, Bearing2_4's multi-domain feature set is input into the network to predict the bearing life. The network hyperparameter Settings are shown in Table 6. The life prediction results are smoothed, in this case, the window length is chosen to be 40, and the life prediction results are shown in Figure 21.

Table 6. Experiment 2 hyperparameter setting.

Hyperparameters	Value
Epochs	100
BatchSize	16
Learning rate	0.00395
Multi-head attention head count	4
Size of the convolution kernel	5/4
The number of convolution kernels	42/42
Optimizer	Adam

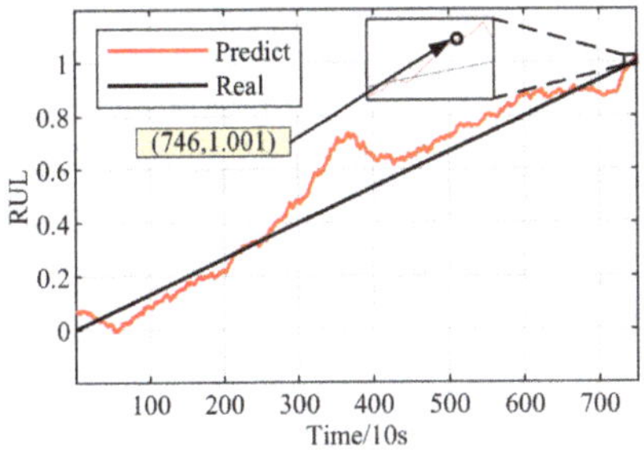

Figure 21. Bearing2_4 Life Prediction Results.

The error index of life prediction results is MAE: 0.0618, MSE: 0.0065. It can be seen that the fitting effect of life prediction is very good, which well reflects the degradation trend of bearings. At the end of the bearing life, the bearing failure is only warned 50 s in advance, accounting for 0.67% of the whole life cycle of the bearing, which accurately predicts the failure of the bearing.

CNN, GRU, TCN, and transformer encoder models are used to compare with the prediction effect of this model, and the comparison effect is shown in Figure 22. It can be seen that although the transformer encoder (blue curve) captures the long-term dependence of the data and predicts the degradation trend well, the prediction accuracy is far from real life, and the prediction curve once seriously deviated from real life. Although TCN (light green curve) is good at capturing local features, it cannot extract global features, so the prediction effect is not ideal. The method in this paper combines the advantages of both and accurately predicts the bearing life. The error metrics are shown in Table 7. Compared with CNN, GRU, TCN, and transformer encoder, the MAE of the proposed method is reduced by 54.79%, 63.26%, 74.44%, and 45.75%, and the MSE is reduced by 72.22%, 82.22%, 92.18%, and 77.18%, respectively. It can be seen that the proposed method is superior to other methods.

Table 7. Experiment 2 error index of multi model prediction results.

Model	MAE	MSE
Method of this paper	**0.0618**	**0.0065**
CNN	0.1367	0.0234
GRU	0.1682	0.0366
TCN	0.2450	0.0832
Transformer encoder	0.1140	0.0285

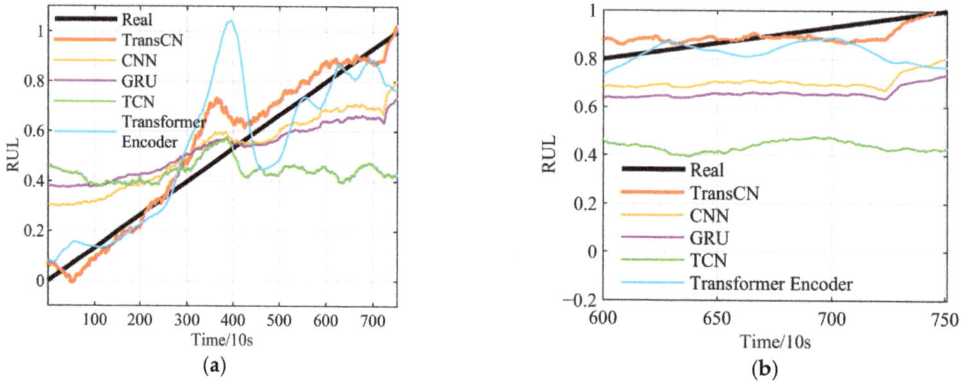

Figure 22. Experiment 2 comparison of multiple model prediction results: (**a**) comparison of the prediction results of the methods; (**b**) comparison results at the end of life.

Combining the life prediction results of working conditions 1 and 2, the life prediction method proposed in this paper can dig out the nonlinear relationship between vibration signals and bearing life and can better predict the life process of rolling bearings.

In summary, the algorithm proposed in this paper is experimentally verified to achieve better performance, accuracy, and generalization of life prediction than many mainstream methods. It provides an effective solution for failure risk management of rotating machinery. We note that in real industrial scenarios, the speed and load of rolling bearings may face random dynamic changes. Such changing conditions may lead to fluctuations in vibration signal features. As a further direction of exploration, the study can consider introducing a dynamic feature extraction module and an online updating strategy to enhance the model's adaptability to variable working conditions and extend the application scope of this study.

5. Conclusions

To further improve the accurate description of the rolling bearing degradation trend and life process, a TransCN model is proposed in this paper. By improving the encoding layer of the transformer and integrating it with the multi-scale convolution module, the accuracy of rolling bearing life prediction is effectively improved, and the following conclusions are drawn:

(1) A new dual-layer self-attention mechanism is proposed to improve the transformer encoding layer to obtain the global features of rolling bearing time series. Through the concatenation of self-attention mechanism, the long-term trend of the time series can be effectively captured, and complex global dependencies and dynamic patterns can be modeled, providing a more comprehensive time representation for the model.

(2) A two-layer same convolutional module structure is designed to capture local features of different scales in the fatigue evolution of rolling bearings and integrate them, which enhances the local feature extraction capability of the model, improves the computing efficiency of the model, and reduces the computing resource requirements.

(3) The tests under different working conditions show that the life prediction results of TransCN fit the actual life curve, and the end of the bearing life is in good agreement, which proves that the model has a certain generalization ability. Compared with the mainstream life prediction model, the life error index is generally reduced by more than 45%. The results of this paper can provide a more accurate theoretical basis for the operation and maintenance of rotating machinery equipment.

Author Contributions: All authors were involved in the conceptualization and design of the study. Data processing and analysis, and model construction were performed by Z.L., Z.W., X.L., and Y.Y. The first draft of the manuscript was written by Z.L. and revised by Z.W. All authors commented on a previous version of the manuscript. All authors have read and agreed to the published version of the manuscript.

Funding: This research was funded by the National Natural Science Foundation of China (Grant No. 52165065), the Yunnan Province Science and Technology Program of China (Grant No. 202401AT070346), and the Yunnan Provincial Major Science and Technology Special Program (Grant No. 202402AC080005).

Data Availability Statement: Data are contained within the article.

Acknowledgments: The author sincerely thanks IEEE for providing the PHM 2012 bearing data set and anonymous reviewers for their constructive suggestions and comments on this paper.

Conflicts of Interest: The authors declared no potential conflicts of interest concerning the research, authorship, and/or publication of this paper. And all the data used in this paper are real and valid.

References

1. Huang, G.; Lei, W.; Dong, X.; Zou, D.; Chen, S.; Dong, X. Stage-Based Remaining Useful Life Prediction for Bearings Using GNN and Correlation-Driven Feature Extraction. *Machines* **2025**, *13*, 43. [CrossRef]
2. He, M.; Li, Z.; Hu, F. A Novel RUL-Centric Data Augmentation Method for Predicting the Remaining Useful Life of Bearings. *Machines* **2024**, *12*, 766. [CrossRef]
3. Chen, B.; Smith, W.A.; Cheng, Y.; Gu, F.; Chu, F.; Zhang, W.; Ball, A.D. Probability distributions and typical sparsity measures of Hilbert transform-based generalized envelopes and their application to machine condition monitoring. *Mech. Syst. Signal Process.* **2025**, *224*, 112026. [CrossRef]
4. Cai, S.; Zhang, J.; Li, C.; He, Z.; Wang, Z. A RUL prediction method of rolling bearings based on degradation detection and deep BiLSTM. *Electron. Res. Arch.* **2024**, *32*, 3145–3161. [CrossRef]
5. Fan, Z.; Li, W.; Chang, K.C. A bidirectional long short-term memory autoencoder transformer for remaining useful life estimation. *Mathematics* **2023**, *11*, 4972. [CrossRef]
6. Ferreira, C.; Gonçalves, G. Remaining Useful Life prediction and challenges: A literature review on the use of Machine Learning Methods. *J. Manuf. Syst.* **2022**, *63*, 550–562. [CrossRef]
7. Lei, Y.; Li, N.; Guo, L.; Li, N.; Yan, T.; Lin, J. Machinery health prognostics: A systematic review from data acquisition to RUL prediction. *Mech. Syst. Signal Process.* **2018**, *104*, 799–834. [CrossRef]
8. Chen, B.; Cheng, Y.; Allen, P.; Wang, S.; Gu, F.; Zhang, W.; Ball, A.D. A product envelope spectrum generated from spectral correlation/coherence for railway axle-box bearing fault diagnosis. *Mech. Syst. Signal Process.* **2025**, *225*, 112262. [CrossRef]
9. Lei, Y.; Li, N.; Gontarz, S.; Lin, J.; Radkowski, S.; Dybala, J. A model-based method for remaining useful life prediction of machinery. *IEEE Trans. Reliab.* **2016**, *65*, 1314–1326. [CrossRef]
10. Tian, Z.; Liao, H. Condition based maintenance optimization for multi-component systems using proportional hazards model. *Reliab. Eng. Syst. Saf.* **2011**, *96*, 581–589. [CrossRef]
11. Singleton, R.K.; Strangas, E.G.; Aviyente, S. Extended Kalman filtering for remaining-useful-life estimation of bearings. *IEEE Trans. Ind. Electron.* **2014**, *62*, 1781–1790. [CrossRef]
12. Zhu, J.; Chen, N.; Shen, C. A new data-driven transferable remaining useful life prediction approach for bearing under different working conditions. *Mech. Syst. Signal Process.* **2020**, *139*, 106602. [CrossRef]
13. Guo, R.; Li, H.; Huang, C. Operation stage division and RUL prediction of bearings based on 1DCNN-ON-LSTM. *Meas. Sci. Technol.* **2023**, *35*, 025035. [CrossRef]
14. Yang, L.; Jiang, Y.; Zeng, K.; Peng, T. Rolling Bearing Remaining Useful Life Prediction Based on CNN-VAE-MBiLSTM. *Sensors* **2024**, *24*, 2992. [CrossRef] [PubMed]
15. Ma, P.; Li, G.; Zhang, H.; Wang, C.; Li, X. Prediction of remaining useful life of rolling bearings based on multiscale efficient channel attention CNN and bidirectional GRU. *IEEE Trans. Instrum. Meas.* **2024**, *73*, 1–13. [CrossRef]
16. Wen, L.; Su, S.; Li, X.; Ding, W.; Feng, K. GRU-AE-wiener: A generative adversarial network assisted hybrid gated recurrent unit with Wiener model for bearing remaining useful life estimation. *Mech. Syst. Signal Process.* **2024**, *220*, 111663. [CrossRef]
17. Lea, C.; Vidal, R.; Reiter, A.; Hager, G.D. Temporal convolutional networks: A unified approach to action segmentation. In Proceedings of the Computer Vision–ECCV 2016 Workshops, Amsterdam, The Netherlands, 8–10 and 15–16 Octorber 2016; Proceedings, Part III; Springer International Publishing: Berlin/Heidelberg, Germany, 2016; pp. 47–54.

18. Vaswani, A. Attention is all you need. In Proceedings of the Advances in Neural Information Processing Systems, Long Beach, CA, USA, 4–9 December 2017.
19. Tang, X.; Xi, H.; Chen, Q.; Lin, T.R. Rolling bearing remaining useful life prediction based on LSTM-Transformer algorithm. In *Proceedings of IncoME-VI and TEPEN 2021: Performance Engineering and Maintenance Engineering*; Springer International Publishing: Cham, Switzerland, 2022; pp. 207–215.
20. Zhou, Z.; Liu, L.; Song, X.; Chen, K. Remaining useful life prediction method of rolling bearing based on Transformer model. *J. Beijing Univ. Aeronaut. Astronaut.* **2021**, *49*, 430–443.
21. Wei, Y.; Wu, D. Conditional variational transformer for bearing remaining useful life prediction. *Adv. Eng. Inform.* **2024**, *59*, 102247. [CrossRef]
22. Deng, L.; Li, W.; Zhang, W. Intelligent prediction of rolling bearing remaining useful life based on probabilistic DeepAR-Transformer model. *Meas. Sci. Technol.* **2023**, *35*, 015107. [CrossRef]
23. Zhang, M.; He, C.; Huang, C.; Yang, J. A weighted time embedding transformer network for remaining useful life prediction of rolling bearing. *Reliab. Eng. Syst. Saf.* **2024**, *251*, 110399. [CrossRef]
24. Li, S.; Jin, X.; Xuan, Y.; Zhou, X.; Chen, W.; Wang, Y.-X.; Yan, X. Enhancing the locality and breaking the memory bottleneck of transformer on time series forecasting. *Adv. Neural Inf. Process. Syst.* **2019**, *32*, 5243–5253.
25. Xu, X.; Zhou, J.; Weng, X.; Zhang, Z.; He, H.; Steyskal, F.; Brunauer, G. A novel evidence reasoning-based rul prediction method integrating uncertainty information. *Reliab. Eng. Syst. Saf.* **2024**, *250*, 110250. [CrossRef]
26. Ke, Z.; Lin, B. Nonlinear characteristic measure of gearbox faults and their category identification with TWSVM. *J. Vib. Shock.* **2018**, *37*, 179–184.
27. Liu, X.; Luan, X.; Zhao, J.; Xiao, B.; Sha, Y. Aircraft engine rolling bearings based on comprehensive dynamic screening Fault feature extraction method. *J. Aerosp. Power* **2024**, *40*, 20240210-1–20240210-12.
28. Zhan, H.; Li, H.; Xiao, X. Prediction of Remaining Useful Life of Rolling Bearings based on Transformer Encoder. In Proceedings of the 2023 IEEE 7th Information Technology and Mechatronics Engineering Conference (ITOEC), Chongqing, China, 15–17 September 2023; Volume 7, pp. 355–359.
29. Li, X.; Li, J.; Zuo, L.; Zhu, L.; Shen, H.T. Domain adaptive remaining useful life prediction with transformer. *IEEE Trans. Instrum. Meas.* **2022**, *71*, 1–13. [CrossRef]
30. Nectoux, P.; Gouriveau, R.; Medjaher, K.; Ramasso, E.; Morello, B.; Zerhouni, N.; Varnier, C. PRONOSTIA: An experimental platform for bearings accelerated degradation tests. In Proceedings of the IEEE International Conference on Prognostics and Health Management, PHM'12, Denver, CO, USA, 20–23 June 2012; IEEE Catalog Number: CPF12PHM-CDR; pp. 1–8.

MDPI AG
Grosspeteranlage 5
4052 Basel
Switzerland
Tel.: +41 61 683 77 34

Machines Editorial Office
E-mail: machines@mdpi.com
www.mdpi.com/journal/machines

www.ingramcontent.com/pod-product-compliance
Lightning Source LLC
Chambersburg PA
CBHW061820210926
51850CB00006B/760
9783725849611